GLACIERS

For Debbie
(my wonderful wife)

and for my Mom and my Dad.

GLACIERS

Peter G. Knight

Ice like a flower
Draws colour from arctic skies.

Stanley Thornes (Publishers) Ltd

First published in 1999 by:
Stanley Thornes (Publishers) Ltd
Ellenborough House
Wellington Street
CHELTENHAM
GL50 1YW
United Kingdom

98 99 00 01 02 / 10 9 8 7 6 5 4 3 2 1

A catalogue record for this book is available from the British Library

ISBN 0-7487-4000-7

Typeset by WestKey Ltd, Falmouth, Cornwall
Printed and bound in Great Britain by Martins the Printers Ltd, Berwick upon Tweed.

CONTENTS

ACKNOWLEDGEMENTS

I owe thanks to numerous individuals and organisations for their assistance, direct and otherwise, in preparing this book. Grants for research that provided original material reported in the text were awarded by The Royal Society, the Natural Environment Research Council, the British Geomorphological Research Group and Keele University. The illustrations were prepared with the assistance of Peter Greatbach, Dave Wilde, Richard Burgess, Mal Beech, and Andy Lawrence at Keele University. Specific sections of the text benefited directly from comments, information and reference materials provided by Andy Russell, Ian Fairchild, Ian Willis, Bryn Hubbard, Thomas Schneider, Simon Ommanney, Mark Twickler, Gary Clow and Debbie Knight. For inspiration, support and correction at various times over a number of years, I am grateful to David Sugden, and also to Chalmers Clapperton, Wilf Theakstone, Roland Souchez, Jean-Louis Tison, Reggie Lorrain, J. Campbell Gemmell, Alistair Gemmell, Judith Maizels, Pat Hamilton, Andy Dugmore, Peter Bull, Judith Pallot, Martin Sharp, Hal Lister, Andrew Goudie, Geoffrey Boulton, and Tony Parsons. Figure 5.17 was provided by my friend and field companion Robin Grimes. Excerpt on page 8, from *The Little Prince*, by Antoine De Saint-Exupery, used with permission of Editions Gallimard (copyright 1946), and reprinted by permission of Harcourt Brace and Company, (copyright 1943 and renewed 1971) and by permission of Egmont Children's Books Ltd for editions published by William Heinemann. Excerpts on this page, and pages 93 and 213, from W.H. Auden, *Selected Poems*, edited by E. Mendelson, reproduced with permission of the publishers, Faber and Faber, and with permission of Curtis Brown Limited, original copyright W.H. Auden 1941, 1967 and 1940. Excerpt on page 225 from Mervyn Peake, *Selected Poems*, reproduced with permission of the publishers, Faber and Faber, and with permission of David Higham Associates Limited. Every attempt has been made to contact copyright holders, but the author and publishers apologise if any have been overlooked.

So here he was without maps or supplies,
A hundred miles from any decent town;

W.H. Auden

INTRODUCTION 1

1.1 THIS BOOK

The aim of this book is to provide an up-to-date introduction to the study of glaciers. It is intended to be of value not only to researchers and advanced students, but also to undergraduates and non-professionals with an interest in glaciers. The book is organised so that it can either be read sequentially from start to finish as a review of the subject, or, with the help of the index, used as a source of reference to specific topics.

Glaciology is a relatively young discipline, but one that has developed quickly and continues to do so. The study of glaciers is a huge and expanding field. In recent years, several major leaps in our understanding of glaciers have been heralded as new paradigms in glaciology, and readers who are not intimately involved in the discipline can quickly fall behind new developments. Much of the existing glaciological literature is indigestible to all but the most expert readers, and many of the excellent glacier-related texts that presently exist cover only parts of the glacier story (e.g. Souchez and Lorrain, 1991; Paterson, 1994) or spread their attention far into neighbouring disciplines (e.g. Sugden and John, 1976; Hambrey, 1994; Bennett and Glasser, 1996; Benn and Evans, 1998). This book focuses directly on glaciers, and aims to incorporate discussion of recent developments in glaciology while remaining accessible to a broad readership. The mathematical content of the book is deliberately kept at a low level to facilitate ease of reading; there is no shortage of glaciological literature available for more mathematically inclined readers.

Glaciers and glaciology have focused the attention of a growing international research community for a little more than 150 years. They are staple components of educational courses at all levels in a range of subjects including Geography, Geology, Earth and Environmental Science and Geophysics. In recent geological history 30% of the Earth's land surface has been covered by glaciers. At present, 10% of the land is permanently ice-covered, and a further 10% is permanently frozen. About 50% of the land is covered by snow and ice in the northern-hemisphere winter. More than 75% of the world's fresh water is contained in glaciers, which provide irrigation water for some of the most densely populated areas of the world. The development potential of large areas of our planet is dependent on our ability to mitigate glacial hazards, to harness glacial resources, and to penetrate the glacial barrier to the immense terrestrial and marine resources of the polar regions. The whole global environment is dominated by the planet's glacial character, and an understanding of the Earth's surface is impossible without an understanding of glaciers. Recently, popular scientific attention has been drawn to global environmental issues. Changing climate and sea-level have become major concerns, and a profusion of recent literature has focused on Quaternary environments and environmental change. A major component of past and future environmental change is the planet's glacier system, and glaciers are now recognised as key players in the drama of the global environment. Modelling of the global environment therefore requires a detailed understanding of the characteristics and behaviour of glaciers. A major goal of glaciology must be to close the gap between the simplifications that fuel much of the environmental modellers' output and the reality of how glaciers behave.

If nature had not been invested with glaciers, it would have taken a bold writer of science fiction to invent them, and it is unlikely that they would have been imagined to be either as complex, or as important to the planet, as in fact they are. This book examines the physical characteristics of glaciers, their role in different parts of the global system, and the interest shown in them by the scientific community.

1.2 TYPES OF GLACIER

A glacier is a body of ice originating on land by the recrystallisation of snow or other forms of solid precipitation and showing evidence of past and present flow.

(Meier, 1964)

Ice exists in many different forms and contexts in the geographical environment, but not all ice masses are properly referred to as glaciers. Meier's (1964) definition, at the head of this section, indicates two key criteria: glaciers are made of ice formed from the accumulation of precipitation; and glaciers flow. Thus, for example, lake ice or sea ice formed by freezing of a water surface is not a type of glacier; so there is no glacier at the North Pole. Similarly, ground ice formed by segregation and freezing of water in the earth is not a type of glacier. Even using this restrictive definition, however, it is possible to identify many different types of glacier. Attempts at glacier classification have been made for almost as long as glaciers have been scientifically studied. Over one hundred years ago, Rink (1877) made a distinction between the Greenland glaciers and what he referred to as glaciers of the 'temperate zone'. Types of glacier have been defined in a variety of different ways, and there is a classification scheme to suit most purposes. However, inconsistencies in terminology between different systems of classification can hamper scientific communication. The value of classification in general can be questioned, since any individual glacier could be fitted into a variety of alternative categories depending on the classification scheme chosen. Different parts of the same glacier might fall into different categories, and there are transitional forms between the categories of most classifications. Each glacier is effectively unique. Nature can provide an exception to any scheme of classification. Nevertheless, classifications are valuable in providing a shorthand description of glaciers, and in identifying shared characteristics or differences that have implications for aspects of glacier study. Most classifications have focused on three main issues: thermal characteristics, dynamic characteristics and morphological characteristics. These are considered in the following paragraphs.

Lagally (1932), Ahlmann (1933, 1935, 1948) and Court (1957) made early contributions to a classification that is still widely in use to differentiate between glaciers with different thermal characteristics. In the so-called Lagally–Ahlmann classification the key distinction is between glaciers in which the ice is at the pressure melting point throughout the body of the glacier (warm or temperate glaciers), glaciers where the ice is below the melting point throughout (cold, high-polar or high-arctic glaciers), and glaciers where most of the ice is cold but the upper portion of the glacier is at the melting point for part of the year (transitional, sub-polar or sub-arctic glaciers). More recent contributions have identified problems with this simple classification. For example, the geographic and stratigraphic distribution of temperature zones in transitional or sub-polar glaciers can occur in a variety of configurations, and processes operating at the bed of warm or temperate glaciers generate temperature variations that cause areas of the bed periodically to fall below the melting point (Robin, 1976; Goodman *et al.*, 1979). Most glaciers are in fact polythermal, which is to say that they have different thermal characteristics in different parts. Nevertheless, recognition of the thermal characteristics of glaciers, or of specific parts of glaciers, is valuable in predicting rheology, basal sliding, geomorphology and hydraulic activity. The thermal classification is widely employed, and thermal characteristics are used especially often in referring to the base of a glacier. Basal thermal regime is particularly important to glacier behaviour; cold-based and warm-based glaciers exhibit quite different characteristics in basal processes such as sliding and erosion.

Glaciers can be classified in terms of their dynamic characteristics, and referred to as active, passive, or dead (Embleton and King, 1968). In recent literature, this terminology has not been widely adopted in a formal sense, although informal usage persists. Active glaciers experience a strong throughput of ice, usually associated with a strong mass balance gradient (Chapter 3). Where the supply of ice and the rate of movement are low, glaciers can be termed passive, and when the ice is stationary, for example in the terminal area of a surging glacier immediately following a surge, it can be referred to as dead. As in the thermal classification, problems can arise in applying the dynamic classification to whole glaciers, as different parts of the same glacier may be in different dynamic states.

Morphological classification is based on the size and shape of glaciers, and their relationship to the topography over which, or through which, they flow. A basic distinction can be drawn between glaciers which are constrained by topography (such as cirque glaciers, valley glaciers and small ice fields), and those which are not (such as ice caps and ice sheets). Within these classes, glaciers are differentiated by shape, size and setting. For example, a distinction is sometimes drawn on the basis of scale between ice caps (less than 50,000 km^2) and ice sheets (more than 50,000 km^2) (Armstrong *et al.*, 1973). Classifications of this type have been summarised by Embleton and King (1968) and Sugden and John (1976). As with most classifications it is possible to find examples of glaciers which do not fit comfortably into this morphological scheme. A particular problem is that many glaciers which are commonly

Figure 1.1 The summit ice cap of the volcano Cotopaxi, Ecuador.

referred to as ice caps are in fact not unconstrained by topography. For example, mountain ice caps such as those which picturesquely adorn the peaks of many mountains are unconstrained only in that they entirely envelop their surroundings (Figure 1.1). The pattern of ice flow within these mountain ice caps, and their surface profiles, are controlled not purely by the rheology and internal dynamics of the ice mass, as would be true of a wholly unconstrained glacier, but primarily by the form of the underlying topography. This type of glacier might more properly be referred to as an ice carapace. A more valuable morphological classification could be derived by distinguishing between ice sheets, which derive their characteristics largely from ice flow properties and internal dynamics, and ice caps, which are largely conditioned in their flow regime and surface morphology by the underlying topography. Because terminology has grown up over many years and classifications have been produced in different contexts, a wide range of glacier types are now referred to in the literature. Some of the most frequently used terms are identified and defined below.

Ice sheets are large ice masses (sometimes defined as larger than 50,000 km^2) that submerge underlying topography and develop surface profiles and flow patterns that are largely controlled by ice flow properties rather than by topographic constraints. The Greenland and Antarctic ice sheets are the only ones that exist at present, but others have existed during former glacial periods, including the Laurentide ice sheet that covered the northern part of North America and was larger than the present-day Antarctic ice sheet. The two sections of the Antarctic ice sheet, East and West, are sometimes referred to as separate ice sheets although they coalesce to form a single ice mass. Marine ice-sheets are based on bedrock that lies predominantly below contemporary sea-level. They drain mainly by ice streams that feed into the ocean or into attached ice shelves. The West Antarctic ice sheet is the only marine ice sheet existing at present. Ground-based ice sheets rest on bedrock that is above contemporary sea-level. Ice sheets may incorporate ice domes and various types of outlet glacier. Ice domes form the central parts of ice sheets, where snow accumulates and from which ice flows outwards. The Greenland ice sheet, for example, comprises a large northern dome and a smaller southern dome. Outlet glaciers are extensions from the periphery of the sheet where the thinning ice becomes constrained by local topography, and can take any of the wide range of forms adopted by valley glaciers (described below). The terms glacier lobe and ice-sheet lobe are applied to outlet glaciers that are as wide as they are long, and to any identifiable prominence or bulge of an ice sheet or glacier margin. Lobes tend to extend into low-lying areas while higher ground forms embayments or re-entrants between the lobes. Many terrestrial ice-sheet margins comprise long sequences of lobes and embayments interspersed with outlet glaciers. Ice streams are linear zones of faster-flowing ice within an ice sheet, often corresponding to overdeepened troughs in the subglacial topography and draining a large proportion of the ice sheet's total discharge (Chapter 7).

The term ice cap is applied to several related types of glacier. The term is sometimes applied to ice masses that are equivalent to ice sheets but somewhat smaller (sometimes defined as smaller than 50,000 km^2). For example, Sugden and John (1976) argued that an ice mass engulfing Wales or Svalbard would be an ice cap rather than an ice sheet simply by virtue of its size. Many ice caps that exist at present occupy upland areas, and do not extend far beyond the periphery of their highland source. Examples include Vatnajökull, in southern Iceland, which is Europe's largest existing ice cap at about 8400 km^2. Ice caps such as this are fundamentally different from ice sheets in that they do depend on local topography for their existence, and their extent and form are constrained by the uplands on which they rest. Like ice sheets, ice caps may incorporate ice domes and outlet glaciers. Mountain ice caps form on individual mountain summits and extend some way down the mountain side, often terminating in series of outlet glaciers. They are sometimes referred to as ice carapaces or summit glaciers and typically comprise steep, thin ice. An ice field occurs where ice accumulates in an upland area between mountain peaks but does not reach sufficient thickness to overwhelm the topography. Ice fields have generally flat or undulating surfaces as opposed to the domed surfaces of ice caps, but ice fields and ice caps probably represent stages in a continuum of forms. Typically, ice fields drain through cols in the surrounding mountains, supplying ice to outlet glaciers. The term transectional glacier is sometimes applied to glaciers that occupy an upland area and supply outlets in several directions but are not sufficiently large to cover the mountains or form an ice cap. Transectional glaciers are responsible for forming transfluent breaches or cols in upland areas. The definitions of transectional glaciers and ice fields overlap somewhat.

Niche glaciers or cliff glaciers are very small glaciers lying in shallow hollows on mountain sides (Figure 1.2). They sometimes develop on very steep slopes, and are sometimes fed by avalanches and ice falls from higher up the slope. Very small glaciers of this type are sometimes referred to as glacierettes. Cirque glaciers, also sometimes referred to as corrie glaciers or cwm glaciers, are small glaciers that occupy topographic hollows on mountainsides and enlarge them into deep basins. If the cirque glacier grows beyond the size of its hollow, ice can flow out to form a larger valley glacier.

Valley glaciers are glaciers confined between valley walls and terminating in a narrow tongue. Alpine valley glaciers are supplied by ice flowing out of mountain cirques into valleys below. Outlet valley glaciers are supplied by ice from an ice sheet, ice cap or ice field when ice extends down a valley away from the edge of the parent ice mass. Valley glaciers tend to occupy pre-existing valleys, and thus tend to meet at valley confluences. Where a glacier is fed by snow or ice from avalanches, and the main part of the glacier is disjointed from the accumulation area, it is known as a rejuvenated, reactivated, or reconstituted glacier. When several valley

Figure 1.2 The smallest scale of glaciation: niche glaciers and ice aprons in the Grand Tetons, Wyoming, USA.

glaciers coalesce as tributaries to a main trunk, a compound valley glacier is formed, often marked by medial moraines indicating the boundaries between individual units. When a small tributary glacier joins the main trunk from a hanging valley, and the ice from the tributary remains suspended in the upper portion of the trunk glacier, the resulting unit within the compound glacier is referred to as an inset glacier. When a valley glacier extends beyond the end of the confining valley walls into an area of lower relief, the ice can splay outwards in the shape of a fan. Glaciers terminating in these fan-like lobes are called piedmont glaciers. Valley glaciers that terminate in the sea, such as those that flow into fjords, are sometimes referred to as tidewater glaciers. The floating terminus of Jakobshavn Isbrae, west Greenland, rises and falls about 3 m with the tide (Lingle *et al.*, 1981).

Ice shelves are formed by the seaward extension of glaciers to form floating tongues, which coalesce to form floating ice masses of considerable horizontal extent attached to the coast. The large Ross and Ronne ice shelves that extend from the Antarctic ice sheet cover areas the size of Spain and are up to 2000 m thick. The point at which the ice becomes afloat is referred to as the grounding line. Where ice shelves become locally grounded on the sea-bed, flow of the shelf is disturbed by what is known as an ice rise if the ice is diverted around the obstruction or ice rumples if flow continues over the obstruction. Ice rises play a major role in regulating ice discharge from the Antarctic ice sheet through to the ice shelves (MacAyeal *et al.*, 1987) and the shelves act as a buttress against advance of the glaciers that feed them. Ice shelves of a different type can be formed by the growth of thick sea ice which remains perennially landfast. These are referred to as sea-ice ice shelves, as distinct from glacial ice shelves. Composite ice shelves include elements of both types. Once established, ice shelves can grow by accumulation of snow *in situ* at the surface, as well as by input of ice from land-based feeder glaciers or the freezing-on of seawater from below. Ice shelves are in hydrostatic equilibrium with the water in which they float, so their melting would not contribute directly to a sea-level rise. Ice shelves ablate by a combination of calving and bottom-melting, and they commonly terminate in cliffs produced by the calving of icebergs.

Icebergs are not a type of glacier, but fragments of a glacier that have broken off or 'calved' into water (Figure 1.3). They are composed of freshwater glacier ice, not sea ice, and they float in sea water with about 90% of their bulk submerged. In freshwater, which has a lower density than saltwater, they float with slightly more of their bulk submerged. Icebergs are classified according to shape and size. Tabular icebergs are formed by the calving of large sections of a floating glacier tongue or ice shelf, so that the surface of the berg is the former glacier surface. The largest bergs are those which originate from fragments of major ice shelves. For example, in September 1986 a 13,000 km^2 section of the Filchner ice shelf, equivalent to the area of Northern Ireland, broke away to form three tabular icebergs. The separation occurred along fracture lines that had been recognised since 1956, and major calving events such as this seem to occur periodically on a time scale of several decades. Fragments from the largest of the 1986 Filchner bergs, designated A24, drifted as far north as 36°S (Vaughan, 1993a). In 1995 and 1996 the break-up of the Larsen ice shelf in the Antarctic Peninsula produced several new large bergs, one of which measured 78 × 37 km and was 200 m thick. Large tabular icebergs that calve into the Arctic Ocean are sometimes referred to as ice islands.

Rock glaciers are features of periglacial environments, and there is some dispute as to whether or not they are a type of glacier according to our earlier definition. Some authors have argued that there is no relationship between glaciers and rock glaciers, and that rock glaciers are morphological expressions of

Figure 1.3 A small iceberg in Disko Bay, near Jakobshavn, Greenland.

creeping mountain permafrost: slope talus features incorporating substantial amounts of interstitial ice (e.g. Haeberli, 1985, 1989; Barsch, 1988). Others have regarded rock glaciers as very debris-rich glaciers (e.g. Whalley, 1979; Johnson and Lacasse, 1988). Elconin and LaChapelle (1997) provided a detailed description of the morphology, structure and flow of Fireweed rock glacier in the Wrangell Mountains of Alaska and found that both glacial and periglacial processes were involved in the feature's accumulation. Lliboutry (1990, p. 125) argued that: 'the geographical study of rock glaciers as an extreme case of glacier fluctuations, as an indicator of favourable mass balances in the past, or of past surges, would be much more rewarding than to consider them as a mere case of standard permafrost, or of creeping regolith.' Giardino *et al.* (1987) brought together a number of papers that provide further information on these issues.

Many of the different types of glacier described above can be considered as elements in a continuum of forms. For example, niche glaciers can develop into corrie glaciers which overspill to form valley glaciers, which in turn might feed piedmont lobes or ice shelves. Ice fields might grow into ice caps, which in turn might grow into ice sheets. Transitions between different types of glacier can thus be recognised both over distance and through time. Kuhn (1995) discussed the continuous size spectrum of glaciers and identified fundamental morphological differences between small and large glaciers. He pointed out that smaller glaciers have larger thickness-to-length ratios (Table 1.1), and that

Table 1.1 Typical thickness to length ratio of different glacier types (after Kuhn, 1995)

Size	*Thickness*	*Length*	*Thickness/length*
Ice sheets	1 km	1000 km	0.001
Valley glaciers	100 m	10 km	0.01
Cirque glaciers	10 m	100 m	0.1

Table 1.2 Accumulation and ice-thickness values for different sizes of glacier (after Kuhn, 1995)

Size	*Accumulation* *($m\ a^{-1}$)*	*Thickness* *(m)*	*Accumulation/thickness* *(a^{-1})*
Antarctic ice sheet	0.1	1000	10^{-4}
Valley glacier	1	100	10^{-2}
Avalanche cone	10	10	10^{-0}

the ratio of glacier thickness to annual accumulation, which defines a characteristic time scale indicative of the time it would take a glacier to disappear or reform in changing climate conditions, is much larger for large glaciers than for small ones (Table 1.2)

1.3 GEOGRAPHY OF GLACIERS

'Geographies,' said the geographer, 'are the books which, of all books, are most concerned with matters of consequence.'

Antoine de Saint-Exupery: *The Little Prince* (1945)

The spatial organisation of natural phenomena is important. The location of a glacier exerts a strong control on its character, and the distribution of glaciers reveals a lot about the factors that control glaciation. Changes through time in the distribution of glaciers reflect changing environmental conditions, and the record of glacier fluctuations can tell us about the dynamics both of the environment and of glaciers. At a smaller scale, glacier morphology and the distribution of features in a glacier reflect a distribution of processes and environments within the glacier. Glaciological information is embodied in the spatial expressions of glacial phenomena. Spatial variations provide clues, and a geographical approach to glaciology is defined by its use of these clues.

The distribution of glaciers is determined by the environmental controls on their initial formation and subsequent survival, which in turn follow from the relationship between glaciers, the hydrological cycle and climate. The main control is the precipitation, survival and accumulation of snow; so glaciation is controlled principally by climate. Regional climate is part of global climate and largely related to latitude. The elevation at which snow is likely to accumulate at specific latitudes is defined by the regional snowline. Local climate is determined by a range of additional factors, including altitude, aspect, topography and continentality. At a local scale, topographic suitability for the accumulation of snow plays an important role in the location of glaciers. Steep slopes, sharp pinnacles and windward or equator-ward aspects do not favour accumulation of snow. Once established, glaciers can alter local topography and climates. In the case of large ice sheets the increase in surface elevation caused by the presence of the ice sheet tends to increase the proportion of precipitation falling as snow and to decrease the surface temperature, both of which serve to reinforce the development of the glacier. Oerlemans and van der Veen (1984) went so far as to suggest that the Greenland ice sheet survives at present only because of the modifications it has made to its environment, and that if the ice sheet were removed, conditions at ground level would not then support its regrowth. Other investigators (e.g. Letréguilly *et al.*, 1991) have contested this opinion, but there is no doubt that once an ice sheet is established, local conditions for its survival can persist even if the regional climate returns to its pre-glacial state.

At present, glaciers cover about $15.8 \times 10^6\ km^2$, or about one-tenth of the Earth's land surface. The distribution of this cover is shown in Table 1.3 and Figure 1.4. The vast majority of the world's glacier ice (~91% of the volume and 84% of the area) is in the Antarctic ice sheet.

Table 1.3 Global distribution of glaciated areas

Continent	*Region*	*Area (km²)*	
Antarctica	Subantarctic islands	7,000	
	Antarctic continent	13,586,310	
	Total		13,593,310
North America	Greenland	1,726,400	
	Canada	200,806	
	USA	75,283	
	Mexico	11	
	Total		2,002,500
Asia	Russia and former Soviet states	77,223	
	Turkey, Iran and Afghanistan	4,000	
	Pakistan and India	40,000	
	Nepal and Bhutan	7,500	
	China	56,481	
	Indonesia	7	
	Total		185,211
Africa	Africa	10	
	Total		10
Europe	Iceland	11,260	
	Svalbard	36,612	
	Scandinavia	3,174	
	Alps	2,909	
	Pyrenees	12	
	Total		53,967
South America	Argentina and Chile	23,328	
	Peru and Ecuador	1,900	
	Bolivia, Colombia, and Venezuela	680	
	Total		25,908
Australasia	New Zealand	860	
	Total		860
Total			**15,861,766**

Most of the rest is in the Greenland ice sheet (8% of the volume, 12% of the area). All the world's other glaciers combined account for only about 4% of the global ice area and less than 1% of the global ice volume. However, it is interesting to note the range of locations in which glaciers occur. The distribution ranges from pole to equator, from sea-level to the highest mountain ranges, and from continental interiors to small oceanic islands. Australia is the only continental mainland not to have glaciers at present, and there is evidence of former glaciation even there. At different times in the past, as climates and the positions of continents have changed, most parts of the Earth's surface have been subject to glaciation.

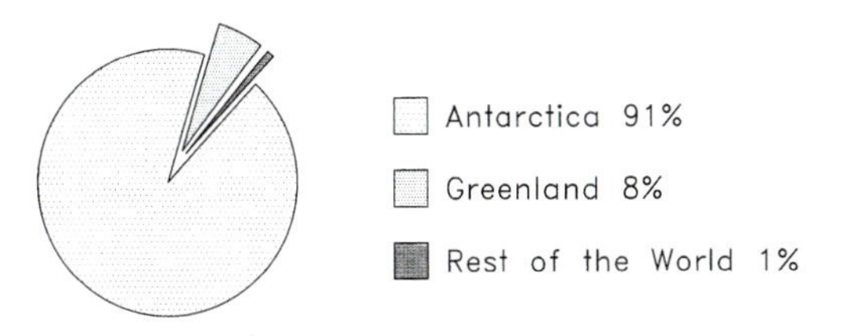

Figure 1.4 Distribution of glaciated areas between continents.

1.4 CONCLUSION

Glaciers come in a huge range of shapes and sizes. Different glaciers, and even different parts of the same glacier, can have a variety of different thermal, hydrological and dynamic characteristics. Glaciers occur in locations ranging from the poles to the equator, and most parts of the world have experienced the direct effects of glaciation at some time in the past. Glaciers currently occupy less of the planet than they have done in geological history, but nevertheless exert a profound influence on the global environment. Our developing understanding of glaciers will play an important role in our understanding of the global environmental system.

GLACIERS AND THE GLOBAL SYSTEM 2

2.1 GLACIERS AS PART OF THE GLOBAL ENVIRONMENT

When I explain to my final-year students the various skips and loops in their lecture programme, I like to say that reality is a bundle, not a list. The world is not organised as a linear narrative. No single element of the physical environment can be analysed comprehensively without reference to the system as a whole, and the physical environment as a whole cannot be understood without reference to its component parts. Therefore, although this book is about glaciers and not about the whole global environment, we need to begin by seeing glaciers as part of a broader environmental system. This chapter outlines some of the principal relationships between glaciers and other components of the environment. The aim is to provide a relatively simple environmental context for what follows in the rest of the book, and the references in the text should provide readers with a route into more specialised literature on the topics that are discussed.

Glaciers affect, and are affected by, many elements of the global system, and play a central role in the operation of that system. Evidence from both present-day and former environments demonstrate the interaction of glaciation with climatic events at a range of time scales; with geological phenomena ranging from geomorphology and sedimentology to tectonics and volcanism; with the ocean/atmosphere circulation and the hydrological cycle; and with biotic activity. The periodic shift between glacial and non-glacial conditions in global history is one of the key controls on the planet's environment.

As only about 0.1% of the world's population live in glaciated regions, there is a tendency for most of us to view glaciers and glaciation as something exotic or exceptional. We tend to see glaciers as external to our 'normal' conception of landscape; as isolated phenomena periodically imposed as an external force onto the landscape of an area. Those of us who live in presently non-glaciated areas think of glaciers as long ago or far away. However, glaciers play an integral role as parts of our landscape in a variety of different ways, and at a range of scales. This applies not only to locations where glaciers presently exist, but also to areas where they have existed in the past, and to areas well beyond the limits of contemporary glaciation. Glaciers affect both glaciated and non-glaciated areas at a global scale through their interaction with climate, sea-level, the hydrological cycle and the ocean/atmosphere circulation. We can also recognise the influence of glaciers that existed in the past on presently non-glaciated areas. Because glaciers in the past have been more extensive that at present, many modern glacier-free landscapes preserve a record of former glaciation in landforms, sediments and drainage patterns. Continuing postglacial isostatic rebound in many formerly glaciated areas, and the survival in the geological record of glacigenic sediments from the Precambrian period as much as about 2.5×10^9 years ago, testify to the historical legacy of glaciation. Glaciers that existed even in the distant past continue to have an impact on both landscape and human activity.

Existing glaciers have a direct effect on both local and more distant environments. At the immediately local scale glaciers obscure former land surfaces, creating a new ice surface with particular topographic and micro-environmental characteristics. Beneath the ice a unique assemblage of geological, geomorphic and hydraulic conditions comes into effect in the subglacial environment, altering the previous surface to varying degrees by subglacial processes. At the regional scale, even small glaciers install into the landscape a whole suite of climatic, geomorphological and hydrological conditions that control surface processes not only in the vicinity of the ice but also, through influences on sediment supply and fluvial regime, downstream of the ice throughout any drainage basin inhabited by glaciers. The effects of glaciers extend through hydrological, geochemical and sedimentological effects into the oceans and atmosphere at a global scale. The global distribution of loess, and the presence of iceberg-rafted glacial sediments in the deep oceans, both testify to the range of glacial processes beyond the limits of glaciation. Subglacial, ice-marginal, proglacial and paraglacial environments (Chapter 9) form a time-transgressive geographical system linking geomorphology, sedimentology and hydrology. It is a pervasive and persistent influence that superimposes sequences of effects, and preserves a record of environmental phenomena in the landscape. Whatever location or time period we consider, we cannot escape the glacial character of our landscapes.

Earth surface processes and environmental conditions are ultimately controlled by solar, geothermal and gravitational energy. Differences in solar heating of the Earth's surface control major components of the global environmental system by driving ocean/atmosphere dynamics and the hydrological cycle. Glaciers can be regarded as one component in this system, and in models of the global environment the cryosphere ranks alongside the ocean and atmosphere as one of the principal realms of environmental activity and influence. Barron *et al.* (1989) argued that the 'hydrological cycle should be of primary interest in studies of future global change', and it seems appropriate to begin our exploration of the environmental context of glaciers with that topic.

2.2 GLACIERS AND THE HYDROLOGICAL CYCLE

Glaciers are a part of the hydrological cycle. The nature of the hydrological cycle is strongly dependent of the extent of glaciation, and many of the characteristics of glaciers can be traced to, and best understood in terms of, their role as part of the global hydrological system. Glaciers represent a land-surface water store of fluctuating size, and a slow transport mechanism for throughput of water from the atmosphere, via the land, back to the ocean. The properties of glaciers are controlled by parameters within the hydrological system. For example, glaciers fluctuate in response to changing rates of water input and output, and they inherit physical and chemical characteristics from their atmospheric and oceanic water sources. In return, there is a direct feedback from glacier conditions to sea-level, ocean chemistry, groundwater systems and atmospheric composition. Because of this close linkage between atmosphere, oceans and glaciers, the characteristics of glaciers can reflect disruptions to the hydrological cycle and hence provide a record of environmental change.

There is about 1.45×10^9 km^3 of water in the whole global hydrological system (Lvovitch, 1970). Of that, about 97.4% is held in the oceans, and 2%, or about 2.8×10^7 km^3, exists as snow and ice (Hoinkes, 1967). At present, there is about 3.2×10^7 km^3 of freshwater in the global hydrological system, of which about 77% is in the form of snow and ice. During glacial periods in the past, ice volumes approximately trebled; ice sheets have sometimes contained more than 90% of the freshwater, and almost 6% of the total water store. The volumetric significance of glaciers in the hydrological cycle is therefore clear (Figure 2.1).

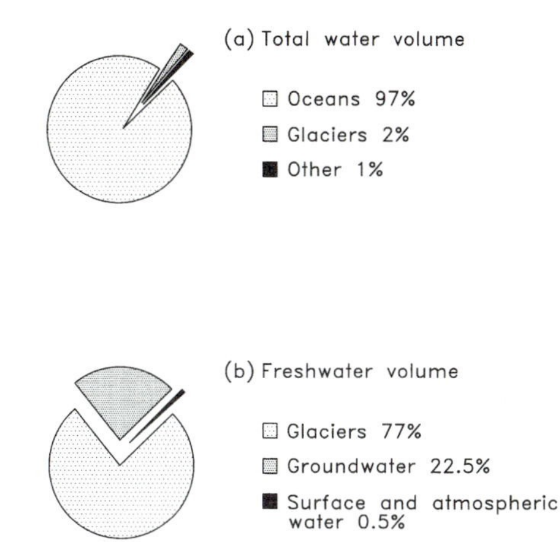

Figure 2.1 The volumetric significance of glaciers in the hydrological cycle. (a) Proportions of the world's total water volume in different locations. (b) Proportions of the world's fresh water in different locations. Data from Martinec (1976).

Water held in glaciers is transferred back to the rest of the hydrological system by ablation. At present about 3000 km^3 of water is released from glaciers annually (Kotliakov, 1970). During deglaciation at the end of the last ice age about 14,000 BP, meltwater pulses of up to 16,000 $km^3 a^{-1}$ occurred (Fairbanks, 1989; Bard *et al.*, 1996) (Figure 2.2). The water that is released from glaciers is transferred into surface streams, into groundwater, directly into the ocean, or by evaporation and sublimation into the atmosphere. One of the largest components of this return flow is by calving of icebergs from glacier termini; these bergs then melt into the oceans over periods of months or years (Hamley and Budd, 1986). The residence time or throughput rate of water in a glacier depends on the size of the glacier and the rate at which ice flows through the glacier. For large ice sheets, throughput times of hundreds of thousands of years can be inferred from the age of the oldest ice retrieved from the base of the Greenland and Antarctic ice sheets. For small, fast-moving glaciers, the residence time is much shorter; for example, ice takes only 150–200 years to travel the 6 km length of the Athabasca Glacier in Canada (Savage and Paterson, 1963). The average residence time for ice in glaciers has been calculated as about 10,000 years, which compares spectacularly with the average residence time of about 12 days for water in the world's rivers (Martinec, 1976). The length of time for which water taken into the glacial section of the hydrological cycle is withheld from the other parts of the cycle can be extended in several ways. Water can be stored as ice for periods much longer than the normal throughput time for glaciers when special conditions such as the burial of ice beneath sediment lead to the underground preservation of glacier ice after regional deglaciation, a condition referred to as retarded deglaciation (Chapter 8). Water released from glaciers is not necessarily transferred directly back to the ocean: meltwater can be stored temporarily in subglacial or proglacial lakes. Basal meltwater entering subglacial Lake Vostok beneath the Antarctic ice sheet can be held in the lake for tens of thousands of years, and the mean age of water in the lake, since it was deposited as

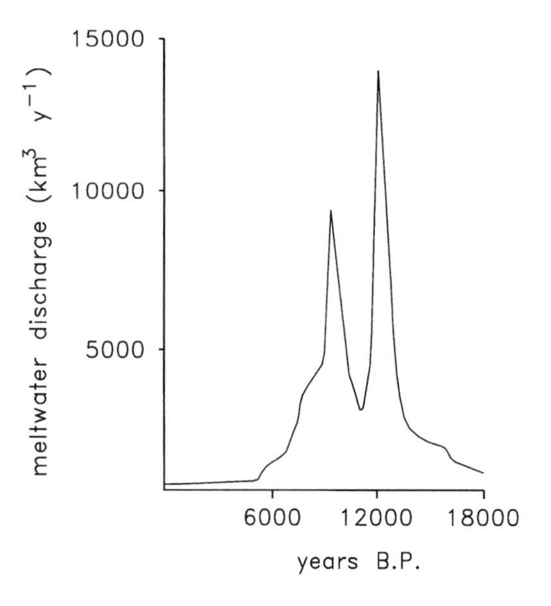

Figure 2.2 Record of meltwater discharge around 18,000 years before present, indicating pulses of high discharge during deglaciation. After Fairbanks (1989).

snow on the ice-sheet surface, is about one million years (Kapitsa *et al.*, 1996).

Major ice-marginal lakes that existed during the last deglaciation held huge volumes of water in temporary storage, and in some cases released the water in major floods. For example, glacial Lake Missoula on the southern edge of the retreating Laurentide ice sheet in North America repeatedly stored and released volumes of up to 2000 km^3 of water (Baker and Bunker, 1985). Glacial Lake Agassiz, which also bordered the retreating Laurentide ice sheet, covered an area of 350,000 km^2, and released floods with peak discharges of around 1 million $m^3\ s^{-1}$ (Teller and Clayton, 1983). Such massive and abrupt releases of water can lead to abrupt rises in sea-level. One flood that drained from the proglacial Lake Agassiz–Ojibway system released 75,000–150,000 km^3 of water, which is equivalent to a global sea-level rise of about 20–40 cm. Meltwater released at the end of the last ice age has been associated with a long-term sea-level rise of 40–50 mm a^{-1} (Bard *et al.*, 1996), but abrupt, metre-scale rises in sea-level occurred with the collapse of the Laurentide and Antarctic ice sheets and the release of huge volumes of subglacial and proglacial water (Blanchon and Shaw, 1995). Sudden meltwater pulses can also be supplied directly from glaciers without temporary proglacial storage. For example, jökulhlaups (Chapter 6), surges (Chapter 7) and Heinrich events (this chapter) can all be associated with sudden production and/or release of water from glaciers into the proglacial zone.

Many of the characteristics of glaciers are controlled by the hydrological cycle, and the hydrological cycle itself changes through time in response to broader environmental changes. For example, during glacial maxima, ocean volume is diminished and ocean chemistry changed. In response, atmospheric water also changes its characteristics and in turn influences the chemistry of snow accumulating on glacier surfaces. Glaciers thus inherit characteristics that record environmental conditions: the chemistry and stable isotope composition of glacier ice provides a record of atmospheric composition, ocean volume and a range of other environmental indicators (e.g. Robin, 1983). The hydrological cycle also controls mass balance and the energy that drives glacier movement. All of these are discussed in more detail in subsequent sections, but are important to note here as illustrations of the importance of the hydrological cycle. It is important also to recognise that glaciers exert control on the hydrological cycle. The presence of glaciers and ice sheets controls the routing of surface water and groundwater; and the effect of glaciers on climate, which will be discussed in the next section, exerts a strong influence on global hydrological conditions.

2.3 GLACIERS AND OCEANS

2.3.1 SEA LEVEL

If we suppose the region from the 35th parallel to the north pole to be invested with a coat of ice thick enough to reach the summits of the Jura . . . it is evident that the abstraction of such a quantity of water from the ocean would materially affect its depth.

Charles MacLaren, *The Scotsman* (13 January 1842)

From what we have already discussed of glaciers and the hydrological cycle, the direct link between glaciers and sea-level is clear. Glaciers store water and delay its return to the ocean, so when glaciers expand, oceans are depleted, and sea level falls. However, the relationship is complex. Sea level changes through time for many reasons, several of which are closely related to the behaviour of glaciers. The two major types of sea-level change are those related to changes in the volume of water in the ocean, and those related to changes in the relative elevation of sections of the earth's crust. Changes associated with changes in water volume have in the past been considered to occur simultaneously over large areas or even globally, and have been called 'eustatic' (Daly, 1934). However, the idea of global eustatic sea-level trends has largely been forsaken in view of regional differences in eustatic effects due to variations in crustal rheology and the shape of the geoid (e.g. Kidson, 1982). Changes associated with changes in crustal elevation relate primarily to isostatic adjustments of the crust under varying amounts of weight or loading. These isostatic variations are confined to the sections of crust affected by the changing load. In considering sea-level change associated with glaciation it is necessary to consider both volumetric change and isostatic change. The main concepts to consider are glacial eustasy, hydro-isostasy and glacial isostasy.

Eustatic response to ice-volume change occurs relatively quickly, as water is added to, or removed from, the ocean basins. The growth or shrinkage of major glaciers is accompanied roughly synchronously by a sea-level response. Fairbanks (1989) and Bard *et al.* (1996) identified specific periods of discernible sea-level rise associated with specific spikes in meltwater production. Variations in ice volume with glacier fluctuation can be translated directly into changes in volume of stored water and hence sea-level equivalents. For example, Table 2.1 shows volumes of former and present-day ice sheets, and the amount of water that their melting would release. If the Antarctic ice sheet were to melt, the water released would be equivalent to 66 m of water distributed across the world's oceans. However, the actual sea-level rise that would occur as a result of this volumetric input is modified by the processes of hydro-isostasy and glacial isostasy. The volume of water released into the ocean basins exerts a weight onto the crustal floor of the basin and depresses the basin floor into the asthenosphere below. Because of the relative densities of the materials involved, the amount of crustal depression is equivalent to about one-third of the depth of extra water, so a 66 m rise following melting of the Antarctic ice sheet would be transformed into a sea-level rise of about 44 m. However, this hydro-isostatic response to loading occurs more slowly than the immediate sea-level change driven by volumetric changes, so sea-level may initially rise upon deglaciation and then gradually fall as hydro-isostatic depression proceeds. At a local scale the picture is further complicated because different sections of an ocean basin can be affected by different depths of water at different times, and different geological areas might exhibit different rheological responses to loading.

Isostatic adjustment also affects continental areas experiencing glaciation. Glaciated continents respond to the weight of glaciers in the same way that ocean basins respond to weight of water. During glaciation, sea levels are low, but glaciated continents are depressed by the weight of glaciers so relative sea level is not as low as that experienced on non-glaciated continents. Upon deglaciation removal of the weight of ice leads to isostatic rebound or recovery of the land. As the isostatic recovery

Table 2.1 The approximate volumes of the ice sheets during the glacial maximum (figures in brackets) and at present, and the equivalent sea-level changes represented by the storage of water in those ice sheets (Based on figures in Flint, 1971)

Ice sheet	*Area (km^2)*	*Volume (10^6 km^3)*	*Sea level equivalent (m)*
Antarctic	12.5 (13.8)	23.5 (26)	59 (66)
Greenland	1.7 (2.3)	2.6 (3.5)	6 (11)
Laurentide	0 (13.4)	0 (29.5)	0 (74)
Cordilleran	0 (2.4)	0 (3.6)	0 (9)
Scandinavian	0 (6.7)	0 (13.3)	0 (34)
Other	0.6 (5.2)	0.2 (1.14)	0.5 (3)
Total	14.9 (43.7)	26.2 (77)	65 (197)

occurs relatively slowly, an initial postglacial sea-level rise resulting from water influx to the oceans is generally followed in glaciated areas by a long-term sea-level fall associated with isostatic recovery. Andrews (1968, 1970) described the long-term recovery of the Hudson Bay area following the disintegration of the Laurentide ice sheet. The area still has 150 m left to rise before isostatic recovery is complete. The sea-level effect of glaciation is yet further complicated by an effect referred to as peripheral bulge. Areas at the periphery of a glaciated area might experience a certain amount of crustal up-warping in response to downwarping elsewhere beneath thicker parts of the ice mass. These areas would thus be uplifted while the ice sheet is present, and lowered upon deglaciation. If these areas were adjacent to the ocean, the bulge would lead to a localised sea-level effect different from that expected from isostatic depression.

Several other glacier-related effects also influence sea level. For example: changes in water temperature can lead to thermal expansion and contraction between glacial and non-glacial periods, resulting in sea-level changes of several metres; the presence of ice sheets on the continents can exert a gravitational effect on the ocean water, causing an apparent sea-level rise at the coasts of glaciated land masses; and the same gravitational effect can affect the distribution of sub-crustal material, altering the shape of the geoid. The overall picture of sea-level response to ice-volume change is thus very complex. Direct addition or removal of water results in a widespread, short-term change in water depth, but this is complicated by regionally variable glacio-isostatic hydro-isostatic, and geoidal responses. Peltier (1987) provided a summary of how different glacial histories could lead to different sea-level effects, and identified areas of the globe where sea-level histories of different types apply.

Not only does glaciation affect sea level, changes in sea level can also affect glaciers. For example, water temperature, circulation patterns and sea level affect the stability of tidewater glaciers and ice shelves. Ice shelves have a buttressing effect that regulates the flow of ice from ice sheets, and weakening of an ice shelf could propagate upstream into the ice sheet so that large areas of land-based ice could become unstable. The West Antarctic ice sheet, grounded below sea-level, could collapse if the stability of its flanking ice shelves were compromised by sea-level rise or by enhanced under-shelf melting (Alley and Whillans, 1991).

2.3.2 OCEAN COMPOSITION AND CIRCULATION

Glaciation affects not only ocean volume and sea level, but also ocean water composition, temperature and circulation. The deepwater circulation of the world's oceans is driven primarily by differences in density between bodies of water, resulting from differences in water temperature and salinity. The salinity and temperature of water in the oceans is closely related to the growth and decay of ice sheets. The growth of ice sheets is accompanied by an overall increase in the salinity of the oceans, while the melting of ice sheets is accompanied by influxes of cold, fresh water to the oceans. These changes are thought to lead to major changes in ocean circulation. Evidence for the effects of glacier fluctuations on ocean water has recently been provided by the discovery of distinctive layers in ocean-floor sediments in the north Atlantic (Heinrich, 1988). These 'Heinrich layers' are rich in ice-rafted debris and poor in foraminifera, and they record decreases in sea-surface temperature and salinity caused by discharges of icebergs from eastern Canada associated with repeated rapid ice-sheet advances (Bond *et al.*, 1992). Paillard (1995) suggested that massive freshwater inputs from the Heinrich events could temporarily stop the normal thermohaline ocean circulation. Seidov *et al.* (1996) modelled the effect of meltwater events on ocean circulation in the North Atlantic and found that the release of meltwater into the ocean caused major decreases in the overturning strength of the North Atlantic salinity conveyor belt and in northward heat transport in ocean water. North Atlantic Deep Water production almost ceased, and the surface circulation in part of the Norwegian and Greenland Seas was reversed, implying that the North Atlantic Current no longer reached the Norwegian Sea. Broeker and Denton (1990) speculated that rapid changes in ocean circulation could be the cause of abrupt climate changes, and Adkins *et al.* (1997) showed that the last interglacial began and ended with abrupt changes in ocean deep-water flow, with transitions occurring in less than 400 years. Ocean circulation changes clearly have important feedback effects on climate, which are discussed in the next section.

2.4 GLACIERS AND CLIMATE

There is a complex two-way relationship between glaciers and climate. By controlling elements of glacier mass balance and temperature, climate exerts control over a range of glacier characteristics including size, thermal and hydrological regime, movement, and geomorphic activity. Glaciers exert control over climate by affecting albedo, the surface energy balance, and the composition and circulation of the atmosphere and oceans. These relationships are closely linked to the hydrological cycle and to the general circulation of the oceans and the atmosphere, and can be identified at a range of temporal and spatial scales.

There is a direct and seemingly obvious climatic control on glaciation. The formation, development and survival of glaciers requires the accumulation of snow, which in turn depends on temperature and precipitation. Glaciers develop and expand in response to increased snowfall or decreased temperatures. Climatic control of glaciation can be recognised at a long-term global scale in the oscillation between glacial and interglacial periods; at the regional scale in the advance and retreat of glaciers over periods of decades and centuries; and at the short-term local scale in seasonal ice-margin fluctuations (Chapter 8). However, this relationship is not entirely simple. A glacier's response to climate is based on a wide range of climatic variables, including temperature, precipitation, cloud cover, wind speed and humidity. Both the seasonal distribution and the absolute values of these variables are important controls on glacier behaviour, and the rate at which glaciers respond to changes in climate can be complicated further by glacier geometry and dynamics; different glaciers can react in different ways. The response of glaciers to climate change thus involves several elements and can be difficult to predict. For example, a rise in global temperature which might lead to increased ablation at ice-sheet margins could also lead to increased snowfall in ice-sheet interiors. Data from ice cores in central Greenland show that ice accumulation is relatively low in cold periods and high in warmer periods (Alley *et al.*, 1993). Global warming could, perhaps counter-intuitively, thus lead to ice-sheet expansion.

Glaciers are not entirely at the mercy of climate; they also exert some control over it. The growth and decay of large ice sheets disrupts the global energy budget and affects ocean and atmosphere dynamics both directly and indirectly. Harrison *et al.* (1992) modelled atmospheric circulation for full glacial conditions (about 18,000 BP) and showed that the Atlantic Westerly jet strengthened and shifted south, the Icelandic low shifted south, and the subtropical anticyclone strengthened and shifted north. The European ice sheet engendered a glacial anticyclone over Europe. Repenning (1990) argued that glacial erosion of the Chugach and St Elias Mountains in northwestern North America between 2.5 and 2.0 million years ago allowed relatively dry Pacific westerlies to extend across Canada, reducing the encroachment of moist Atlantic and Arctic air, and contributing to ending the continental glaciation to the east. The impact of glaciers on climate has for several decades been central to theories of the initiation of ice ages based on catastrophic feedback effects. For example, Wilson (1964) hypothesised that periodic expansion of the Antarctic ice sheet could lead to the growth of ice shelves and a consequent increase in the albedo of 25×10^6 km^2 of ocean from 8% to 80%, decreasing heat input to the earth by 4% and reinforcing ice-sheet growth. Potential negative feedbacks can also be identified. For example, ocean warming leading to ice-shelf bottom melting will produce cold meltwater that will reduce ocean temperature. Recent theories, discussed below, have focused on the possibility that variations in production of glacier meltwater can influence climate by changing the pattern of ocean circulation.

The relationship between glaciers and climate is non-linear, with some changes occurring very abruptly. This might be due to non-linearity in the climate system itself or to non-linearity in glacier dynamics. Abrupt changes in climate have been revealed by data from the GRIP and GISP2 ice cores in central Greenland, which have shown that climate can shift dramatically over periods of as little as a few years (Dansgaard *et al.*, 1993; GRIP members, 1993). Glacial and interglacial periods prior to the Holocene have been characterised by intense climate instability, with periodic abrupt switching between warmer and colder conditions. These rapid climate shifts, sometimes referred to as 'Dansgaard–Oeschger events', have been interpreted as a 'flickering' between two alternative stable states, occurring at intervals of 1000–3000 years. and involving extremely rapid reorganisations in global circulation (Taylor *et al.*, 1993b). Alley *et al.* (1993) argued that the extreme rapidity with which accumulation rates in Greenland increased in association with warming events at the end of the last ice age (doubling in as little as 1–3 years) implies that the changes recorded in the ice cores occurred in response to some threshold or trigger in the North Atlantic climate system.

Recent models of climate change have also recognised instability and non-linearity in glacier behaviour. A central part of these new models is a periodic fluctuation of ice sheets associated with massive discharges of icebergs and water. Heinrich (1988) reported a series of layers in north Atlantic sediments that are rich in ice-rafted debris and poor in foraminifera. The layers record decreases in sea-surface temperature and salinity caused by short-lived but massive discharges of icebergs as a result of repeated rapid advances of ice sheets in eastern North America. Each of these Heinrich layers, deposited at intervals of 5000–10,000 years between 14,000 and 70,000 BP, marked the end of a period of gradual cooling and was followed by an abrupt climate warming (Bond *et al.*, 1992, 1993). However, the timing of the events is inconsistent with Milankovitch climate forcing, and modelling by Oerlemans (1993) suggested that the Heinrich events are unlikely to reflect a direct response of the Laurentide ice sheet to climate cooling. The periodic instabilities that lead to Heinrich events might be attributed to instabilities in ice sheet dynamics and to feedback between ice-sheets, oceans and atmosphere. Hughes (1992a) invoked periodicity of flow in ice streams to explain ice-sheet instability, and referred to a new paradigm of climate change based on unstable ice-sheet dynamics. Others (e.g. MacAyeal, 1993; Clark, 1994) have invoked instability of ice-sheet flow over deforming sediment. MacAyeal (1993) referred to 'binge–purge' oscillations of the Laurentide ice sheet on a periodically mobile substrate as the cause of the Heinrich events, envisaging a surge-like alternation between periods of accumulation over an immobile bed and periods of evacuation over a mobile bed, the alternation between the two states triggered perhaps by a critical ice thickness. Atmospheric dust concentrations may also play an important role. Atmospheric aerosol content was high during glacial episodes, and has been invoked as a cause of regional cooling (e.g. Harvey, 1988), but Overpeck *et al.* (1996) suggested that peaks in atmospheric dust concentrations found to occur immediately prior to glacial terminations could have caused regional warming of as much as 5°C, and that this dust-induced warming could have been involved in triggering the large climate shifts associated with Heinrich and Dansgaard–Oeschger events. Broeker (1994) provided a convenient review of evidence connecting Heinrich events and climate change.

The impact of the Heinrich events on ocean temperature and salinity illustrates the potential impact of glacier instability on climate. Paillard (1995) suggested that massive freshwater inputs from the Heinrich events temporarily stopped the thermohaline circulation, enhancing the temperature contrast between high and low latitudes. Immediately after the events, the circulation restarted and heat transfer from low latitudes resumed, engendering abrupt warming in high latitudes. This warming in turn affected the ice sheet. Paillard suggested that the flickering Dansgaard–Oeschger events revealed in the ice-core record could be secondary oscillations in climate in response to the Heinrich events, driven by heat transfer via ocean circulation. This theory underlines the potential importance of the coupling between ice-sheet dynamics and oceanic thermohaline circulation in the complex structure of climate history. However, the existence of rapid ocean-surface changes independent of glacier changes on land has suggested the need to look for causes other than ice-sheet instability. At the time of writing this book, our understanding of the ice/ocean/climate system is far from complete.

Because they reflect aspects of the global system, glaciers can be used as tool to reconstruct conditions from former periods or to predict future changes. Ice surviving from earlier periods provides a record of accumulation rates, temperatures and the composition and circulation of ocean and atmosphere. Geological evidence provides a record of former ice extent. If the climate–glacier relationship is clearly understood, prediction of future glacier conditions in changing climates will be possible (US Department of Energy, 1985). For example, Loutre (1995) modelled the collapse of the Greenland ice sheet in response to increased levels of atmospheric CO_2 over the next 5000 years. However, modelling remains inconclusive because of poor understanding of ice and climate dynamics. A major problem for modellers is the complexity of interactions between elements of the physical system in which glaciers are involved.

The role of CO_2 in the interaction between atmospheric and ground-surface processes is a good example of this complexity. There seems to be an important relationship between glaciation, weathering processes and atmospheric composition. Climate records such as that provided by the 140,000-year-old Vostok ice core indicate a direct link between atmospheric temperature and CO_2 concentration. On long time scales atmospheric CO_2 levels are controlled by the balance between the rate of volcanic input from the Earth's interior and the rate of output through chemical weathering at the Earth's surface (Raymo and Ruddiman, 1992). Variations in atmospheric CO_2 might therefore be driven by changes in Earth surface weathering processes. Kump and Alley (1994) argued that continental ice sheets reduce chemical weathering significantly, partly because large areas of ice sheets are frozen to their beds, inhibiting chemical processes. However, certain glaciological conditions, especially during deglaciation, can lead to high chemical weathering rates, and Gibbs and Kump (1996) reported observations from a range of sources that suggest that modern glacial environments have chemical weathering rates of up to 4 times the global mean. Proglacial environments around the margins of warm-based and melting glaciers present a potentially very active chemical weathering environment: abundant fresh, comminuted rock fragments; copious water; recently exposed fresh rock surfaces; and an atmospheric supply of CO_2 (Collins *et al.*, 1996; Gibbs and Kump, 1996; Sharp, 1996). It has been suggested that glacially driven chemical weathering could be a significant factor in carbon cycling and climate change on glacial/interglacial time scales (e.g. Sharp *et al.*, 1995). Tranter (1996) considered the role of glacial meltwater as a sink for atmospheric CO_2 during glacial/interglacial transition, and found that chemical weathering processes associated with glacial runoff at the time of maximum runoff during deglaciation about 10,000 years ago could have made a significant contribution to the recorded ~50 ppm decrease in atmospheric CO_2 at that time. The feedback loops between surface processes, atmospheric composition and climate remain incompletely understood, and pose something of a chicken-and-egg problem. Long-term climate changes might be tied to tectonic mountain-building cycles that periodically engender enhanced montane erosion and weathering, hence inducing drawdown of CO_2 from the atmosphere and consequently climate change (Raymo *et al.*, 1988; Molnar and England, 1990).

2.5 CONCLUSION

Glaciers are an important part of a linked global system involving global energy sources and sinks, the hydrological cycle, the atmospheric and oceanic circulation, climate, crustal rheology and sea level. The system is internally complex and poorly understood, rich in feedback loops and non-linearities. Glaciers both drive and are driven by elements of this system, and can by their characteristics and behaviour give us insight into the dynamics, history and possible future of the physical environment. Successful modelling of the global environment requires an understanding of the role of glaciers in the environmental system, of glacier dynamics and of the glacial response to environmental inputs. The remainder of this book explores the properties and characteristics of glaciers as far as they are known.

FORMATION AND MASS BALANCE OF GLACIERS 3

In order to qualify as a glacier, an ice mass must have (1) an area where snow or ice usually accumulates in excess of melting and (2) another area where the wastage of snow or ice usually exceeds the accumulation, and there must be (3) a slow transfer of mass from the first region to the second.

(Meier, 1964)

3.1 CONDITIONS OF GLACIER FORMATION AND SURVIVAL

Glaciers are created when snow accumulates on the ground from year to year without completely melting away each summer. As the snow gets thicker it is transformed by compression and recrystallisation into ice. The essential requirements for the formation of glaciers are the precipitation of snow, its accumulation on the surface, and its survival through the summer melt season until the next season's snowfall. At low temperatures only a small amount of snowfall is required to support glaciers, while glaciers in warmer environments require higher seasonal supplies of snowfall. The Antarctic ice sheet testifies to fact that in areas with very low ablation, survival of even small amounts of precipitated snow can support major glaciers. The controls

Table 3.1 Factors influencing accumulation of snow and the development of glaciers

Factors affecting glaciation	*Conditions encouraging glacier development*
Precipitation	High Dominantly winter
Temperature	Low mean Low summer temperature
Insolation	Low
Wind	Low
Humidity	High
Altitude	High
Latitude	High
Continentality	Maritime situation
Aspect	Poleward Lee-side
Gradient (relief)	Not very high
Accumulation area	Large

on glacier formation are thus essentially climatic, primarily related to the amount and seasonal distribution of the precipitation and melting of snow. These climatic controls are in turn influenced by a range of environmental variables at both local and regional scales. Table 3.1 lists the main factors that tend to encourage or discourage snow accumulation and glacier formation.

On a broad scale, regional climate related to altitude and latitude determines a regional snowline below which snow does not accumulate. This indicates altitudes above which glaciation is probable at different latitudes. This altitude is modified locally by other effects. For example, aspect is very important; the local snowline is usually lower on poleward-facing aspects because less solar radiation is received, and lower on windward slopes because of precipitation-shadow effects. Ohmura and Reeh (1991) found that the surface topography plays an important role in determining regional precipitation patterns and accumulation over the Greenland ice sheet. The airflow over most of Greenland is dominated by southerly and south-westerly winds, and the precipitation distribution is marked by a south-to-north decrease in precipitation, with a pronounced precipitation shadow in the north-east of Greenland and a belt of high precipitation midway up the western (windward) flank of the ice sheet.

The lowest boundary of climatic glaciation can be represented by the position of the so-called equilibrium line on glaciers, above which the perennial accumulation of snow and ice exceeds the loss. The climate prevailing at the equilibrium line altitude (ELA) can be thought of as just sufficient to maintain the existence of the glacier or to support the development of one (Ohmura *et al.*, 1992). Changes in climate can cause changes in the altitude of the equilibrium line. Oerlemans and Hoogendoorn (1989) calculated that a 1K temperature change would lead to an ELA change of 130 m for typical conditions in the Alps. Bradley and Serreze (1987) reported mass balance measurements from two small ice caps on Ellesmere Island which lay entirely below the local equilibrium line. Out of equilibrium with contemporary climate, these ice caps were predicted to disappear within 100–200 years.

Once glaciers have formed, their existence can modify the local climate in a variety of ways. The ice lowers local temperatures; surface albedo is increased; local and regional circulation and precipitation patterns can be changed. For example, the existence of the Greenland ice sheet raises the topographic surface by about 2000 m above the level of the preglacial land surface. This reduces surface temperature, establishes orographic precipitation and circulation, causes a higher proportion of the precipitation to fall as snow and reduces summer melting. The ice sheet thus establishes its own self-sustaining microclimate. Oerlemans and van der Veen (1984) suggested that the level of the ELA at about 1500 m in Greenland is such that if the ice sheet were to disappear it would not reform, as the land surface would be below the climatic glaciation level. By contrast, Letréguilly *et al.* (1991) modelled the characteristics of the ice sheet under different hypothetical climates, and suggested that the ice sheet would reform from bedrock even under present-day conditions. However, their model did not take account of all the influences on accumulation, such as changes in precipitation due to orographic effects. Because the precipitation and accumulation of snow are affected by so many variables, and because the existence of the ice sheet has such a profound effect on general circulation and climate, it is difficult to predict accurately what could happen if the ice sheet were to disappear. In Antarctica the ELA is much lower, and there is less doubt that the Antarctic ice sheet would reform under present conditions. For Greenland, it seems that there are two possible stable conditions under the present climate. One condition is to have no ice sheet; the other is to have a large ice sheet that develops a self-sustaining local climate. For the Antarctic, a large ice sheet is presently the only stable condition.

The growth of glaciers, and ultimately the development of large ice sheets, probably occurs differently in different environments. In mountainous environments, glaciation may begin with initial snow-patch and niche-glacier formation, then progress through stages of corrie glaciation, the development of ice fields in between peaks, the growth and coalescence of mountain ice caps into regional ice caps, and the growth of these regional caps into ice sheets. In this scenario early accumulation and glacier formation will occur where snow gathers in sheltered lee-side locations. As the ice mass grows and develops a domed form, precipitation shadowing will lead to higher accumulation on the windward side of the dome and extension of the glacier in that direction. This type of development is sometimes referred to as leeward initiation and windward growth. In lowland areas ice-sheet initiation might take a different form, with widespread snowfields thickening to form domes which coalesce into ice caps. Recent evidence from the ice cores drilled to the base of the Greenland ice sheet at Summit (Chapter 11) suggests that the ice now at the bottom of the ice sheet in that central area formed at the ground surface, and is a remnant of the growing stages or early build-up of the ice sheet (Souchez *et al.*, 1994).

3.2 MASS BALANCE

The ice-fjords of course prove that an enormous surplus of ice is produced in the interior, and that the vast sheet of ice would be liable to expand and spread over the coast regions and their inlets if the ice-fjords did not discharge the surplus.

(Rink, 1877)

The mass balance of a glacier describes the input (accumulation), throughput (transport) and output (ablation) of snow and ice. Specifically it defines the difference between the amount of accumulation and the amount of ablation experienced by a glacier and is hence a measure of the state of health of the glacier. For a glacier that is in a steady state and has a constant volume, the input from accumulation of snow and ice is equal to the output by ablation and the mass balance is zero. If input is greater than output then the mass balance is positive and the glacier is growing. If input is less than output then the mass balance is negative and the glacier is shrinking. Mass balance is an important concept in glaciology because it provides the important link between climatic changes and glacier variations, because glacier dynamics and hydrology are closely related to mass balance changes, and Because many aspects of the mass balance equation are of practical application.

There has been a long history of mass balance studies. Rink (1877) demonstrated a grasp of the principles of mass balance over a century ago when he calculated the area of the Greenland ice sheet by dividing the discharge of its outlet glaciers by an estimate of the mean accumulation. A proliferation of early schemes for describing and quantifying mass balance generated ambiguous and conflicting terminology, but deliberate efforts to produce widely applicable terminology defining key terms resulted in increasing uniformity (Meier, 1962; Anonymous, 1969; IHD, 1970; Mayo *et al.*, 1972). The main elements of the mass balance of glaciers are listed in Table 3.2, and illustrated in Figure 3.1.

Figure 3.1 shows some of the main mass balance terms as functions of time. Note that in American usage the word 'balance' is frequently replaced by the word 'budget'. The accumulation season is the period when the accumulation rate exceeds the ablation rate, and the ablation season is when the ablation rate exceeds the accumulation rate. The difference between the accumulation rate and the ablation rate is the net mass flux, which is positive in the accumulation season and negative in the ablation season. The balance year runs from the start of one accumulation

Table 3.2 Selected mass balance terms

Term (with alternatives)	*Symbol (if common)*	*Definition*
Ablation	a	Loss of ice/snow from the glacier
Ablation area	S_a	Where $b_s < 0$ at the end of the budget year
Ablation rate	a^r	Rate of mass loss
Ablation season		Period when net mass flux is negative ($b^r < 0$), ($a^r < c^r$)
Accumulation	c	Addition of ice/snow
Accumulation area	S_c	Where $b_s > 0$ at the end of the balance year (t_2)
Accumulation area ratio	AAR	Area of S_c divided by area of whole glacier
Accumulation rate	c^r	Rate of mass gain
Accumulation season		Period when net mass flux is positive ($b^r > 0$), ($a^r > c^r$)
Activity index (Energy of glacierization)	E	Function of the change in net balance with altitude across the ELA On Figure 3.2, $(db_0/dz)z'o$
Apparent ablation	a^*	Difference between max. and min. values of b_s: $(c^* - b)$
Apparent accumulation	c^*	Maximum value of b_s
Balance (budget) year	start: t_1; finish: t_2	Time interval between (approximately annual) minima in b_s
Cumulative mass flux	b_s	$\int b^r\, dt$, reaching annual maximum at time t_m
Equlibrium line		Line dividing S_a from S_c (end-of-balance-year snowline). $b = 0$
Equilibrium line altitude	ELA, z'	Altitude of line dividing S_a from S_c. $b = 0$
Firn edge		Limit between glacier ice and firn at end of balance year (t_2)
Firn line (firn limit)		Lower limit of winter snow at the end of the balance year (t_2)
Mass (budget) imbalance	b_i	Difference between net balance and steady state net balance: $b - b_0$
Mass balance (regime)		Relationship between accumulation and ablation of snow and ice
Net balance total	B	Net balance integrated over the area of the glacier: $\int_S b\,dS$, or $\int_{S_c} b\,dS + \int_{S_a} b\,dS$
Net balance	b	Net mass flux integrated through the balance year ($\int_{t1}^{t2} b^r\, dt$), i.e. the overall change in mass by the end of the budget year
Net balance altitude function	$b(z)$	Variation of net balance (b) with altitude (z)
Net mass flux	b^r	$c^r - a^r$
Slush limit (runoff line)		Highest point from which runoff occurs
Specific . . .	(lower case)	Measured at a point, not over an area; Term is usually implicit
Steady state	0	$B = 0$, (B_0)
Total balance rate	B^r	Net mass flux across the area of the glacier $\int_s b^r\, dS = dB/dt$
Total . . .	(upper case)	Integrated over the area of the glacier, not at a single point
Transient snowline		Line separating transient accumulation and ablation areas (Effectively a transient equilibrium line)
Transient . . .		Measured as an instant value, not annual, mean, or total value

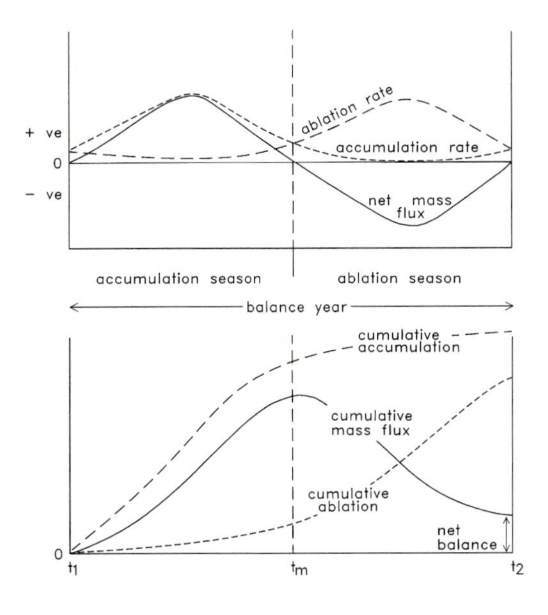

Figure 3.1 Graphical representation of mass balance components.

season to the start of the next, and the difference between the total annual ablation and the total annual accumulation defines the net balance, or the overall annual mass balance of the glacier. These terms can be applied to a glacier as a whole or to a specific location on a glacier. Where the specific net balance can be identified for different levels on a glacier surface, the mass balance can be defined with reference to altitude as a vertical net balance profile (VNBP), as shown in Figure 3.2. Kaser *et al.* (1996) showed how tropical mountain glaciers have distinctively different VNBPs from mid- and high-latitude glaciers: the lack of thermal seasonality induces a continuously positive sensible heat flux in the ablation area, and hence a much sharper VNBP in the ablation zones of tropical glaciers. The net balance is generally negative in the lower parts of glaciers (the ablation area), and positive in the upper parts (the accumulation area), although it may become less positive in the upper areas of large ice masses where precipitation is low and ablation by processes such as wind erosion is high. The boundary between the accumulation and ablation areas, where the specific net balance is zero, is defined as the equilibrium line.

The scheme described above is sometimes referred to as the 'stratigraphic system'. It has been formulated in the context of glaciers where accumulation occurs primarily in the winter season and ablation during the summer season. These are referred to as 'winter-accumulation' type glaciers. However, not all glaciers are of this type. For example, Ageta and Higushi (1984) and Ageta and Fujita (1996) described 'summer-accumulation' type glaciers from the Himalayas and the Tibetan Plateau in which the seasonal precipitation pattern is such that accumulation and ablation are both concentrated simultaneously in the summer. In this situation the stratigraphic system of mass balance analysis that relies on the clear differentiation of summer from winter is not entirely appropriate, and the so-called 'fixed date' system, in which the measurement

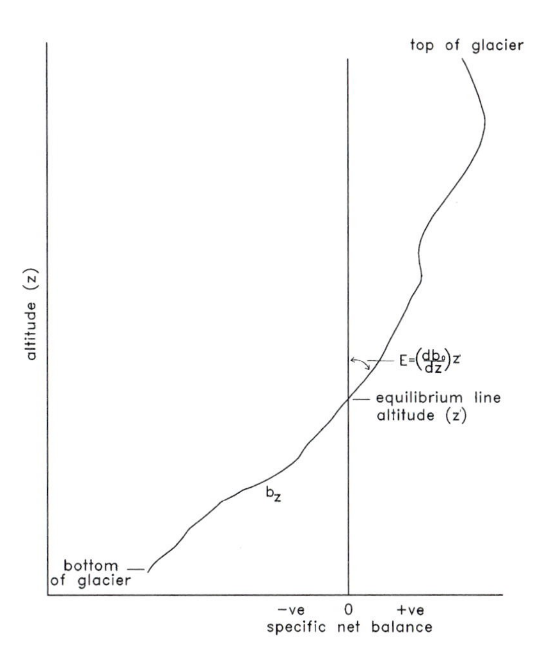

Figure 3.2 Components of a glacier's vertical net balance profile.

year is defined by fixed calendar dates as shown in Figure 3.3, is more useful.

In assessing the general state of a glacier, three parameters that are frequently used are the equilibrium line altitude (ELA), the accumulation area ratio (AAR) and the activity index. Year-to-year variation of the ELA is a good indicator of variation in mass, and of climatic conditions (Meier and Post, 1962; Østrem, 1975). Once a relationship between ELA and mass balance at a glacier is established, it can be used to estimate mass balance on the basis of ELA for periods when mass balance data is unavailable, and to extrapolate existing mass balance series into the past. It is often easier to measure ELA than to make direct mass balance measurements; the ELA can be assessed quickly from satellite or air-photo data. Braithwaite (1984) studied mass balance records from 31 glaciers in North America, Europe and Asia and found a high correlation between mean specific balance and the ELA. Kulkarni (1992) found correlation coefficients of –0.92 and –0.94 between ELA and mass balance data for two Himalayan glaciers. Ohmura *et al.* (1992) provided detailed discussions of climate at the equilibrium line and identified a relationship between climate, ELA and mass balance that depended on the geometry and mass turnover of individual glaciers but nevertheless offered the possibility of predicting mass balance changes from climate-driven ELA changes. Oerlemans (1992) calculated for three glaciers in Norway that a 1K temperature increase would cause ELA rises of 108–135 m, with corresponding mass balance changes of between –0.715 and –1.11 m a^{-1} (area averaged water equivalent) and snout retreat of up to 6.5 km.

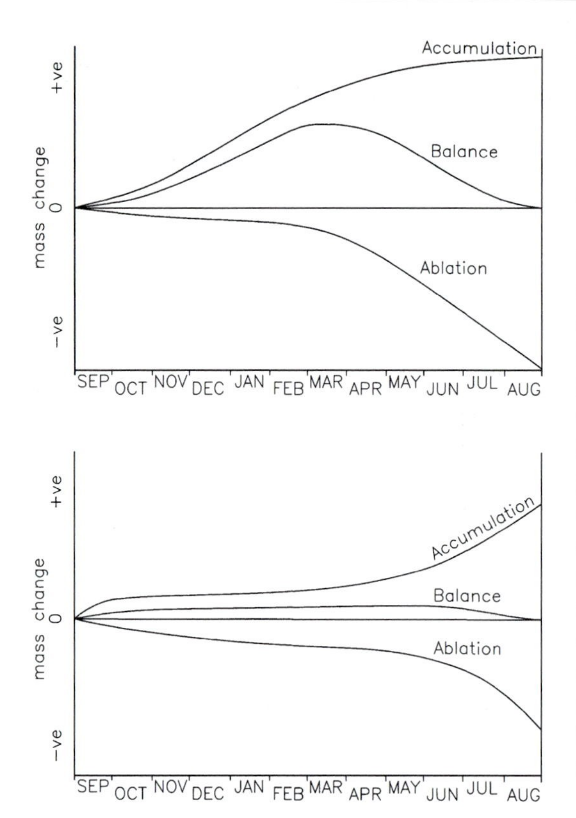

Figure 3.3 Examples of the fixed date system of mass balance analysis, illustrating (top) winter-accumulation type and (bottom) summer-accumulation type glaciers.

The activity index is the sum of the rate of increase of accumulation and the rate of decrease of ablation with altitude. It is a function of the net balance gradient at the ELA, and can be envisaged as the steepness of the transition from ablation to accumulation with distance up the glacier, or the gradient of the line b_z in Figure 3.2. It is the same as the net balance gradient, and is sometimes referred to as the energy of glaciation (Schumskiy, 1946). The activity index is related to changes with altitude in the amount of precipitation falling as snow; surface albedo; air temperature; and the effect of valley walls or extraglacial land surfaces on melting. The activity index is also related to ice residence time. Glaciers with high activity indices tend towards faster throughput of ice.

The accumulation area ratio (AAR) is the ratio of the accumulation area to the total area of the glacier. A range of different values have been measured at different glaciers. Porter (1970) suggested that typical values for modern glaciers were between 0.6 and 0.7. Meier and Post (1962) suggested that glaciers with a zero mass balance (glaciers in equilibrium) would have AARs > 0.5 and generally in the range 0.5–0.8. They found a value of 0.58 for North Cascade Glaciers. The AAR is partly a function of the mass balance: glaciers with unhealthy balances are likely to have low AAR values. Topographic and geometric influences can also be important; for example, the typical equilibrium (zero mass balance) AAR for mountain glaciers in areas of very high relief might be lower than for other glaciers because of the number of glaciers supplied from avalanches rather than from a contiguous accumulation area (Müller, 1980). Kulkarni (1992) found an AAR of 0.44 for a group of Himalayan glaciers. The AAR provides a simple way of reconstructing former ice masses from geomorphological data. If the AAR is estimated on the basis of similar glacier types and environments, the position of the ELA can quickly be plotted on a map of former glacier extent. Porter (1970, 1975) mapped the extent of former glaciers in Pakistan and New Zealand on the basis of geological evidence, and assumed an AAR of ~0.6 to calculate the ELA of the reconstructed glaciers. Pelto (1988) calculated annual balance using a combination of ELA, AAR and activity index methods, suggesting that the answer to Braithwaite's (1984) question, 'Can the mass balance of a glacier be estimated from its equilibrium-line position?' is 'yes' if the activity index and the area–altitude distribution are also known.

The mass balance of different glaciers is controlled by different parameters, although climate is the dominant factor. What controls mass balance in any particular case depends on the specific accumulation and ablation characteristics of the glacier in question. Rogerson (1986) found that, for four small (0.7–1.4 km^2) cirque glaciers in Northern Labrador, mass balance was directly related to the climate record, especially winter snowfall. Local variation between glaciers was determined by other factors including altitude, slope, proximity to cirque wall, height of the cirque wall above the glacier (hence shadiness) and amount of supraglacial debris cover. Letréguilly (1988) compared the mass balance of three glaciers at different distances from the coast in western Canada and found that winter precipitation was dominant in controlling annual variations in mass balance close to the coast but that summer ablation was dominant further inland. Variations in the mass balance of Peyto Glacier (550 km from the coast) were almost entirely related to summer temperature; variations at Sentinal Glacier (30 km) were mostly controlled by winter precipitation; and at Place Glacier (160 km) the two factors were closely matched in importance.

Mass balance parameters can be measured in a variety of different ways. Indirect estimates based on proxy criteria such as those mentioned above provide one source of information. Field measurement of accumulation and ablation can be achieved using stakes, pits, cores and probes. Remote sensing of the glacier surface by satellite or aerial reconnaissance can identify positions of surface features such as the snowline at different times of year. Remote sensing can also measure changes in surface elevation and ice thickness through time. Meteorological and hydrological monitoring can reveal flux and storage of snow and ice. Energy budget models and precipitation data can be used to predict mass balance.

From the point of view of glaciology, the key elements of mass balance are accumulation of snow, the transformation of snow into glacier ice, and ablation, or release of material from the glacier. These processes control all other aspects of the mass balance and hence the general status of the glacier, and will be considered in the following sections.

3.3 ACCUMULATION

The term 'accumulation' is generally taken to include all the processes by which snow or ice is added to a glacier. The main processes of accumulation include: direct precipitation of snow; freezing of liquid water; transport and deposition of snow by wind; deposition of snow or ice by avalanche; precipitation of rime or hoar; and freezing of water to the base of an ice shelf or floating ice tongue.

Direct precipitation of snow accounts for the overwhelming bulk of accumulation in most glaciers. According to LaChapelle (1992), snow consists of ice crystals in the atmosphere which grow large enough to fall and reach the ground. A snow crystal is a single ice particle which has a common orientation of the orderly array of molecules which make up its structure. A snow grain is a mechanically separate particle of snow which may or may not be a single crystal. Recent classifications of snow types has been provided by IAHS (1990) and LaChapelle (1992). About 5% of global precipitation falls in the form of snow (Martinec, 1976). Accumulation is controlled largely by the frequency and intensity of snow precipitation events, which in turn depend partly on air temperature, as moisture transport is limited by the saturation vapour pressure of cold air. There is a general geographical correlation between low temperatures and low accumulation rates (e.g. Fortuin and Oerlemans, 1990).

Freezing of liquid water can include freezing of rain on the glacier surface, or refreezing of meltwater in the snowpack. The latter process makes a major contribution to the ice of many glaciers, creating what is referred to as 'superimposed ice'. Fujita *et al.* (1996) found that about 60% of the summer meltwater produced in the accumulation zone of Xiao Dongkemadi Glacier on the Tibetan Plateau refroze at the snow–ice interface to form superimposed ice, representing about 26% of the total accumulated snowfall. Hooke and Clausen (1982) found that superimposed ice constituted most of the ice at the margin of the Barnes ice cap, Baffin Island.

In cold, windy environments, large amounts of snow can be transported and deposited by wind. In Adelie Land, Antarctica, Wendler (1989) calculated an annual mass flux of wind-blown snow of 6.3×10^6 kg m^{-2} a^{-1}. Kuhn (1981) estimated that wind drift increased accumulation in the firn basins of large glaciers by a factor of two. The formation of areas of bare ice (blue ice zones) close to nunataks in Antarctica may be due to locally reduced snow drifting in wind shadows close to nunataks (Orheim and Lucchitta, 1990).

Snow or ice can be deposited onto glaciers by avalanches. Avalanches can sweep snow from large areas of steep mountainsides into small cirque basins, and Kuhn (1994) estimated that accumulation in such basins was about four times the winter precipitation. Whole glaciers can be maintained by avalanches, and are known as rejuvenated or reactivated glaciers. One limb of Morsarjökull, a valley glacier descending from Vatnajökull ice cap in southern Iceland, is fed entirely by ice avalanching down a cliff face from the ice cap above.

Ice can be precipitated as rime when supercooled water droplets freeze onto the ice surface, or as hoar when ice condenses from vapour. These sources usually provide only negligible amounts of ice, although Linkletter and Warburton (1976) suggested that 5–10% of accumulation on the Ross ice shelf was derived from rime and hoar. Hoar forms not only at the surface but also in crevasses and at depth in the snowpack.

Accumulation occurs dominantly at the glacier surface, and in common usage the term is confined to surface processes. However, accumulation can also occur englacially or subglacially. Mechanisms for englacial and subglacial accumulation include freezing of water within or beneath the ice, and the entrainment of ice formed in subglacial cavities. Ice formed by freezing of water has been found to form basal ice layers tens of metres thick in some glaciers (Chapter 5). Regelation processes involving localised melting and refreezing involve no overall change in glacier mass, and so are not considered as accumulation or ablation.

Ice shelves represent an unusual class of glacier. They are largely fed by inflowing glaciers at the upstream end, but accumulation also occurs by precipitation of snow at the ice-shelf surface. Ice shelves account for 11% of the area of the Antarctic ice sheet but 25% of its precipitation (Jacobs *et al.*, 1992). Accumulation can also occur by freezing of water from the ocean below onto the base of the shelf. The basal ice of many ice shelves is of marine origin. Pedley *et al.* (1988) identified localised freezing of freshwater to the base of George VI ice shelf, Antarctica, where meltwater became trapped between the base of the ice shelf and more saline deeper water. Cold meltwater produced at the base of the shelf close to the grounding line can produce a thermohaline convection cell or 'ice pump' (Lewis and Perkin, 1986) in the water beneath the shelf. Because of the pressure dependence of the freezing point, the rising limb of the cell can cause ice platelets to form and accrete to the bottom of the shelf at more distal locations (Figure 3.4). Oerter *et al.* (1992) described marine ice accumulating at the base of the Filchner-Ronne ice shelf in this way.

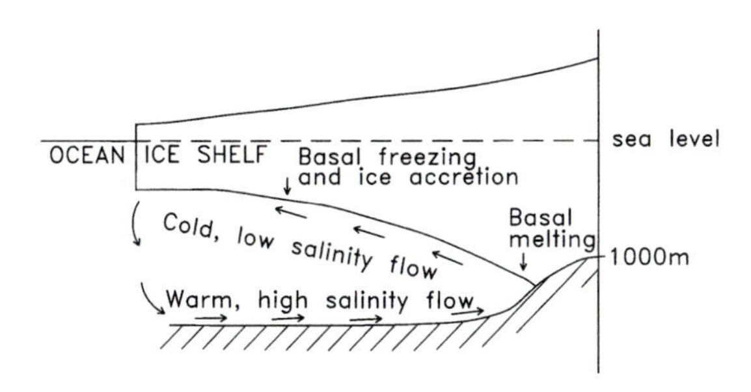

Figure 3.4 A thermo-haline circulation cell beneath an ice shelf causing accretion of marine ice at the shelf bottom. Melting near to the grounding line produces low-salinity, low-density water that rises and flows outwards along the shelf bottom. With decreasing depth the pressure-dependent freezing point rises, and the supercooled water releases ice platelets that accumulate at the shelf bottom. This freezing increases the salinity and hence the density of the water, and also releases latent heat that warms the water. This warm, dense water subsides in a deep counter-current and fuels continued melting close to the grounding line.

3.4 TRANSFORMATION OF SNOW TO ICE

As snow accumulates at the surface, processes of compression and recrystallisation operate to effect its gradual transformation into ice (Meier, 1964; LaChapelle, 1992) (Table 3.3). The density of loose-packed, freshly fallen snow ranges from < 100 to 300 kg m^{-3}, with the snowflakes forming a very open matrix. Several processes serve to convert snowflakes to rounded grains after they land. Breaking and wear occur as the grains impact each other on landing and in drifting over the ground. Fresh, intricately branched snow crystals have a thermodynamically unstable shape. Molecules evaporate from the sharp asperities where vapour pressure is high, and condense back to ice in the re-entrants. Thermodynamic processes tending to reduce surface free energy to a minimum thus result in the rounding of particles and the conversion of large irregular crystals into several smaller ones. This process operates best at close to 0°C and stops altogether below about –40°C. Settling and packing of the rounded snow grains can lead to a bulk density of up to 580 kg m^{-3}. Beyond this density, further compaction occurs as result of recrystallisation and creep as the snow is transformed to firn. Firnification is complete, and the snow fully converted to ice, when all the air spaces between grains are sealed off into isolated bubbles. At this point the ice has a density of about 830–840 kg m^{-3}.

The dominant processes in firnification depend to some extent on whether water is present in the snowpack. Where the snow is dry, the transformation takes place by compaction, involving processes of pressure sintering, whereby intercrystalline bonds are formed. Wilkinson (1988) described these processes for polar firn and glacier ice and found that power-law creep is the dominant process for densification between densities of 500–980 kg m^{-3}. When the snowpack is subject to wetting by seasonal melting, these processes are supplemented by regelation bonding of grains, and the percolation and freezing of

Table 3.3 Characteristics of material during stages of the transformation from snow to ice. Data from Meier, 1964, and Paterson, 1994 after Seligman, 1936

Material	*Definition (characteristics)*	*Density ($kg\ m^{-3}$)*	*Porosity (% void/ice)*	*Air permeability ($g\ cm^{-2}\ s^{-1}$)*	*Grain size (mm)*
Snow	Falling of deposited ice particles (UNESCO, 1970) Porous permeable aggregate of ice grains (Bader, 1962)	10–550	99–67	> 400–40	0.1–5
New snow		10–70			
Damp new snow		100–200			
Settled snow		200–300			
Wind-packed snow		350–400			
Depth hoar		100–300			
Firn	Material (snow, rime, hoar, refrozen meltwater) that has survived to the end of one ablation season and has not yet compacted and recrystallised to the point where it is impermeable to liquid water	400–840	56–8	40–0	0.5–5
Very wet snow/firn		700–800			
Glacier ice	Where interconnecting air passages between grains are sealed and air is present only in bubbles	830–917	8–0	0	1– > 100

water. Braithwaite *et al.* (1994) found that the main control on near-surface firn density in the lower accumulation area of the Greenland ice sheet is the refreezing of meltwater penetrating from the surface. Where large temperature gradients exist within the snowpack (for example, between adjacent layers of snow), a process known as temperature-gradient metamorphism contributes to the firnification process. In this process ice is removed as vapour from one crystal and deposited as ice on an adjacent, colder crystal. This process generally prevails when temperature gradients exceed about 0.1°C cm^{-1}, and substantial portions of the snowpack can be transformed. The crystals produced are referred to as depth hoar, and are the buried equivalent of surface hoar frost, deposited directly from a vapour phase. Depth hoar tends to be structurally weak, and is a major cause of snow avalanches. The rate at which the transformation from snow to ice takes place thus depends on temperature, on the accumulation rate which determines the rate of increase of pressure, and on the presence or absence of water. In extreme cases superimposed ice on the glacier surface can form in a single season when melting and refreezing occurs. Where accumulation is low and the snow is dry, firnification can take thousands of years. Barnola *et al.* (1987) found the firn-ice transition at Vostok, Antarctica, occurred at a depth equivalent to an age of 2500 years.

Stratification of snow and firn is readily seen in pits or cores, from which data regarding the rate and nature of ice formation can be gathered. Scott (1905) observed a visible stratigraphic record in Antarctic firn, and Sorge (1935) recognised differences between summer and winter layers in firn in Greenland. Forster *et al.* (1991) demonstrated that stratification was also measurable by means of radar. A 'standard model' of firn stratification was developed as a result of a large number of studies, and has been widely tested by comparison of pit observations with accumulation measurements and measurements of marker horizons (Benson, 1962; Gow, 1965; Alley, 1988a). Visible stratification is determined by depositional processes and diagenesis in the upper 50–100 mm of the snow pack. Coarse-grained, low-density hoar layers are formed in the autumn when a strong temperature gradient causes upward vapour transport; fine-grained, high-density, wind-packed crusts are formed in winter. Pairs of hoar and wind-crust layers are interpreted to represent one winter/autumn. Goodwin (1991) identified three distinctive seasonal surfaces in firn stratigraphy in Wilkes Land, Antarctica, related to wind and radiation glazes associated with different seasonal climate conditions. A spatially continuous wind-glazed crust formed during the autumn snow-supply hiatus, and correlated well with $\delta^{18}O$ stratigraphy as an annual marker for dating and accumulation studies. Spatially coherent seasonal variations of various kinds including $\delta^{18}O$, chemical impurities and dust offer potential for dating levels in the firn and thereby calculating accumulation rates.

3.5 ABLATION

The term ablation includes all processes by which snow and ice are lost from the glacier. These processes include melting, sublimation, wind erosion, removal by avalanche, and calving.

Melting is the dominant ablation mechanism for most land-terminating glaciers. Melting occurs when snow or ice at the melting temperature is supplied with heat. The heat is supplied primarily by radiation or by heat exchange with the air. Rain falling onto an ice or snow surface can also transfer heat, and local factors such as proximity of rock surfaces on nunataks or valley walls that are heated by the sun might also have an effect. Melting of polycrystalline ice occurs both at externally exposed crystal faces and at crystal boundaries within the polycrystalline mass. The melting point is likely to be lower in the veins than in the crystals, because of the concentration of impurities in the vein network, and because of the molecular curvature of the ice at the veins. An increase in the bulk temperature of the ice, or the penetration of solar radiation into the interior of an ice mass, can initiate intergranular melting. Wolfe and English (1995) found that differences in surface albedo rather than differences in altitude controlled variations in melting in the lower part of Quviagvaa Glacier on Ellesmere Island, and van der Wal *et al.* (1992) demonstrated that variation in ablation over the Hintereisferner glacier tongue was almost entirely due to differences in surface albedo controlling absorption of solar radiation. Braithwaite and Olesen (1990b) found at sites in Greenland in June–August that radiation accounted for about two-thirds of mean ablation, and turbulent fluxes for about one-third. Zwally and Fiegles (1994) found that the amount of surface melting on Antarctic ice shelves and around the margin of the Antarctic ice sheet was largely correlated with regional air temperatures. The difference in ice ablation caused by future climate change will depend on the seasonal distribution of the temperature change, since winter temperatures over the majority of the world's ice are so low that temperature rises of several degrees centigrade would be insufficient to increase ablation significantly. Braithwaite and Olesen (1993) found that only 2% of annual ablation in West Greenland occurred in the 7 months from October to April, and determined that even a 5°C temperature rise would not significantly increase ablation during those months because temperatures were so low anyway. By contrast, a 5°C temperature rise during the summer would lead to a doubling of the annual ablation (Braithwaite and Olesen, 1989). The onset of deglaciation in the Norwegian Sea about 18,000 years ago occurred during a period of rapidly increasing summer insolation (Bard *et al.*, 1990), and Lehman *et al.* (1991) suggested that the retreat of the Fennoscandian ice sheet may have been initiated by increasing insolation.

Surface melting of glaciers can be inhibited if the ice is protected from heat input. Surface debris is a common cause of inhibited melting, and in extreme cases can lead to 'retarded deglaciation', when large amounts of glacier ice survive into a postglacial climate beneath a protective debris cover (e.g. Lorrain and Demeur, 1985; Astakhov and Isayeva, 1988). Yoshida *et al.* (1990) reported 'fossil ice' 1000–1700 years old that had survived by burial beneath a perennial snow patch. Deliberate control of ablation for hydrological engineering has been suggested. A light dusting of low-albedo material such as coal dust onto a glacier surface could increase surface melting and water production by increasing the absorption of solar radiation by the surface.

Melting can also occur at the glacier bed, depending on the basal thermal regime. Heat for basal melting is supplied by geothermal heat flux; by friction between the ice, debris and substrate; and by meltwater. For the Antarctic ice sheet, basal melting is 6–10 times greater than surface melting (sources in Paterson, 1994). The bottom-melting of floating glacier tongues and of ice shelves is largely controlled by the temperature of the water in which they float. Jacobs *et al.* (1992) suggested, contrary to some previous estimates, that bottom melting contributes about 20% to the total ablation of the Antarctic ice shelves, with the bulk of the remainder being due to iceberg calving. Basal melting is especially important during periods of advance when ice has a long residence time and the calving rate is reduced. Rignot (1996) found that basal melting was the dominant ablation process in the floating section of the tongue of Petermann Gletscher, Greenland: 95% of the ice that crossed the grounding line into the floating terminus section melted before it reached the calving front, and about three-quarters of that melting occurred at the base. Jenkins *et al.* (1997) found that basal melting beneath the floating tongue of Pine Island Glacier, West Antarctica, removed ice at a rate of about 12 m a^{-1}, accounting for about half of the total ablation.

Sublimation is the direct transformation of ice to vapour. It is especially important in arid environments and at high altitudes, and can account for all of the surface ice loss from cold polar glaciers. Sublimation accounts for all of the ablation of some glaciers in the Transantarctic Mountains, including Taylor Glacier, where sublimation averages 0.18 m a^{-1} water equivalent at 1000 m elevation (Mercer, 1971; Robinson, 1984). Sugden *et al.* (1995) calculated the survival potential of ice buried beneath a thin soil cover at Taylor Glacier, and found that with saturated soil as little as 1 m might be lost by sublimation over a period of 8 million years. However, with a lower soil humidity, as much as 1000 m of ice could be sublimated in the same time. In arid environments, the high rate of sublimation is such that the latent heat of evaporation occupies a large part of the summer heat balance. For a glacier in the Chinese Tien Shan, Ohno *et al.* (1992) recorded sublimation of 81 mm a^{-1}, equivalent to 12% of the total ablation but consuming 54% of the energy causing ablation. High sublimation rates thus tend to suppress total ablation.

Loss of snow from glacier surfaces by wind erosion, and of snow and ice by avalanching, is spatially very variable but can be locally important. Ice avalanches (Chapter 10) are generated by calving mechanisms described below. Wind erosion is controlled by local wind speed, topographic exposure and snow characteristics. Dry, windy conditions favour erosion. The so-called blue ice zones of Antarctica are snow-free areas on the ice-sheet surface close to nunataks, the most likely origin of which is thought to be either locally reduced accumulation of drifting snow or locally enhanced erosion of surface snow by wind patterns close to the nunataks (Takahachi *et al.*, 1988; Orheim and Lucchitta, 1990).

Calving is the process by which blocks of ice become detached from the margin of a glacier. Calving of icebergs into water is a major component of the ablation of many glaciers (Figure 3.5). For example, calving at present accounts for about 77% (Jacobs *et al.*, 1992) of ablation from the Antarctic ice sheet, and about 56% of ablation from the Greenland ice sheet (Reeh, 1994). At Jakobshavn Isbrae, the ice stream that drains about 6.5% of the Greenland ice sheet, the discharge of ice calving from the snout has been calculated to be as much as 40 km^3 a^{-1} (Pelto *et al.*, 1989) or ~60 × 10^6 tonnes d^{-1} (Echelmeyer *et al.*, 1991). At Glaciar San Rafael, Chile, Warren *et al.* (1995) recorded 7000 individual calving events over 32 days. Calving exerts a strong control on the position of floating

Figure 3.5 A calving front. This section of the ice-sheet margin in West Greenland terminates as an ice cliff from which calving of icebergs occurs. Note the undercutting of the cliff by the lake water, and the propagation of crevasses downwards from the surface. The lake shown here periodically drains subglacially to produce jökulhlaups in the proglacial area (Sugden *et al.*, 1985; Russell, 1989).

ice fronts (Chapter 8) and was a major factor in the retreat of marine and lacustrine margins of northern hemisphere ice sheets during the last deglaciation.

Calving can affect both floating and grounded ice margins and can occur in a variety of different ways (Warren, 1992). The most common mechanisms include: toppling of seracs by propagation of crevasses downwards from the surface; shearing downwards of complete blocks; and detachment and flotation of submerged ice. Calving can involve both tensional and shearing failure in the ice. Hughes (1989, 1992b) discussed calving of slabs from ice walls grounded on land and in shallow water and accurately predicted the calving rate using a model that took bending creep to be the rate-controlling calving mechanism. Shear rupture occurred along near-vertical shear bands in the ice wall, with calving following rupture at a shear stress of about 1 bar. Detachment and flotation of submerged ice has been described by Lingle *et al.* (1993) from Vitus Lake at Bering Glacier. They found fracture to be caused by upward-bending moment induced when the ice surface is lowered below hydrostatic equilibrium. Fracture propagates from the base of the ice, and the iceberg breaks off, floating to the surface. The berg floats higher in the water than the snout of the glacier, which is held down by the tensile strength of the thicker ice upstream or by adhesion to the bed or valley walls. A general calving rate law incorporating all the variables known to be associated with calving rate has yet to be achieved, but a range of controlling factors can be identified. The most important of these seem to include water depth, subglacial runoff, ice thickness, ice cliff height, presence of pre-existing weaknesses such as crevasses in the ice approaching the cliff, water and ice temperature, tidal flexure and buoyancy. Dowdeswell (1989) related rates of iceberg production, and the size of icebergs, to characteristics of glaciers producing the icebergs. In Svalbard he found that surging glaciers produced large numbers of small, debris-rich, icebergs in their surge phase, but few bergs in their quiescent retreating phase. Less active tidewater glaciers often produce larger individual bergs because there is less surface crevassing, although the total amount calved is less than more active glaciers. Some larger glaciers and ice shelves seem to experience short-lived major calving episodes separated by periods of as much as several decades during which calving is negligible.

It has been suggested (Hughes, 1986) that calving could be critical in the unstable decay of ice sheets in a greenhouse scenario. In the so-called 'Jakobshavns effect' a series of positive feedbacks are associated with ice-stream flow. Fast flow causes intense surface crevassing which both enhances calving and allows summer melt to penetrate and lubricate the glacier. This causes increased velocity, which in turn leads both to increased calving and to increased crevassing, increased meltwater penetration and further acceleration. Calving would increase continuously and theoretically the glacier would retreat rapidly as far as the subglacial headwall of the fjord. In glaciers or ice streams that have no headwall the effect could lead to ice-sheet collapse.

It is clear that different glaciers, or different parts of a glacier, will be dominated by different ablation processes. For glaciers of a given type (floating or land-based, for example) the dominant processes depend primarily on the climate. For example, cold arid conditions favour sublimation, while warm conditions favour melting. Seasonal distribution of snow cover is also important, as ablation from a snow surface is only about half that from an ice surface at the same temperature (Braithwaite and Olesen, 1990a). However, the relative importance of different processes varies also with the thermal regime and dynamic characteristics of glaciers as well as with their geographical setting.

3.6 MEASUREMENT AND MODELLING OF MASS BALANCE

Mass balance parameters can be measured directly in the field or by remote sensing, or can be predicted on the basis of physically based or statistical models. They can also be calculated indirectly from climatological and hydrological measurements. Mass balance measurements have been carried out since at least 1882, when a stake network was established on Rhonegletscher. Early researchers employed networks of stakes and pits for measurement of seasonal and annual variations in snow depth and surface melting. Marker horizons of known age such as radioactive layers in the snowpack can also be used to calculate accumulation rates (e.g. Lefauconnier *et al.*, 1994). More recently, airborne and satellite remote sensing and the use of global positioning satellite systems have provided the opportunity to recover data on surface conditions more easily from large areas. However, the precision of remotely sensed data is generally too poor to identify small mass balance fluctuations with confidence, and field measurements cannot be applied rigorously to large complex systems with numerous, inaccessible input and output locations. For example, it is impossible to measure the amount of water transferred to groundwater from subglacial melting, or the discharge of meltwater beneath floating ice margins. In mass balance estimations for ice sheets, errors in estimating accumulation have been improved by the availability of remotely sensed data for inaccessible areas and are generally < 25%. However, errors in estimating ablation are generally > 100% because it is virtually impossible to measure many of the outputs with accuracy (Giovinetto and Zwally, 1995). Consequently it is generally impossible even to be confident at present whether the balance of the ice sheets is positive or negative.

Different approaches to mass balance modelling have developed for glaciers of different size (Braithwaite, 1995a). For small glaciers, mass balance studies typically involve detailed measurement of mass balance parameters in accurately delimited areas over extended time periods that can then be compared with climatic data and extrapolated into past or future environments. For example, mass balance measurements have been carried out continuously since 1945 on the ~3 km^2 Storglaciären in northern Sweden. This is the longest continuous record of mass balance. Surveys are conducted sometimes on a weekly basis by staff from Tarfala research station, which is only 1 km from the glacier, using a sampling grid with as many as 100 points per km^2 (Holmlund *et al.*, 1996a). By contrast, mass balance studies of large ice sheets have generally involved sparse data sets over inaccessible and inadequately mapped areas, and have required a greater degree of theoretical modelling. Several different approaches to modelling mass balance have been taken, including energy balance modelling, degree-day modelling and regression modelling.

Modelling of mass balance is necessarily concerned with the surface energy balance, as melting and accumulation are temperature dependent. Energy balance models are based on the assumption of a glacier surface energy balance of the form:

$$Q_n + Q_s + Q_l + Q_c + Q_m = 0$$

where Q_n is net radiation, Q_s is sensible heat, Q_l is latent heat, Q_c is heat conduction and Q_m is heat used for melting. Energy balance models calculate ablation from climate data. Braithwaite and Olesen (1990b) summarised an energy budget model to calculate ablation in the following form:

$$\text{Abl} = \text{SHF} + \text{LHF} + \text{SWR} + \text{LWR}$$

where Abl = simulated ablation; SHF and LHF = turbulent sensible and latent heat fluxes; and SWR and LWR = short-wave and long-wave radiation fluxes.

Models of this type require detailed information on a variety of physical conditions, such as air temperature, wind speed, vapour pressure, sunshine duration and radiation totals. However, in many cases these data are not available and need to be estimated. An alternative, simpler approach can be based on the evident correlation of many of the factors controlling mass balance (especially ablation) with temperature. The simplest approach is the degree-day model, which relates melt rate to air temperature (e.g. Braithwaite and Olesen, 1989). The degree-day method is based on the sum of temperatures over a period of time, and relates this degree-day figure to characteristics of mass balance. For example, the melting of snow and ice during a period is often taken as proportional to the positive degree-day sum (i.e. the sum of all temperatures above the melting point during that period). The factor linking the positive degree-day sum to ablation is the positive degree-day factor, which is expressed in millimetres d^{-1} $°C^{-1}$. Typical values range from about 5.0 to 8.0 mm d^{-1} $°C^{-1}$. Braithwaite (1995b) discussed the use of positive degree-days in modelling ablation, and argued that the positive degree-day factor is a simplification of complex processes more properly described in an energy balance formulation, and thus is not a universal constant but varies with local energy balance conditions. For example, melting snow will have a lower positive degree-day factor than melting ice, and melting ice will have different factors depending on local, and time-dependent, energy balance regimes. Hall and Weston (1993) showed that neither energy balance nor degree-day models are adequate for characterising ablation. Detailed boundary layer characteristics are also important, and an evolving boundary layer over an evolving ice sheet is likely to produce feedback effects between the air and the ice sheet so that there is more ablation as the ice sheet becomes steeper. Turbulent mixing of warmer air from above into the boundary layer close to the ice surface is greatest where the ice-sheet gradient is steepest, which is near to the margin. This mixing results in enhanced warming of the ice and more ablation close to the margin, and hence further steepening of the surface profile. In accordance with this, Braithwaite and Olesen (1990c) indicated that the Greenland ice sheet is becoming steeper as it retreats. Vincent and Vallon (1997) also questioned the validity of simple correlations between local climate and mass balance in modelling. They pointed out that glacier surface conditions influencing albedo are critical in determining the climate/mass-balance relationship, and that models ignoring this would generate false mass balance reconstructions. Surface energy balance and heat budget models are discussed further in Chapter 4.

Regression models are based on the extrapolation of observed relationships between mass balance and climate parameters into periods when no direct balance measurements are available. Chen and Funk (1990) described regression formulae linking specific glaciers to local climate station records. For example, comparing balance data from the Rhone Glacier with climate data from Reckingen climatic station, they found that:

$$b = 6320 + 1.094P - 601T$$

where b is specific net balance (mm water equivalent), P is annual precipitation (mm), and T is mean summer temperature (°C). Models of this type are sometimes referred to in the literature as PT models (for precipitation and temperature). Once such a relationship is established on the basis of a calibration period, it can be used to estimate balance figures from climate data outwith that period and to estimate the relative importance of temperature and precipitation to mass balance.

An important aspect of mass balance modelling with respect to glacier fluctuations and sea-level change is the sensitivity of a glacier to climate change. Models consider either static sensitivity or dynamic sensitivity. Static sensitivity involves the ratio of the change of mass balance to the change in temperature, but ignores time-dependent changes in glacier geometry. Dynamic sensitivity takes into account the time-dependent change in glacier characteristics concomitant with mass balance change, including the warming or cooling caused by ice-mass reduction or growth. Jóhannesson (1997) used a degree-day mass balance model coupled to a dynamic glacier model to predict the disappearance within 200 years of two of the major outlet lobes of Hofsjökull in Iceland.

3.7 MASS BALANCE OF SPECIFIC GLACIERS

Notwithstanding the problems of measurement that were mentioned earlier, many attempts to estimate the mass balance of the ice sheets have been made, especially because of the potential impact of ice-sheet reduction on sea level. For large ice sheets, however, the imprecision of existing data is such that it is generally impossible even to be confident whether the balance of the ice sheets is positive or negative. Recent estimates have suggested that the Antarctic ice sheet is broadly in balance, with both accumulation and outflow in the region of 2000 $km^3 a^{-1}$. However, some accumulation records suggest that recent increases in accumulation following a minimum around the year 1960 might be responsible for a positive imbalance of 5–25%. This would be equivalent to an annual sea-level fall of 1.0–1.2 mm a^{-1} (Morgan *et al.*, 1991). Zwally *et al.* (1989) and Zwally (1990) reported thickening of the Greenland ice sheet by about 23–28 cm a^{-1} on the basis of satellite altimetry, which could also indicate increased accumulation and a positive balance. Tables 3.4 and 3.5 show some of the estimates that have been made of the balance of the Greenland and Antarctic ice sheets. Modelling by Gregory and Oerlemans (1998) suggested that climate warming during the period 1990–2100 will lead to shrinkage of the Greenland ice sheet to an extent equivalent to about 76 mm of sea-level rise, but that the same warming will lead to growth of the Antarctic ice sheet by a similar amount.

For smaller glaciers, it is more feasible to measure mass balance with some confidence, and substantial amounts of data are becoming available with the recent initiation of global glacier monitoring initiatives. Even where records are not continuous, long-term trends can be identified by combining data sources of different types from different technological eras. For example, Jacobs *et al.* (1997) compared 1993 Landsat data with 1961 photogrammetry to identify an average recession rate of at least 4 m a^{-1} around the 183 km southern perimeter of Barnes ice cap, Baffin Island. A number of glaciers in Europe and North America have continuous mass balance records stretching back several decades. The longest series of measurements come from Storglaciären in Northern Sweden, which Holmlund (1988) found to be close to a steady state. Many records reveal a general pattern of negative mass balance (mass loss) over recent years, although, in the northern hemisphere especially, the picture is quite heterogeneous (Haeberli *et al.*, 1989). For example, Armstrong (1989) found that Blue Glacier, Washington, was in approximate equilibrium with present climate, having a mixture of positive and negative balance years during a 30-year study period, while Krimmel (1989) showed that the nearby South Cascade Glacier had a consistently negative balance over approximately the same period. Reports of rising ELAs and shrinking glaciers are widespread, especially with regard to sensitive mountain glaciers. For example, Aellen (1995) and Böhm

Table 3.4 Some estimates of the mass balance of the Greenland ice sheet ($km^3 a^{-1}$ water equivalent)

Balance component	*US Dept of Energy (1985)*	*Bauer (1955)*	*Huybrechts et al. (1991)*	*Weidick (1975)*	*Benson (1962)*
Total accumulation	500 ± 100	446	600	500	500
Iceberg calving	205 ± 60	215	318	205	215
Melting	295 ± 100	315	282	295	272
Total ablation	500 ± 160	530	600	500	487
Balance	~0 ± 260	–84	0	0	+13

Table 3.5 Estimate of the annual mass balance of the Antarctic ice sheet, (From Jacobs *et al.*, 1992)

Balance component	*Gt* ($\times 10^{12}$ *kg* a^{-1})[a]	
Accumulation on grounded ice	1528	
Accumulation on ice shelves	616	
Total accumulation		2144
Iceberg calving	–2016	
Ice-shelf melting	–544	
Surface runoff	–36	
Subglacial runoff	–17	
Total ablation		–3513
Balance		**–469**

[a] A measurement of –469 Gt would be equivalent to 1.3 mm a^{-1} sea-level rise if it all came from grounded ice.

(1995) both found a history of recent mass loss in the European Alps. Ames and Hastenrath (1996) and Hastenrath and Ames (1995a,b) found that glaciers in the Cordillera Blanca of Peru were in negative balance and likely to disappear within half a century. Kaser and Noggler (1996) reported glacier recession in the Ruwenzori Range, and Hastenrath (1984, 1989, 1992) has catalogued progressive retreat of glaciers in equatorial East Africa. Dyugerov (1996) reported negative mass balance for the past few decades for glaciers in the Caucasus and Tien Shan. Climatic amelioration need not lead to negative mass balance: at Wolverine Glacier, Alaska, Mayo and March (1990) noted how increases in both air temperature and precipitation in the period 1968–1988 were accompanied by an increase in accumulation, but no increase in ablation, as the warming only affected temperatures below about –5°C. There are also many reports of recent negative balance and retreat of Antarctic ice shelves. Rott *et al.* (1996) reported the rapid disintegration of part of the Northern Larsen ice shelf. In January 1995 4200 km^2 of ice broke away suddenly, in a period of a few days, following a long period of gradual retreat associated with regional warming. It may be that there is a critical threshold of retreat beyond which the shelf becomes unstable and retreats catastrophically, in the manner of some fjord glaciers (Chapter 8). It has been suggested, however, that the retreat of ice shelves and other contemporary changes in the configuration of the Antarctic ice sheet are not occurring in response to recent climate change but are continuing adjustments to the end of the last ice age (Alley and Whillans, 1991).

3.8 CONCLUSION

Glaciers can be viewed as a system of input, throughput and output of mass. The mass balance of a glacier is largely driven by environmental (climatic) controls, and has a major influence on the glacier's characteristics and behaviour. Measurements of mass balance can be used as an indicator of a glacier's 'health', and as a proxy measure of climate change, but oversimplification of the complexity of the controls on mass balance can lead to misinterpretation of the evidence they present. Even major and sustained periods of negative mass balance can arise from cyclical dynamic instabilities, such as surging or ice-shelf calving, that do not necessarily reflect climate change. Mass balance is intimately linked with glacier dynamics as well as with climate, and mass balance offers a convenient conceptual link between the broad environmental context of glaciation and the details of glacier behaviour.

MATERIAL, CHEMICAL AND THERMAL PROPERTIES OF GLACIERS

4

4.1 DIFFERENT TYPES OF ICE

Ice can occur in a variety of different crystalline forms, controlled by pressure and temperature. These different forms of ice are usually designated by roman numerals as ices I–IX. The phase diagram in Figure 4.1 shows the conditions in which the various forms exist. The form of ice with which we are familiar, and which occurs naturally in the geographical environment, has a hexagonal crystal structure and is known as ice Ih. It is to this form of ice that we refer throughout this book unless otherwise specified. Ice Ih exists at pressures up to 200 MPa and at temperatures down to about −100°C. Below −100°C there can exist a cubic-crystalline modification of ice I, designated Ic, that is metastable with ice Ih. Other forms of ice include an amorphous vitreous form of ice that can exist below about −160°C and a number of high-pressure ices that exist above 200 MPa. These have more densely packed atomic structures than ice Ih, and different physical properties. For example, at a pressure of 20,000 MPa, ice VII can exist at temperatures up to about 440°C. These high-pressure ices do not exist naturally on Earth, but are thought to exist on other bodies within the solar system, including some of the moons of Saturn and Jupiter (e.g. Poirier, 1982; Krass, 1984). Ice of materials other than water also occurs naturally on other planets, including ices of CO_2, CH_4 and NH_3. Table 4.1 shows the basic characteristics of the different forms of (H_2O) ice. Further details were given by Fletcher (1970) and Kamb (1972).

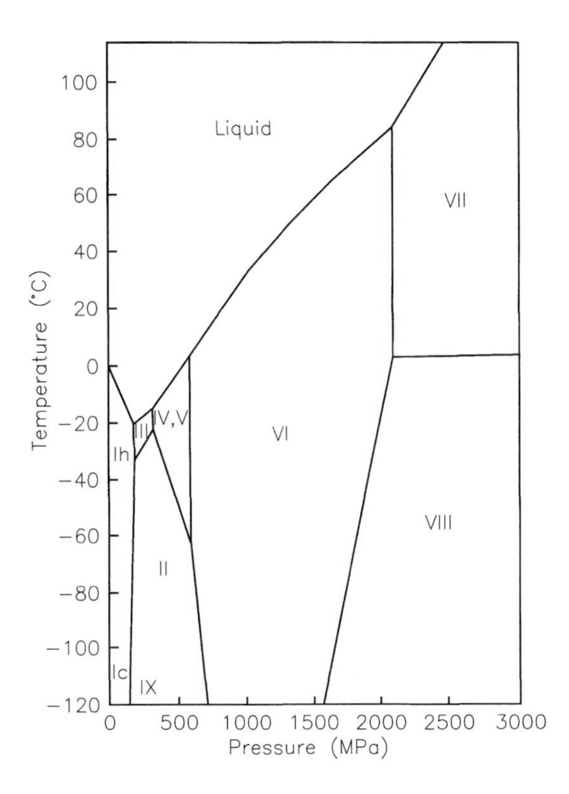

Figure 4.1 Phase diagram for ice.

4.2 STRUCTURE OF ICE

4.2.1 ATOMIC STRUCTURE

The basic constituent of ice is water (H_2O). The water molecule comprises one oxygen atom and two hydrogen atoms. The oxygen atom shares

Table 4.1 Characteristics of different forms of ice (data from Fletcher, 1970; Lock, 1990)

Ice	*Density ($g\ cm^{-3}$)*	*Crystal structure*	*Environment (see also Figure 4.1)*
Ih	0.92	Hexagonal	Stable at Earth surface conditions < 0°C
Ic	0.92	Cubic	Metastable with ice Ih below –100°C
vit.	0.94	Vitreous	< –160°C. (Transforms to Ic ≥–160°C)
II	1.17	Rhombohedral	High pressure
III	1.14	Tetragonal	High pressure
IV	1.28	Monoclinic	High pressure. Metastable with ice V
V	1.23	Monoclinic	High pressure
VI	1.31	Tetragonal	High pressure
VII	1.50	Cubic	High pressure
VIII	1.50	Cubic	High pressure
IX	1.14	Tetragonal	High pressure

the single electron of each of the hydrogen atoms and the molecule is commonly visualised as a regular tetrahedron with positive charges in two of the corners and negative charges in the other two. In ice, the H_2O molecules are joined to one another by hydrogen bonds whereby each hydrogen atom lies between two oxygen atoms. Each oxygen atom lies at the centre of a regular tetrahedron, with adjacent oxygen atoms at each of the four corners of the tetrahedron separated by a distance of 0.276 nm. This arrangement results in an atomic lattice made up of layers of hexagonal rings of atoms, or what Fletcher (1970) called 'a stack of crinkled molecular sheets'. These layers lie in the basal plane of the crystal, perpendicular to the main crystallographic axis (the c-axis or optic axis). This atomic structure is very open, with almost enough space between the atoms to accommodate interstitial water molecules. This open structure accounts for the low density of ice-I compared with liquid water and with the high-pressure ices. When the temperature of ice rises above melting point the hydrogen bonds begin to break down and the lattice begins to collapse. This leads to an increase in density on melting. As the temperature in the liquid rises, more bonds break and density continues to increase up to a temperature of 4°C, at which point expansion due to molecular agitation overcomes the effect, and the density of water decreases with temperature from 4°C upwards.

4.2.2 POLYCRYSTALLINE ICE

Ice in glaciers is made up of many individual crystals, and is hence termed polycrystalline. Many behavioural similarities exist between polycrystalline ice and metals, and metallurgical terminology and theory is frequently applied to ice crystallography. This includes the habit of referring to crystals as 'grains' (Alley *et al.*, 1986a). Crystals in glacier ice form initially from compression and recrystallisation of snow (described in Chapter 3), but crystal growth and recrystallisation continue after the initial formation of the ice (Duval and Lorius, 1980; Duval, 1985). In a low-strain environment such as that in the accumulation area of an ice sheet, crystals grow steadily with age. Crystal boundaries migrate in response to pressure differences caused by surface tension related to the curvature of crystal boundaries. The overall effect is for crystal boundaries to migrate towards the centres of smaller crystals, consuming them to the benefit of larger ones. Ice cores have revealed a good correlation between depth (age) and crystal size (Duval and Lorius, 1980; Lipenkov *et al.*, 1989). This correlation breaks down where cores penetrate ice formed during glacial periods, in which crystals are generally much smaller, indicating a slower growth rate. This slow crystal growth in ice-age ice could be due to the low surface temperature at the time of ice deposition (Petit *et al.*, 1987; Alley *et al.*, 1988), or to the presence of dissolved impurities in the ice, which has been shown to inhibit crystal growth (Alley *et al.*, 1986a,b). Small crystals can also result from the subdivision of existing crystals by polygonisation, or the nucleation and growth of new crystals. Crystal size is also dependent on stress. High stress and high strain favour nucleation of new crystals at a greater rate than the growth of existing ones. High strain zones such as shear bands are characterised by very small crystals.

The disposition or orientation of crystals within a polycrystalline ice mass is referred to as the fabric of the ice (Rigsby, 1960). Fabric is measured in terms of the orientation of the c-axes or optic axes of the crystals. Fabrics take a variety of forms, ranging from a random distribution to a strong single maximum in which all the grains are oriented in the same way. Ice fabric exerts an important control on ice viscosity (Azuma and Mae, 1988; Shoji and Langway, 1988; Budd and Jacka, 1989). A 'soft' fabric, in which most of the crystals are favourably oriented for deformation with respect to the prevailing stress, can deform an order of magnitude more rapidly than a 'hard' fabric, where many crystals are unfavourably oriented. Ice fabrics are largely determined by the deformational history of the ice, and reflect cumulative strain and stress (Azuma and Higashi, 1985; Alley, 1988b). Fabrics change in response to changing dynamic conditions by both crystal rotation and recrystallisation. Crystals are rotated during deformation into a preferred orientation with respect to the prevailing stress. For example, unconfined vertical compression, such as might occur beneath the summit of a model ice sheet, produces a strong single-maximum fabric in which the crystals are strongly vertically oriented. A similar vertical single maximum occurs in response to simple shear, such as might occur at the base of an ice sheet, where the basal planes of the crystals rotate into alignment with the direction of shear parallel to the bed, rendering the c-axes near vertical. Other stress configurations produce different rotation fabrics. Dynamic recrystallisation leads to the formation of new grains with their basal planes favourably oriented for basal glide in response to the prevailing stress. For example, in unconfined uniaxial compression or extension, recrystallisation produces a girdle fabric in which crystal orientations are clustered around the strain axis but separated from it by an angle of 45°. Specific fabrics can thus be interpreted as evidence of flow conditions. Paterson (1994) provided a brief review of some of the field and laboratory investigations that have explored the relationship between stress configurations, total strains and fabric. Crystal fabrics do not require long periods to develop. Pfeffer (1992) measured crystal fabrics upstream and downstream of the limit of the 1982/3 surge at Variegated Glacier, and found that the surge had imposed a new fabric on ice it affected. Rigsby (1960) observed complete rearrangement of fabric in fine-grained ice at 0°C within a month. Fabric adjustment may occur more quickly in fine-grained ice, because grain boundaries form nucleation sites for new grains.

Crystals in polycrystalline ice are surrounded by a network of interconnecting grain boundaries (Nye and Frank, 1973; Raymond and Harrison, 1975). This vein network is described more fully in Chapter 6 (Figure 6.5). Water can exist in the vein network even many degrees below the normal freezing point, both because of microscopic curvature of the ice/water interface and because vein water contains much of the soluble impurity content of the ice (Harrison and Raymond, 1976; Mulvaney *et al.*, 1988; Nye, 1989, 1991; Mader, 1992b). The vein network plays an important role in the behaviour of polycrystalline ice. The interconnectedness of the vein network controls the permeability of ice, and the presence of water in the vein network can affect the creep rate of the ice. Impurities and bubbles in the veins can inhibit rates of grain boundary migration and crystal growth. Because liquid in the veins can be much richer in impurities than the ice itself, and because of the curvature of the ice/liquid interface in the veins, the local melting point in the veins is lowered. Melting of a polycrystalline ice mass thus begins at grain boundaries. This explains why melting of glacier ice is often accompanied by the disaggregation of the ice into its individual grains. Water inclusions in the vein network and within ice crystals are discussed further in Chapter 6.

4.2.3 GLACIER ICE

The ice to be found in glaciers is somewhat different from that found typically in the physics laboratory or textbook. Glacier ice includes: liquid water at grain boundaries and in intracrystalline inclusions; solid particles ranging from microscopic aerosols to large rocks; dissolved impurities; air bubbles and dissolved gases. Its behaviour is affected by these inclusions, and the physics and chemistry of glaciers are not the same as the physics and chemistry of pure ice. Properties critical to glacier behaviour, such as density, viscosity and melting temperature, all vary from site to site, depending on the local characteristics of the ice. Some of these properties are discussed further in the following sections.

4.3 PHYSICAL PROPERTIES OF ICE

Ice can be considered either at the scale of individual crystals or as a polycrystalline mass. We can also differentiate between pure ice and ice with gases, solutes and particulate inclusions like the ice we find in glaciers. The characteristics of ice in glaciers depend on properties relevant to all these different cases: (i) properties derived from the atomic/crystalline structure of ice as a material; (ii) properties based on relationships between crystals; (iii) properties derived from the effect of impurities within the ice.

Ice is a remarkably strong and extraordinarily hard solid (Goodman *et al.*, 1981). It is often thought of as a relatively weak material only because most observations and measurements of ice have necessarily been made at temperatures close to the melting point. At 0°C the resistance of ice to elastic elongation is about two-thirds that of lead and about one-twentieth that of steel. However, if comparisons with other materials are made in terms of temperatures normalised to the melting temperature, and stress normalised to the shear modulus, ice is one of the strongest of materials (Poirier, 1982). Some of the characteristics of ice at the scale of individual crystals have already been discussed in the previous section. Table 4.2 lists some of the physical properties of ice.

Many of the properties of ice are environmentally variable. The density of ice varies with temperature and pressure, in a range of about 1.4%. At 0°C and atmospheric pressure the density is 0.9167 g cm^{-3}, and the maximum density in the natural environment is about 0.9295 g cm^{-3} (Schumskiy, 1978). The melting temperature varies with pressure. At atmospheric pressure the melting temperature is 0°C. With increasing pressure (p) the melting temperature (phase equilibrium) (θ_m) falls according to the Clausius–Clapeyron equation (Schumskiy, 1978):

$$d\theta_m/dp = -7.28 \times 10^{-3}\ °C\ kg^{-1}\ cm^{-2}$$
$$= 7.52 \times 10^{-3} °C\ atm^{-1},$$

reaching −2.91°C under a pressure of 400 kg cm^{-2}. This is equivalent to a melting point reduction of about 2.5°C beneath an overburden of 3000 m of ice. Gagnon and Gammon (1995) found that the flexural strength of iceberg and glacier ice increased with decreasing temperature and with increasing intragranular bubble density.

Table 4.2 Some physical properties of ice (data from Drewry, 1986; Fletcher, 1970; Paterson, 1994; Schumskiy, 1978)

Properties of ice		*Value*	*Units*
Young's modulus		9.1	GPa
Poisson's ratio		~0.36	
Shear modulus		3.4	GPa
Bulk modulus		8.3–11.3	GPa
Yield strength (elastic limit)		85	MPa
Fracture toughness		0.2	MN $m^{-3/2}$
Toughness		0.003	kJ m^{-2}
Creep activation energy		6.07×10^4	J mol^{-1}
Density (glacier ice)		0.84–0.917	g cm^{-3}
Density (pure ice)		0.916–0.93	g cm^{-3}
Volume coefficient of thermal expansion		1.53×10^{-4}	K^{-1}
Albedo (pure ice, average)		0.41	
Latent heat of fusion	0°C	334	kJ kg^{-1}
	−10°C	285	kJ kg^{-1}
	−20°C	241	kJ kg^{-1}
Latent heat of vaporisation		2800	kJ kg^{-1}
Heat capacity		37.7	J mol^{-1} K^{-1}
Thermal diffusivity at 0°C		2.1×10^{-6}	m^2 s^{-1}
Thermal conductivity at 0°C		2.51	W m^{-1} K^{-1}

4.4 DEFORMATION OF ICE

The flow law relating strain rate to stress, temperature and ice properties is to glaciology what the Holy Grail was to chivalry – an important goal but one that may not be attainable.

(Alley, 1992, p. 245)

For glaciology, an important property of ice is the way in which it deforms, or strains, under stress. The relation between stress (the applied force) and strain (the deforming response) is commonly referred to as the flow law. Deformation in ice can be accomplished by a variety of different mechanisms (see Goodman *et al.*, (1981), for a useful review), but in situations most frequently relevant to glaciers, deformation of individual ice crystals occurs primarily by 'dislocation glide' along the basal plane. This process can be envisaged simplistically as the layers of hexagonal atomic rings moving across each other like a stack of playing cards. Dislocations are defects in the crystal structure that facilitate the movement of planes of atoms across each other. In ice, deformation occurs most easily by the movement of dislocations in the basal plane of the crystal. Glide along other orientations is 100–1000 times harder.

When a stress is applied to a single ice crystal the crystal immediately deforms elastically by a limited amount; if the stress is removed, the ice recovers. If stress is applied continuously and the 'elastic limit' or 'yield strength' is reached, then permanent plastic deformation, or 'creep' begins. Creep then continues for as long as the stress is applied. In polycrystalline ice, the rate of deformation in response to a constant stress changes through time as the stress continues to be applied. This can be represented as a 'creep curve' as shown in Figure 4.2. When stress is first applied, a limited elastic deformation occurs. If stress continues, then elastic deformation is replaced by 'primary' or 'transient' creep. During primary creep, the strain rate decreases through time. This is because crystals with different orientations begin to interfere with each other and inhibit deformation. After a certain

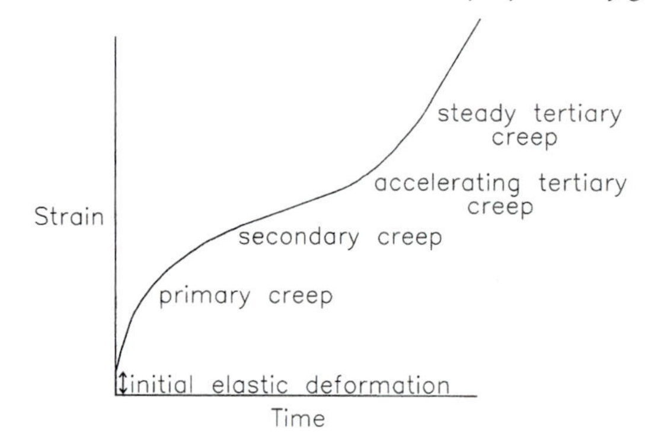

Figure 4.2 Creep curve for ice showing the strain response over time to a constant applied stress.

time, however, recrystallisation and reorientation of crystals into positions favourable for glide begins to offset this effect. The minimum strain rate reached before reorientation leads to creep enhancement is the 'secondary creep rate'. If stress continues, then favourable crystal fabric, multiplication of dislocations and development of microcracks lead to 'accelerating tertiary creep' as the strain rate increases through time. Ultimately, the ice reaches a maximum strain rate in 'steady tertiary creep', and strain continues at that rate for as long as the stress is applied.

While single crystals can deform most easily by slip on the basal plane, deformation of polycrystalline aggregates involves additional processes such as grain boundary migration, intergranular adjustment and recrystallisation (Barnes *et al.*, 1971). The steady creep rate for polycrystalline ice depends primarily on the applied stress and the characteristics of the ice. Glen (1955) measured the deformation of laboratory ice samples in response to stress and arrived at an empirical relationship that formed the basis for what is sometimes referred to as Glen's Flow Law (Nye, 1953; Glen, 1958). The relationship is often expressed in the form:

$$\varepsilon = A\tau^n$$

where ε = strain rate; A = a rate factor reflecting hardness of the ice; τ = stress deviator; n = power-law constant, often given as = 3. The hardness parameter A is exponentially related to temperature:

$$A = A_0 \exp\left[-\frac{Q}{kT}\right]$$

where A_0 = flow-enhancement factor that depends on fabric, grain size, impurity etc. (not temperature); Q = activation energy for creep; k = Boltzmann's constant; T = absolute temperature; τ = stress deviator.

A steady increase in temperature will thus lead to an accelerating increase in deformability. At melting point a stress of 1 bar would lead to a strain of 30% a^{-1}, at −15°C the same stress would lead to a strain of only 3% a^{-1}. This relationship is strictly only accurate below about −10°C, but can be used for higher temperatures with adjustment of Q. The value of A is also related to hydrostatic pressure, but the effect is very small. Paterson (1994) reports mean values of A from a range of sources, ranging in different environments from 4.9×10^{-16} s^{-1} kPa^{-3} at −10°C to 68×10^{-16} s^{-1} kPa^{-3} at 0°C.

The value of the power law constant n, and indeed the question of whether it is a constant, has been widely discussed (Alley, 1992a). For dislocation glide, which is probably the dominant creep mechanism in glacier ice, a value of 3 is theoretically valid, and experimental results confirm that value. However, for the range of situations that exist in glaciers, values of n other than 3 might also be possible. At very low stress, values less than 3 might apply, and at very high stress, values greater than 3. Weertman (1973b) found that values of n derived from a range of sources varied between $n = 1.5$ and $n = 4.2$, with an average of $n = 3$. If $n = 3$, then the strain rate is proportional to the third power (the cube) of the applied stress. A doubling of the applied stress would then be reflected in an eightfold increase in the strain rate. This sensitivity of deformation to small changes in strain is very important to mechanisms of glacier movement.

Deformation of ice in glaciers is sensitive to a range of factors such as water content, impurity content, crystal size and dirt inclusions that effectively contribute to the flow-law (hardness) parameter *A*. These are discussed in the context of glacier movement in Chapter 7.

4.5 ICE COMPOSITION AND CHEMISTRY

Ice in glaciers includes not just water but also a range of solid, dissolved and gaseous impurities. Ice and snow in different conditions or environments can therefore have unique chemical signatures. The composition and chemistry of glacier ice can be used both to indicate the origin and history of the ice, and also to reconstruct the environmental conditions under which snow was precipitated and ice formed. Major issues frequently considered in glaciological literature include gas in the ice, other soluble and particulate impurities, and stable and radioactive isotopes.

4.5.1 GAS IN ICE

Ice can include gas either as bubbles, or dissolved in the ice, or in the form of clathrate hydrate. When firn is transformed to ice at the surface of a glacier, air is sealed within the ice (Chapter 3). Because snow remains permeable to air and airborne particles until it is transformed into ice, the age of the trapped air reflects the date at which the transformation of firn to ice was completed, not the date of deposition of the snow. Depending on accumulation rates and transformation processes at different locations, air can be up to 3000 years younger than the ice in which it is trapped (Schwander and Stauffer, 1984). The gas concentration of bubbly glacier ice is about 0.1 g cm^{-3}, compared with about 0.0001 g cm^{-3} for bubble-free ice. At the firn/ice transition, air accounts for about 10% of the volume of ice. Total gas content of glacier ice in cold glaciers reflects a number of controls, including pore size at the firn/ice transition, atmospheric pressure and temperature. Since pressure and temperature are both related to altitude, there is some correspondence between total gas content and altitude. Herron and Langway (1987) derived palaeo-elevations for Camp Century and Dye 3 from gas content and paleotemperature records. However, other factors such as seasonal temperature variation and the influence of impurity content on pore volume are such that gas content of ice in ice cores is not a perfect surrogate measure of former ice-surface elevation.

The composition of the gas reflects atmospheric composition at the time of sealing, and measurement of gases such as CO_2, N_2O and CH_4 in ice cores shows changing atmospheric composition through time (e.g. Pearman *et al.*, 1986). However, in warm glaciers the presence of water affects the gas composition. The gas composition of air dissolved in water is very different from the composition of the atmosphere because some gasses dissolve much more readily than others. Water in a glacier-surface snowpack is therefore commonly enriched in CO_2 relative to atmospheric values. This contributes to the chemistry of the glacier ice when the water freezes, and preferential diffusion of different gas components upon freezing has a further effect on the composition of the ice. Diffusion of O_2 is faster than diffusion of the other common gases in ice, so O_2 is preferentially expelled upon freezing, leading to a depletion of O_2 from the ice. Further changes in the chemistry of the ice occur with the flow of meltwater through the glacier. When meltwater is derived from the surface it tends to be saturated with atmosphere-derived gas, and has little impact on the gas composition of the ice through which it flows. Meltwater derived from internal melting of glacier ice, however, is relatively gas free, reflecting the gas composition of the ice. As this gas-poor water flows through the ice it is able to dissolve gases from the ice, and remove the gas via the intergranular drainage network, effectively leaching the gas from the ice. Weiss *et al.* (1972) identified a 20-fold depletion in gas composition of ice between the accumulation and ablation zones of the Aletsch Glacier, which could lead to the progressive transformation of bubbly ice into bubble-poor ice along a flow-line. Battle *et al.* (1996) reviewed some of the major processes that influence the composition of air in firn close to the surface of a glacier, and listed the following:

1. The upper ~10 m can be mixed convectively by surface wind stress.
2. As the equilibrium concentration for component species of the gas mixture increases with depth in the firn, heavier gases and isotopes become progressively enriched with depth.
3. Molecular diffusion allows atmospheric changes to be communicated to firn air, but the extent to which this operates decreases with depth, so the air at any given depth is a weighted mean of atmospheric compositions over some interval of time.
4. Some molecules, including O_2, are preferentially excluded from air bubbles trapped in the ice because their shape allows them to escape along molecular-sized channels during bubble close-off. These gases then diffuse upwards through the firn, possibly altering the composition of the firn air en route.
5. Thermal gradients in the firn cause an enrichment of heavier species and isotopes in its colder regions.

In addition, Severinghaus *et al.* (1998) described how rapid temperature change can cause fractionation in gas isotopes in unconsolidated snow, producing a signal that is preserved in the trapped air bubbles as firnification proceeds. They used this phenomenon to demonstrate the rapidity of warming in Greenland at the end of the Younger Dryas.

Gas bubbles often grow smaller with depth and cease to appear below a certain depth in a glacier. This can be because the gas dissolves into the ice with increasing pressure, such that oxygen or nitrogen molecules occupy positions within the ice lattice. It can also occur because the gas forms clathrate hydrates (Miller, 1969). These are atomic structures where voids or 'cages' in the ice lattice can contain molecules of oxygen or nitrogen. Clathrate hydrates begin to form when the pressure in the bubbles exceeds what is termed the 'dissociation pressure', and have been observed to appear below 727 m in the Byrd ice core, below 1099 m in the Camp Century core and below 1092 m in the Dye 3 core. In these circumstances, gas could be released from melting ice even though no bubbles were visible in the ice.

4.5.2 PARTICULATE AND SOLUBLE IMPURITIES IN ICE

Particulate and soluble impurities in glacier ice mainly reflect: the composition of snow and rain falling on the glacier surface; dry fallout of atmospheric materials onto the surface; and the influence of subglacial materials on the basal layer.

Insoluble microparticles reaching a glacier surface through the atmosphere reflect atmospheric turbidity. Most of the microparticle content recovered from ice cores is in the size range 0.1–2 μm, and is generally referred to as 'dust'. In core records, dust events generally reflect periods of desertification, volcanic activity, or increased entrainment of terrestrial material due either to increased windiness or to enhanced sediment supply. Cyclic variations in concentration of microparticles in cores seem to be annual features reflecting seasonal climate characteristics, but long-term changes in microparticle concentration also reflect transitions between glacial and interglacial periods (Petit *et al.*, 1981; Thompson and Mosley-Thompson, 1981). Glacial periods seem to be characterised by windy, arid conditions with copious terrestrial dust sources, and ice formed in these conditions is commonly characterised by high microparticle content. Variations in particulate impurities are often accompanied by variation in soluble impurities, and both of these variations affect the rheological characteristics of the ice (Chapter 7). Both particulate and soluble impurities can be derived from volcanic eruptions. Cores drilled through glaciers commonly reveal layers of volcanic tephra that reflect fallout from past eruptions (Chapter 11). There is also an extraterrestrial component to the particle content of glacier ice. Cosmic dust and micrometeorites have been reported from several locations (e.g. Langway, 1970; Maurette *et al.*, 1987). Nickel, iridium and platinum are characteristic components of extraterrestrial input. Other impurities, such as lead, derive from anthropogenic pollution.

The ionic composition of ice has been used in a variety of glaciological contexts. The solute load of the ice derives from the chemical composition of the precipitation input, although post-depositional photochemical reactions (Conklin *et al.*, 1993) and the chemical action of water in the snowpack (Koerner, 1970, 1997), at the bed, or within the ice. Cations of Na and K are primarily atmospheric in origin, derived, for example, from sea salt via precipitation. Mg and Ca, referred to as earth alkali cations, are generally associated with mineral weathering, and hence commonly with ice frozen from meltwater at the bed. Cationic composition can be used to distinguish between basal and non-basal ice and to identify processes of water flow near the bed (e.g. Souchez and Lorrain, 1991). Anionic components such as Cl^-, SO_4^{2-} and NO_3^- are also relevant. Legrand and Delmas (1984) demonstrated that chloride in polar snow results mainly from the deposition of sea-salt aerosol (as NaCl), particularly in coastal areas. Sea-salt chloride accounts for ~95% of total chloride in snow at Dolleman Island in the Antarctic peninsula (Mulvaney and Peel, 1988) and 50–60% at the South Pole (Legrand and Delmas, 1984). Sulphate derives primarily from sea-salt sources and from volcanic sources. The rate of deposition at the surface of sulphate derived from volcanic sulphur in the atmosphere is dependent on photochemical oxidation in sunlight, and hence seasonally variable. Anthropogenic activity might also contribute to the sulphate input. Goto-Azuma *et al.* (1997) identified seasonal peaks in ion concentration in snow pits on Agassiz Ice Cap, Ellesmere Island, and suggested that they were not substantially affected by post-depositional processes. However, as previously mentioned in the context of water percolation through the vein network, the chemical record preserved in ice from precipitated snow can be disturbed by water flow. For example, Aristarain and Delmas (1993) measured sulphate (SO_4^{2-}, nitrate (NO_3^-), chloride (Cl^-), sodium (Na^+), potassium (K^+) and ammonium (NH_4^+) concentrations through a shallow (13.17 m) core from the Patagonian ice cap and found that some soluble impurities were washed out by percolation. This leaching process was explored experimentally by Souchez *et al.* (1973) and Souchez and Tison (1981) who demonstrated selective flushing of ions, with alkali metals being more easily removed than alkali earth metals. Koerner (1997) argued that melting could lead to misinterpretation of all of the well-established ice-core proxies, including stable isotopes, solutes and insoluble microparticles.

Close to the base of a glacier, the chemical composition of the ice is strongly conditioned by interactions between the ice and the bed. The chemical consequences of interaction of glaciers with material at their beds is discussed in Chapter 5.

4.5.3 STABLE ISOTOPES IN ICE

The analysis of stable isotopes in glaciers has assumed a position of some significance both in glaciology and in environmental reconstruction in recent decades. Ice is made of water, which is made of atoms of hydrogen and oxygen, but water contains different isotopes of hydrogen and oxygen in different proportions in different circumstances. Stable isotope glaciology uses the isotopic composition of ice and water in glaciers to infer its origin and history. Research has focused on two main areas. First, the preservation of an isotopic signature reflecting environmental conditions in glacier ice formed from precipitated snow provides an opportunity to reconstruct former climatic environments. Second, the alteration of that precipitation signature by processes within the glacier provides the opportunity to identify glacial processes and conditions. This section covers some basic issues in stable isotope glaciology; its application is considered in more detail in sections on ice cores and basal ice in later chapters.

Isotopes are atoms of the same element with different numbers of neutrons and hence different atomic weights. Thus the three isotopes of oxygen, ^{16}O, ^{17}O and ^{18}O, are chemically alike but have different weights and hence different physical properties. Likewise the three isotopes of hydrogen: ^{1}H, D (deuterium, ^{2}H) and ^{3}H (tritium). Water molecules can occur with different combinations of hydrogen and oxygen molecules, and some (e.g. $H_2{}^{18}O$) are heavier than others (e.g. $H_2{}^{16}O$). Naturally occurring water (or ice) contains a mixture of isotopes. The reference standard against which water samples are commonly compared (Standard Mean Ocean Water – SMOW) comprises for oxygen: 99.756% ^{16}O, 0.039% ^{17}O and 0.205% ^{18}O. Naturally occurring hydrogen comprises 99.98% ^{1}H, 0.02% D (^{2}H), and only tiny amounts of ^{3}H. In stable isotope analyses ^{17}O and ^{3}H are not considered. The isotopic composition of a sample is measured as the ratio of heavy isotopes to light isotopes (that is, $^{18}O/^{16}O$ and D/H) in terms of difference from SMOW, given as $\delta^{18}O$ or δD, where:

$$\delta^{18}O = \frac{^{18}O/^{16}O(\text{sample}) - {}^{18}O/^{16}O(\text{smow})}{^{18}O/^{16}O(\text{smow})} \times 1000$$

$$\delta D = \frac{D/H(\text{sample}) - D/H(\text{smow})}{D/H(\text{smow})} \times 1000$$

Values of δ are given in parts per thousand (‰), so for example a sample with $\delta^{18}O$ = +10‰ would be enriched in ^{18}O (depleted in ^{16}O) by 10 parts per thousand (1%) relative to SMOW. Positive δ values indicate an isotopically heavy sample and negative values a light sample, relative to SMOW.

Several natural processes, notably evaporation, condensation and freezing, result in isotopic fractionation, or the separation of heavier from lighter isotopes in a sample. It is the operation of these processes as part of the hydrological cycle that imparts a characteristic isotopic signature to glacier ice. The importance of the hydrological cycle in controlling the isotopic composition of glaciers is discussed further in Chapter 11 (ice cores). It is convenient (if not strictly accurate) to imagine fractionation in terms of a greater willingness for the heavier isotopes to remain in or move into lower energy states, and an eagerness on the part of the lighter isotopes to jump up to, or remain in, higher energy states. In evaporation the vapour pressure of the heavier isotopes is lower than that of the lighter isotopes, so the vapour is relatively depleted in the heavier isotope relative to the source water. In an equilibrium condition water vapour is depleted by about 10‰ in ^{18}O and about 100‰ in D relative to water. As evaporation progresses, and light isotopes are preferentially removed, the remaining water in the source reservoir becomes progressively enriched in heavy isotopes.

In condensation and in freezing the reverse is true: the liquid condensate or the frozen fraction of a sample is isotopically heavier than the source vapour or liquid, and as the condensation or freezing of the sample continues, the shrinking reservoir becomes increasingly depleted of heavy isotopes. In the case of freezing, which is relevant to later discussions on basal ice formation, the first part of a reservoir to freeze in a closed system will produce ice which is isotopically slightly heavy relative to the parent water, enriched by about 3‰ in ^{18}O (Craig, 1961). However, this initial depletion of heavy isotopes from the parent water means that subsequent freezing will be drawing on an isotopically lighter parent. Thus each successive freezing fraction produces isotopically lighter ice and leaves behind an increasingly δ-negative reservoir. When most of the reservoir has been depleted, the last fractions to freeze will produce the most isotopically negative ice. In a closed system the total isotopic composition of the ice must be equivalent to the initial water, and although the first fractions to freeze were isotopically heavy, and the last fractions isotopically light, if the ice were completely remelted the isotopic composition of the meltwater would be unaltered from the initial reservoir. In reality, however, closed-system total freezing does not always occur. When part of a glacial water source freezes, and part escapes, the fractionation effect of the freezing can be preserved significantly in both the refrozen ice and the escaping water. Fractionation does not occur on melting of ice as the rate of molecular diffusion through ice is very slow.

Considering fractionation of both hydrogen and oxygen isotopes together (co-isotopic analysis), it is possible to identify specific isotopic characteristics or signatures associated with specific processes. Craig (1961) identified the so-called precipitation effect, whereby samples of meteoric water (including unaltered glacier ice derived from compaction of snow) conform to the linear relationship:

$$\delta D = 8\,\delta^{18}O + 10$$

Samples of meteoric water or glacier ice will lie along this 'meteoric water line' of gradient 8 when plotted on a $\delta D/\delta^{18}O$ graph (Figure 4.3). By contrast, samples of ice formed by freezing from a liquid will lie along a quite different 'freezing slope' when plotted on a $\delta D/\delta^{18}O$ graph. Jouzel and Souchez (1982) demonstrated that for a closed system, where no water is lost or added, the gradient of the freezing slope depends on the initial isotopic composition of the liquid admitted to freeze, such that:

$$S = \frac{(\alpha - 1)}{(\beta - 1)} \times \frac{(1000 + \delta_i D)}{(1000 + \delta_i{}^{18}O)}$$

where α and β are the fractionation coefficients for D and ^{18}O respectively, (1.0212 and 1.00291; Lehmann and Siegenthaler, 1991), and ($\delta_i D$, $\delta_i{}^{18}O$) is the isotopic composition of the initial liquid. When the initial liquid is SMOW, $\delta_i D/\delta_i{}^{18}O = 0$, so the slope $S = (\alpha - 1)/(\beta - 1) = 7.29$.

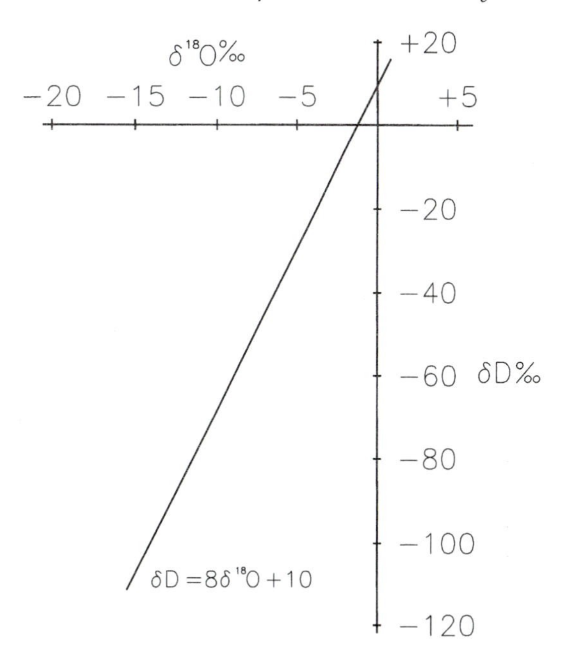

Figure 4.3 $\delta D/\delta^{18}O$ graph showing the relationship typical of precipitation and unaltered glacier ice.

Freezing more negative parent waters produces lower gradient freezing slopes. If the liquid admitted to freeze has the isotopic characteristics of meteoric water ($\delta D = 8\,\delta^{18}O + 10$), then the point at which the freezing slope intersects the precipitation slope gives the isotopic composition of the parent water. If water is frozen in successive stages in a closed system, then the first fractions to freeze (up to ~0.67 of the sample) will be enriched in heavy isotopes relative to the parent water, and the later stages depleted. For an open system where the reservoir of water is continually replenished by water of like isotopic composition, Souchez and Jouzel (1984) showed that the freezing slope is defined by:

$$S = (\alpha / \beta) \times [(\alpha - 1) / (\beta - 1)] \times [(1000 + \delta_i D) / (1000 + \delta_i{}^{18}O)]$$

Under certain circumstances, the refrozen ice can be isotopically indistinguishable from meteoric glacier ice. This can happen, for example, when the water replenishing the reservoir is isotopically lighter than the original water

(Souchez and de Groote, 1985), or when different meteoric parent waters contribute sequentially to an accretion of basal ice (Knight, 1987). Sharp *et al.* (1994) pointed out that when basal ice forms from a range of different parent waters, the empirically determined slope may be no more than a statistical artefact. Hubbard and Sharp (1995) adopted a model where both the isotopic composition of the parent liquid and the proportion of the liquid that freezes were variable over time in an open system. In this model the possible isotopic character of the frozen fraction is defined by an envelope of isotopic signatures, such that any sample of ice falling within the envelope could be produced by refreezing. On that basis, possible refrozen ice can be identified even though the samples do not define a clear co-isotopic freezing slope.

Souchez and Lorrain (1991) have provided a useful review of co-isotopic analysis.

4.5.4 UNSTABLE ISOTOPES IN ICE

Glacier ice can include radioactive isotopes. Radioactive layers caused by fallout of radionuclides such as ^{90}Sr onto glacier accumulation areas have resulted from several incidents, including nuclear bomb tests and the 1986 Chernobyl nuclear accident (e.g. Pourchet *et al.*, 1988). These layers can be used as reference horizons for dating ice cores and measuring accumulation rates. Naturally occurring unstable isotopes such as ^{32}Si and ^{210}Pb have also been used in glaciological research. For example, ^{210}Pb is produced naturally in the atmosphere and accumulates in glaciers via precipitation. The average concentration of ^{210}Pb in precipitation is of the order of five disintegrations per minute per litre. ^{210}Pb decays with a half-life of 22 years to ^{210}Bi and with a half-life of 5 days to ^{210}Po, which itself decays with a half-life of 138 days to the stable ^{206}Pb. Where no melting occurs measurement of the decay of ^{210}Pb and its daughter products can be used to date ice up to 150 years old, and can also therefore be used for reconstructing short-term ice dynamics. For example, Piciotto *et al.* (1967) dated ice samples from a glacier in the Austrian Alps to reveal a pattern of ageing with distance down-flow from the equilibrium line that was consistent with expected flow patterns. Similarly, Nijampurkar *et al.* (1985) found a systematic decrease in ^{210}Pb activity along a longitudinal profile down the surface of Changme-Khangpu Glacier (India), and used that to infer flow rates and ages of ice.

4.6 IMPORTANCE OF TEMPERATURE TO GLACIER CHARACTERISTICS

The temperature of a glacier is one of the most important parameters controlling its character. Glacier morphology, glacier behaviour and the impact of glaciers on their environment are all thermally controlled. The heat budget at the glacier surface is a direct interface between ice and climate in the global system, and glacier surface temperatures play a crucial role in determining the character of the transformation of snow to ice. Ice temperature exerts a direct control on ice rheology, on the presence of water, on mechanisms of movement, on geomorphic processes and on the physical and chemical characteristics of the ice. Temperature distribution through a glacier can be used as an analytical tool: temperature-depth profiles can be used to model former ice-sheet conditions and to test paleotemperature histories derived from data such as $\delta^{18}O$ measurements. Both absolute temperature and the temperature relative to the melting point are important. The melting point itself varies with pressure and with the chemistry of the ice and of the water in the vein network. Glaciers are frequently defined in terms of their thermal state as a shorthand statement of their overall character. In temperate glaciers the ice is at the melting point, water can flow through and under the ice, basal sliding can occur and the ice is relatively soft for deformation. In cold glaciers the ice is below the melting point, hydrological processes are suppressed, sliding is limited and the ice is relatively stiff. In polythermal glaciers, transitions between different thermal zones can be marked by major changes in glacier characteristics.

4.7 CONTROLS ON GLACIER TEMPERATURE

Temperatures in glaciers are controlled by heat sources and sinks at the surface, at the base and in the interior of the ice, and by transfer of heat through the ice. Table 4.3 summarises some of the key parameters, which are discussed in the following paragraphs.

4.7.1 SURFACE TEMPERATURE CONTROLS

. . . the changes of temperature in the different seasons must little affect the deeper masses of the ice, the mean temperature of which must exceed that of the air. During the whole summer all the heat cast upon the surface is employed in melting, and, being absorbed by the water, it is carried down in a latent state through the ravines. The cold of winter, on the other hand, will in some degree be kept off owing to the snowy covering being a bad conductor.

(Rink, 1877)

The temperature of the upper surface of a glacier, and of the snow that accumulates at the surface, is determined mainly by local climate and meteorological conditions, primarily related to elevation, latitude and continentality. The effect can be modelled, and the key parameters identified, as a surface energy balance of the type that was discussed in Chapter 3. A simple formulation for an ice surface equates temperature and phase changes at the surface with the net radiation flux, turbulent heat flux, and the effect of freezing externally derived water:

$$k\,(dT_{ice}/dt) + B = Q(1-\alpha) + L_1 - L_O + H + L_E + L_F$$

where k = heat capacity of ice; dT_{ice}/dt = change in ice temperature through time; B = snowmelt or refreezing of meltwater; Q = incoming short-wave radiation; α = albedo; L_1 = incoming long-wave radiation; L_O = long-wave radiation emitted from the surface; $Q(1-\alpha) + L_1 - L_o)$ = net radiation flux; H = sensible heat transfer by turbulence (convection, conduction, etc.); L_E = latent heat transfer by evaporation (+ve) or condensation (–ve); L_F = latent heat transferred

Table 4.3 Factors controlling glacier temperature

Control on temperature	*Relevant parameters*
Energy flux between ice and atmosphere	
Incoming short and long wave radiation	Solar constant, atmospheric absorption, cloudiness
Reflected and emitted radiation	Albedo, surface characteristics
Turbulent heat flux	Temperature, wind and humidity profile close to ice/atmosphere boundary
Sensible heat in accumulating ice	Snow temperature, accumulation, surface heat flux
Energy flux between ice and substrate	
Geothermal energy	Geothermal flux, substrate conditions
Dynamic heat sources	
Frictional heating	Basal shear stress, sliding velocity
Strain heating	Ice deformation
Advection	Horizontal ice velocity
Energy released or absorbed by phase changes	
Evaporation, condensation, freezing, melting, sublimation	Temperature and pressure fluctuations, proximity to critical temperature.
Heat transfer by horizontal advection	
Meltwater – sensible/latent	Meltwater production and output
Ice – sensible/latent	Horizontal ice velocity, ice temperature

by freezing of precipitated rain (or other externally derived water).

The left side of this equation is the change in ice temperature combined with melting or freezing at the surface. The right side is the energy flux between the ice surface and the atmosphere, which is primarily dependent on solar radiation, on surface albedo (which determines how much of that radiation is effective) and on turbulent energy transfer. Turbulent transfer involves convection and conduction between the atmosphere and the surface, combined with latent heat transfer by evaporation and condensation at the surface. Heat transfer by these processes increases with increased atmospheric turbulence. Clarke *et al.* (1987), Colbeck (1989) and Clarke and Waddington (1991) described 'wind pumping' that can heat the firn by frictional dissipation of energy as air is forced through firn both vertically and horizontally in response to surface air-pressure variations associated with turbulence. Albedo (the reflectivity of the surface) is determined by a range of factors including density, texture and colour of the surface. There is a close relationship between mass balance and albedo as different glacier surface materials have different albedos (Table 4.4). An example of the possible linkage between albedo and accumulation is provided by the so-called 'blue ice' zones at the surface of Antarctic ice sheet. These are areas where the surface of the ice sheet is bare of snow due to local topographic controls. The albedo of bare ice is ~0.56, while the albedo of adjacent snowfields is ~0.8. The blue ice areas therefore absorb more radiation, such that ambient temperature in blue ice areas can be 3°C warmer than the snow-covered areas (Bintanja, 1995). Phase changes are critical in controlling ice temperature on the left side of the equation, and in the latent heat transfers on the right of the equation. For example, the percolation and refreezing of meltwater transfers both sensible and latent heat into the surface layers of the ice. Refreezing of 1 g of meltwater in a snowpack releases sufficient energy to raise the temperature of 160 g of snow by 1°C. When the surface is below freezing point, rainwater falling onto the surface also supplies heat as it freezes. Condensation of 1 g of water at the surface releases enough heat to melt about 8 g of ice.

Table 4.4 Albedo of glacier surface materials (data from sources in Paterson, 1994)

Surface	*Range of albedo*
Dry snow	80–97
Melting snow	66–88
Firn	43–69
Clean ice	34–51
Dirty and debris-covered ice	10–83

The temperature at the glacier surface has a direct effect on ice temperatures in a surface layer about 10 m thick. In this layer, temperatures fluctuate seasonally as heat conduction transfers surface temperature fluctuations into the ice. Ice temperature fluctuates less than surface conditions, as several stabilising processes operate within the heat budget. For example, increased wind speeds increase evaporation and hence cool the ice, but also increase turbulent heat transfer into the ice. Increasing solar radiation increases surface temperature, but also increases evaporation and long-wave radiation from the surface, leading to cooling. If there is a seasonal snow cover it serves to moderate surface temperature fluctuations as they are transmitted into the glacier, because the snow protects the ice surface from the winter cold, and melts in spring so the cold is not transmitted downwards. At the bottom of this surface layer, at a depth of about 10 m, the temperature is steady. If the surface temperature never rises above the melting point, and no water is produced, then the 10 m temperature is about equal to the mean annual surface temperature. Below this depth, ice temperatures reflect former surface temperatures being carried downwards into the glacier and being modified by subsurface temperature effects. Conduction from the surface makes only a small contribution to ice temperature below the 10 m level.

4.7.2 INTERNAL TEMPERATURE CONTROLS

Temperature in the interior of the glacier is determined mainly by the following factors.

1. The temperature of the accumulated snow and ice brought down from the surface by vertical advection.
2. The accumulation rate. The accumulation rate determines the rate at which accumulating firn is carried down into the ice. A high accumulation rate carries cold down into the ice most effectively, so temperatures within the glacier and at the bed tend to be inversely related to accumulation rate (e.g. Robin, 1955; Sugden, 1977).
3. Conduction of heat into the ice from the surface of the glacier. Seasonal variations penetrate 10–15 m, but long-term (glacial/interglacial) cycles could penetrate much deeper.
4. Conduction of heat into the ice from the base of the glacier.
5. Horizontal conduction of heat from adjacent parts of the glacier.
6. Horizontal advection of ice. The temperature of ice at any point in the glacier is affected by the temperature of ice that flows into that position from up-glacier. High accumulation rates, low snow temperature and rapid flow velocity all favour rapid replacement of warmed ice by cold fresh ice from upstream. In the ablation zone, horizontal advection can introduce warmer ice from up-glacier.
7. Energy released or consumed by melting or refreezing.
8. Energy released by ice deformation. Strain heating tends to be concentrated near base because most deformation occurs there. Funk *et al.* (1994) suggested that a basal layer of ice warmed by strain heating may be an important mechanism of fast ice-stream flow.
9. Penetration of radiation close to the surface.

The primary controls on interior ice temperature are thus, indirectly, the controls on temperature in the surface layers of ice, moderated by internal thermal effects and heating from the base.

4.7.3 BASAL TEMPERATURE CONTROLS

The heat flux at the base of a glacier is determined mainly by the following factors.

1. Geothermal heat input.
2. Heat generated by friction of basal sliding.
3. Heat released by freezing of water or consumed by melting of ice.

As early as the eighteenth century, Gruner (1760) had suggested that the melting of glaciers along their lower surface was due largely to the internal heat of the earth. Geothermal heat flux into the base of the ice varies from place to place, being greater on oceanic crust and near volcanically active areas. Measurements of geothermal heating beneath glaciers are relatively few, but values probably range from about 20 to 90 mW m^{-2}. Budd and Young (1983) estimated a value of 50 mW m^{-2} for Camp Century, Greenland; and Firestone *et al.* (1990) gave a value of 41.7 mW m^{-2} for Summit, Greenland. Realistic values would be sufficient to melt up to ~2 cm of ice in a year. Waddington (1987) provided a useful review of data on geothermal heat flux beneath ice sheets, and suggested that values of steady geothermal flux could vary by as much as 50% over spatial and temporal scales relevant to ice sheet modelling. Frictional heat generated by basal sliding is proportional to the sliding velocity and the basal shear stress. For typical glaciers it provides an amount of energy of the same order of magnitude as the geothermal heat flux. Transfer of latent heat by freezing of water or consumed by melting of ice can be associated with closed-system processes such as regelation sliding and the heat pump effect (Chapter 7) or with open-system effects such as the movement of ice and water between areas of different basal thermal regime. In addition, sensible heat can be imported or exported by flow of water into or out of an area of the bed.

The relative importance of these controls varies between different locations. The possibility of heat exchange on melting or freezing depends on the basal temperature being close to the melting point. Where the bed is below the melting point, the basal temperature depends mainly on the geothermal energy input and heat of friction; where the bed is close to the melting point, heat released by melting of ice or absorbed by freezing of meltwater is also important. In polythermal glaciers water flow between different thermal zones at the bed can provide major heat sinks and sources at certain sites. The possibility also exists of superheating water above the melting point in some subglacial stores, and of supercooling water below the freezing point. For example, Strasser *et al.* (1996) described water being supercooled as it flowed uphill in an over-deepening near the terminus of Matanuska Glacier, Alaska. As the water flows uphill out of the hollow, the pressure melting point rises faster than the water temperature increases, and ice begins to crystallise within the supercooled water. Zotikov (1986) elaborated alternative thermal conditions for glaciers with warm and cold beds, and with rock and till substrates. The nature of the substrate is important, for example in controlling water flow and affecting geothermal flux to the base of the ice. Heat exchange can be complicated by circulation and storage of water in a permeable substrate. Where the bed is rock, the energy supplied to the base of the ice is equivalent to the geothermal heat flux; but where the substrate is permeable, some of the heat supplied by the geothermal flux can be employed in phase changes within the substrate, so the heat supplied to the base of the ice is less than the geothermal flux.

A simple description of controls on the basal heat flux, neglecting vertical and horizontal advection of ice, can be given by:

$$q_\lambda = q_{fr} + q_o - (q_r + q_w)$$

where q_λ = heat flux at the base; q_{fr} = heat generated by friction; q_o = geothermal heat; q_r = heat on melting/freezing; q_w = heat transfer through meltwater advection.

4.8 THERMAL REGIME AND SPATIAL VARIATIONS IN TEMPERATURE

Different glaciers exhibit different thermal characteristics. Some glaciers are below the melting point throughout, and are referred to as 'cold'. Others are at the melting point throughout, and are referred to as 'warm' or 'temperate'. Between these two extremes are many other variations. For example, many glaciers are at the melting point except for a surface layer that is subject to seasonal temperature fluctuations, and some glaciers are cold except for a layer of temperate ice near to the bed. Many glaciers exhibit areal variations in temperature, with some parts being at the melting point and some parts being cold. Glaciers with several different thermal zones are sometimes referred to as 'polythermal'. Both vertical and horizontal temperature distributions exert important controls on glacier behaviour. Temperature distributions can in many cases be predicted from the controls listed in the preceding paragraphs, although the interaction of different parameters makes modelling difficult. Waddington (1987) provided a review of the contemporary status of ice temperature models, and concluded that the major limiting factor in thermal modelling was likely to be boundary data. Temperatures have been measured directly in many glaciers, and measurement and modelling of temperatures at the surface, in the interior and at the base of glaciers have indicated substantial spatial variations in thermal conditions.

4.8.1 VERTICAL TEMPERATURE PROFILES

The temperature of the upper 10–15 m of a glacier is affected by seasonal variations that propagate from the surface. The magnitude of the seasonal variation decreases with depth, and at the bottom of this 10–15 m layer the temperature is close to the mean annual surface temperature, unless heat is released by refreezing of percolated meltwater. Below that level, the effects of internal and basal heating lead to a general increase in temperature with depth. If geothermal energy were the only heat source, and the ice were stationary, temperature would increase steadily with depth at a rate equivalent to the geothermal gradient (line A on Figure 4.4). A geothermal flux of 50 mW m^{-2} would lead to a temperature gradient through the ice of 0.024 K m^{-1}. For a 1000 m ice mass this implies that if the surface temperature was −24°C, the temperature at the base would be 0°C. If the surface temperature was warmer than −24°C, the thermal gradient would be less, so a heat excess at base would be available for melting. Strain heating, concentrated near to the bed, increases the warming effect towards the base (line B on Figure 4.4). Accumulation at the ice surface and transport of surface ice downwards into the body of the glacier has a cooling effect such that the temperature increase with depth is less

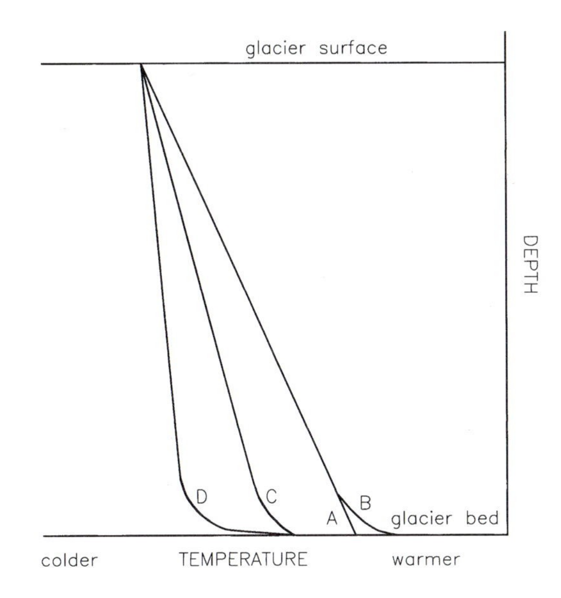

Figure 4.4 Vertical thermal gradients through a glacier in four different theoretical situations (lines A, B, C, D). See text for explanation.

pronounced. The greater the accumulation rate, the less is the temperature difference between the base and the surface of the ice (line C, Figure 4.4). For a moving glacier, horizontal advection of ice from up-glacier has an important effect. At progressively lower levels in the ice column, ice is derived from locations higher in the accumulation area, where snow originally accumulated at lower temperatures. Horizontal advection of this ice thus introduces a negative aspect to the vertical temperature gradient (cooling with depth) which reduces the overall warming tendency. If horizontal advection is pronounced, portions of the ice column can be marked by temperature reduction with depth. However, high rates of horizontal advection might be associated with high strain or basal sliding, which would generate additional heat close to the bed, resulting in a profile like line D in Figure 4.4. Further complications are added by long-term climate changes and changes in glacier thickness or surface elevation, which alter the temperature of snow accumulating at the surface. For example, Blatter (1987) reported that the vertical temperature profile through White Glacier, Axel Heiberg Island, was marked by a temperature minimum 100–150 m below the surface as a result of climatic warming since 1880.

Temperature profiles that have been measured do reflect many of the general characteristics described above, with local conditions determining specific features. Figure 4.5 illustrates profiles that have been measured at various locations. Robin (1983) presented a review of data from a range of sources, and Paterson (1983) reviewed some of the methods by which these profiles have been measured.

4.8.2 VARIATIONS IN TEMPERATURE PROFILE ALONG A FLOW LINE

The vertical temperature profile through an ice sheet varies with distance along a line from the centre to the margin. Dahl-Jensen (1989) modelled these variations, making some simplifying assumptions (horizontal bed, steady-state

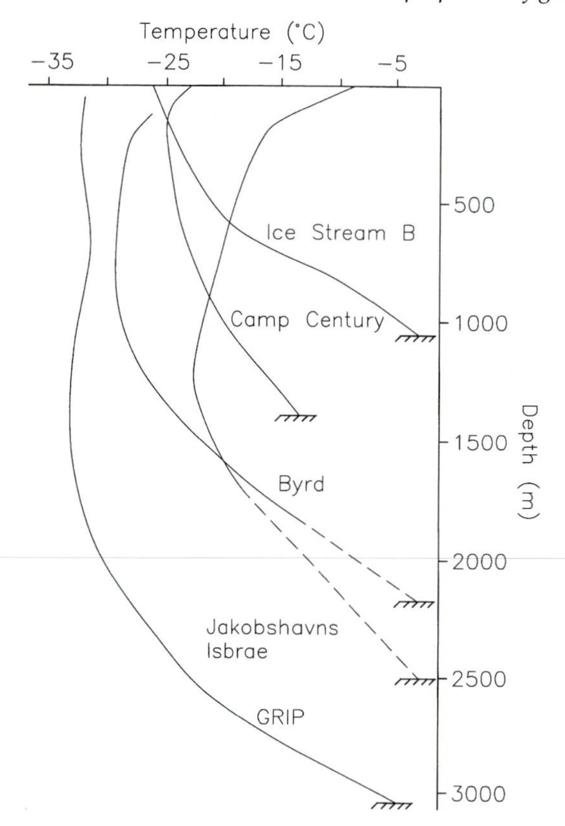

Figure 4.5 Vertical thermal gradients observed in a number of glaciers. Data from Robin (1983) and Johnsen *et al.* (1995)

two-dimensional flow, constant thermal conductivity and geothermal heat flux, and idealised longitudinal variations in elevation, surface temperature, and mass balance). The model showed that basal temperature increased with distance from the top of the flow line towards the margin as surface temperature and strain heating increased (Figure 4.6). There was an intermediate zone of basal melting and temperate basal ice, and a peripheral zone where the basal temperate layer thinned to zero as strain heating reduced in the thinner ice close to the margin. Horizontal advection produced a temperature minimum in the mid-section of the profile that accentuated down-glacier into the ablation zone, where it persisted in the upper section of the profile until removed from the ice by surface ablation. Measured flow-line variations in vertical temperature distributions reflect the general characteristics of this model, but with complications arising from local conditions that differ from the model's simplifying assumptions. For example, real glaciers generally do not have horizontal beds, do not vary in thickness and surface temperature steadily with distance from the centre-line, and are not necessarily in a steady state.

4.8.3 HORIZONTAL TEMPERATURE DISTRIBUTIONS AND BASAL THERMAL REGIME

Variations in thermal characteristics at the bed of a glacier are especially important, as the basal thermal regime plays a dominant role in

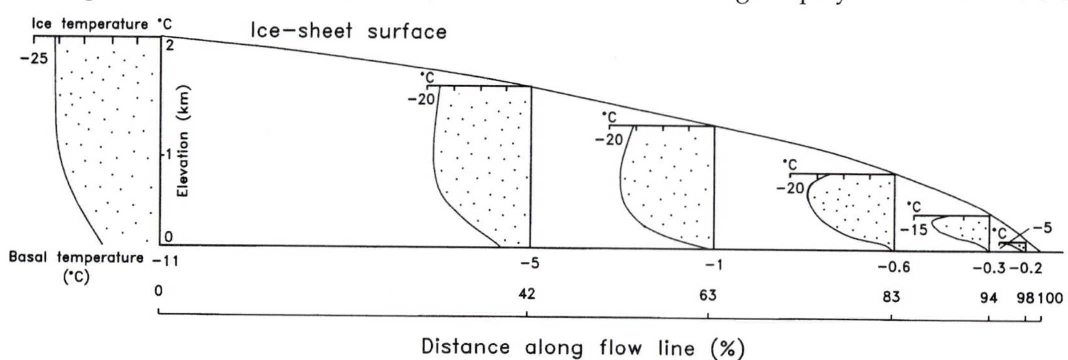

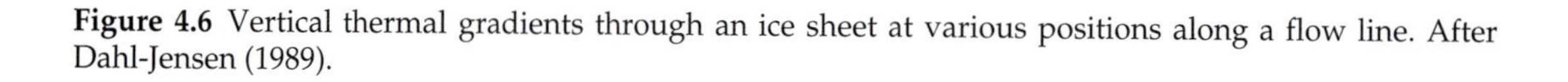

Figure 4.6 Vertical thermal gradients through an ice sheet at various positions along a flow line. After Dahl-Jensen (1989).

glacier movement and in the impact of glaciers on landscape. The temperature at the glacier bed is determined by the input of heat to the bed and the ability of the ice to conduct the heat away from the bed. Three thermal situations can arise, each associated with different basal characteristics (Weertman, 1961):

1. More heat is provided than can be removed. In this situation net melting of ice occurs at the base, and the base of the glacier is at the melting point.
2. The amount of heat provided at the base is approximately equal to the amount that can be removed. In this situation there is a balance between melting and freezing, and the base of the glacier is at the melting point.
3. The amount of heat provided is less than that which could be removed. In this case one of two situations can develop. Either the glacier bed is frozen, or water arriving from elsewhere freezes in this 'cold' zone and in doing so releases enough heat to maintain ice at melting point.

Boulton (1972) elaborated these conditions into a four-fold model of basal thermal regime. He identified possible thermal zones as follows: (A) net basal melting; (B) a balance between melting and freezing; (C) meltwater freezing to the bed maintains the bed at the melting temperature; (D) below the melting point. These different thermal zones can exist simultaneously in different parts of a glacier, and both large ice sheets and smaller glaciers exhibit distinct zones of different basal thermal regime. A warm base is favoured by thicker ice, high surface temperatures, light accumulation rates, high basal velocity and high geothermal heat flux. The opposite conditions favour a cold base. Many glaciers terminating in cold environments are marked by a cold fringe around their margins, at which meltwater derived from warmer basal zones in the interior may freeze to the bed beneath thinner, slower moving ice close to the margin. These cold-fringed glaciers are sometimes referred to as 'sub-polar'. Similar features can result in different ways. For example, Storglaciären in Sweden is a temperate valley glacier with a perennially cold surface layer 20–70 m thick in the ablation area. Where the ice is thin close to the margin this surface layer makes contact with the bed to create a cold rim 100–200 m wide around the periphery of the glacier where the ice is frozen to the bed (Holmlund *et al.* 1996b).

For large ice sheets, several numerical models of the distribution of thermal conditions have been produced. Budd *et al.* (1970) developed a model that related basal temperatures to ice thickness, surface temperature, accumulation rate, velocity, surface warming rate and the basal temperature gradient. Sugden (1977) applied that model to reconstruct the basal thermal regime of the Laurentide ice sheet at its maximum. Using input data from known or presumed characteristics of the ice sheet's dimensions and surface conditions, his reconstruction indicated four roughly concentric zones of different thermal conditions from centre to periphery:

1. a zone of warm-melting beneath the centre of the ice sheet;
2. a zone of warm-freezing;
3. a cold-based zone;
4. a peripheral zone of warm-melting.

Models of this type have important implications for the distribution of geomorphic processes beneath ice sheets (Chapter 9).

4.9 CONCLUSION

The characteristics and behaviour of glaciers are determined to a large extent by the properties of the material from which they are made, so an understanding of the properties of ice is the necessary basis of a sound understanding of glaciers. However, glaciers comprise more than ice alone. Huge, long-lasting masses of polythermal, polycrystalline glacier ice, incorporating particulate debris, gas, solutes and water under self-generated thermal and pressure gradients, are complex physical and chemical systems. The physical, chemical and thermal properties of glaciers do exert a control on glacier behaviour, but are also controlled by it. Glacier dynamics, thermal regime, hydrology, chemistry and rheology are all complexly interrelated. The interrelationship between the physical, chemical and dynamic properties of ice and of glaciers remains a major area in which our understanding and modelling of glacier behaviour is imperfect.

STRUCTURE AND MORPHOLOGY OF GLACIERS 5

5.1 INTRODUCTION

Glaciers come in all shapes and sizes: from continental ice sheets that engulf substantial portions of the planet to tiny ice aprons that shelter precariously on precipitous mountainsides; and from high-altitude tropical ice caps to floating polar ice shelves. However, the physical properties of ice are approximately constant between glaciers, and the major environmental controls on glacier behaviour – such as gravity, precipitation and ablation – are ubiquitous. Therefore, glaciers of different types share many characteristics, and we can discuss glaciers in terms of a common basic anatomy. This chapter will consider first the overall shape of glaciers, and then explore the different parts or zones of which a glacier is composed.

5.2 GROSS MORPHOLOGY

5.2.1 PROFILE AND PLAN

The overall shape, or gross morphology, of a glacier can be considered in terms of its profile and its plan. Both are controlled by a combination of (i) the distribution of accumulation and ablation, (ii) the subglacial topography and (iii) ice rheology and dynamic forces.

The plan form of a glacier depends primarily on how far ice flows in any direction from its point of origin before it ablates. A theoretical ice sheet with a circular accumulation area in an otherwise isotropic environment would be expected to be circular in plan as the ice would flow the same distance in all directions before ablating. If ice flowed faster in one direction than another (for example, because of different temperature or bed material), the ice would travel further in that direction before ablating, so the ice sheet would extend further in that direction. If more ice was supplied from one part of the accumulation area than others, perhaps because of topographic control on the distribution of snowfall, then the part of the ice sheet supplied from that area would extend further than other parts. If ice flow was funnelled in a particular direction by local topography, the ice sheet would extend further in that direction. If the ice descended down a steeper slope in one direction, moving more abruptly into warmer elevations, quicker ablation would limit the ice extent in that direction. For a large ice sheet, differences in latitude between different aspects could have a similar effect; the sheet would be expected to extend furthest in a poleward direction.

The profile of a glacier is also determined by the relationship between ablation and distance travelled, which is in turn largely dependent on flow dynamics. For example, surging glaciers and ice-sheet lobes resting on highly mobile deforming beds tend to have lower (flatter) profiles than slowly moving glaciers on beds with a high shear stress. The gross morphology of a glacier thus depends on internal ice rheological characteristics, on mass balance conditions, and on external environmental controls.

5.2.2 THE INFLUENCE OF ICE RHEOLOGY AND DYNAMIC PROCESSES

The way in which ice moves is a major control on glacier morphology. Glacier movement is discussed more fully in Chapter 7. The main processes are (i) deformation of the ice under its own weight and (ii) movement of the glacier across its substrate. Both of these styles of movement occur in response to stress, or force applied to the ice. The driving stress for glacier movement is generally approximated by the weight of the ice and its surface gradient, given as:

$$\tau = \rho g h \sin \alpha$$

where τ = shear stress; ρ = density of ice; g = force of gravity; h = thickness of ice; and α = ice surface gradient.

In simple modelling approximations it is common to consider a glacier as a 'perfectly plastic' body that does not start to deform or move until a certain critical stress is reached. Above this 'yield stress', deformation or movement increases suddenly. This is not strictly an accurate description of how ice flows, but is a useful approximation if we are trying to incorporate the effects of all the different movement mechanisms into a general picture of ice flow. In this approximation, the surface profile of a glacier is dependent on the yield stress. For example, a glacier slides more easily over a wet bed than over a frozen bed, so the yield stress at which sliding begins is higher with a frozen bed than with a wet bed; in other words, a higher stress is required to initiate motion. The variables that control the driving stress are ice thickness and surface gradient, so a frozen-bed glacier would be expected to attain a greater thickness, or a steeper surface slope, before the yield stress is reached and movement is initiated. Once the yield stress is reached, any further tendency towards an increase in stress will be accommodated by movement of the glacier, so the stress will be limited at the yield stress and the glacier will maintain a thickness and steepness in equilibrium with that yield stress. Thus glaciers with frozen beds will tend to be thicker and/or steeper than warm-based glaciers. Likewise, colder ice is more viscous than warm ice, so the yield stress for internal deformation is higher for colder ice. Cold glaciers are thus likely to attain greater thickness and steepness than warm glaciers. In theory, a glacier thus adjusts its profile by responding to applied stress in such a way that the stress throughout the glacier is equal to the yield stress: $\tau = \tau_0$. The yield stress for a glacier depends on friction or resistance to movement both internally and at the bed. Lower resistance equates to a lower yield stress, and hence to lower ice thickness and/or surface gradient. Floating ice with virtually zero basal shear stress has very low surface gradient. Yield stress values of about 50–150 kPa seem to be realistic for alpine glaciers, and 0–100 kPa for ice sheets. For a glacier 250 m thick, with a surface gradient of 2°, assuming ρ = 0.9 g cm^{-3} and g = 9.8 m s^{-2}, $\tau = 0.9 \times 9.8 \times 250 \times \sin 2 = 77$ kPa. If bed conditions beneath such a glacier changed such that it behaved as if the yield stress were lower than 77 kPa, the glacier would flow faster, and the thickness and/or gradient would decrease until the driving stress matched the yield stress. A glacier 200 m thick, in equilibrium with a yield stress of 50 kPa, would have a surface gradient of ~1.5°. Surface gradient and ice thickness are thus part of a dynamic equation, $\tau = \rho g h \sin \alpha$, and are inversely related in an equilibrium condition. For a constant stress, an increase in ice thickness would be expected to accompany a decrease in gradient, and vice versa. For a change in stress, (for example, in a non-equilibrium glacier that is growing or shrinking), adjustments in surface profile would be predicted. An increase in stress would be expected to lead to an increase in gradient and/or thickness of the ice. If the ice volume remains constant, changes in glacier profile result in changes in the areal extent of the glacier. The mutual adjustment of profile and driving stress has important implications for glaciology. Glacier surface profiles can be used to calculate stress conditions and hence infer probable bed conditions and flow states for both existing and former glaciers. The low surface gradient of lobes of the former Laurentide ice sheet suggest low stress conditions, and have been cited to support hypotheses of deforming bed conditions and high-magnitude subglacial floods. Shoemaker (1992) suggested that low-relief ice-sheet lobes resulted from episodic subglacial floods occurring as water sheets hundreds of kilometres wide. He suggested that the Michigan Lobe of the former Laurentide ice sheet witnessed subglacial water sheets in excess of 6 m thick, causing release of basal shear stress and elongation of the lobe by tens of kilometres. In surging glaciers, the change of stress conditions between pre-surge state (thicker, steeper ice: high stress) and post-surge state (thinner, lower-gradient ice: low stress) can contribute to surge initiation and termination.

The equilibrium profile of a simply approximated theoretical glacier is a thus parabola controlled by the actual stress generated in the glacier and the yield stress required for motion to occur (Figure 5.1). The ice thickness adjusts so that $\tau_b = \tau_0$ throughout, so if the yield stress is known, the thickness and gradient can be predicted for any position between the margin and the centre (Nye, 1952a). Various models of different types have attempted to characterise this theoretical parabolic profile, but comparisons between modelled and measured profiles have indicated that the use of a single term to define a universal profile is unrealistic. Conditions, and hence profiles, vary substantially between different glaciers and even between different parts of a single glacier. A major problem for modellers is that the controls on flow within a glacier are more complex than can be easily accommodated within a conveniently solvable mathematical approximation. Mathematical models based on approximations of ice dynamic properties have frequently ignored some of the primary controls on glacier behaviour (such as basal water, or impurities in ice) and have made barely tenable assumptions, such as that the ice sheet rests on a flat bed or is in stable equilibrium condition. Others, including a class known as finite-element models, have taken greater account of differences between sections of an ice mass. Finite-element models have achieved some realistic results that stand up to comparison with measured profiles (Reeh, 1988). For models of existing ice sheets, it is easy to test the model output against reality. However, for models of former ice sheets this is less feasible. For models of former ice sheets there are, on the whole, too many unknown elements for the models to be reliably constrained. A great deal of glacier modelling has assumed glacier morphology to be controlled almost exclusively by ice rheology, and topographic influences have often been neglected entirely. The literature is full of parallel-sided slabs resting on rigid horizontal beds and conforming to perfectly plastic parabolic profiles. One of the exciting developments in the last few years has been an increasing awareness that modellers will have to take on board the reality of different subglacial landscapes and materials in order to model the behaviour of realistic ice masses. The mechanical properties of 'realistic' glacier materials, including polythermal mixtures of ice, debris and water, are poorly known, but must play a key role in determining the dynamic and hence morphological properties of

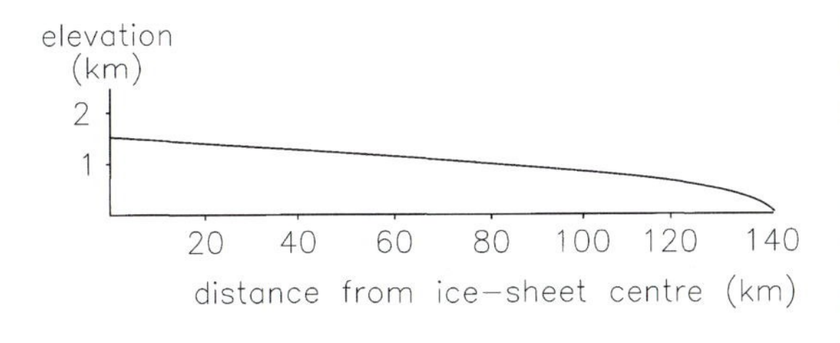

Figure 5.1 Theoretically modelled parabolic form of ice-sheet profile.

ice sheets. Lawson (1996) found that the relationship between the behaviour of clean and debris-laden ice under uniaxial compression is complex and variable, partly because of the temperature dependence of pressure-melting processes. She concluded that the effects of a debris-rich basal layer are likely to be highly variable in space and time. This is an important issue, and one that seems to warrant further work in order to supply modellers with critical information. Ice-sheet modelling is a major area of ongoing progress in glaciology, albeit one fraught with difficulties. Hindmarsh (1993) provided a convenient review of some of the main issues involved in modelling ice-sheet dynamics.

5.2.3 TOPOGRAPHIC CONTROL

In ice sheets that entirely overwhelm the underlying landscape, the flow direction is not strongly controlled by topographic features. However, the shape of ice sheets is often limited by the position of the coastline, as the coastline marks the position at which rapid and efficient ablation by calving can begin. Ice sheets generally do not extend far beyond the grounding line, except in confined embayments where they form ice shelves. Ice shelves do not extend far into open sea before calving. The coast thus exerts an ultimate topographic constraint. Glaciers other than ice sheets do not completely overwhelm the subglacial topography; and they experience a substantial degree of topographic control both on ice-flow patterns and on glacier morphology. Topographic control of glacier form is most clearly recognisable in cirque glaciers where the ice is confined to a closed topographic depression, and in valley glaciers where the ice is confined between valley walls. Alpine valley glaciers form dendritic glacier networks following the pattern of pre-existing river valleys. The influence of topography on planform is well illustrated where valley glaciers extend beyond their confining upland topography and expand as piedmont lobes onto flatter terrain, or where outlet glaciers from ice fields or ice caps spill through narrow gaps between areas of higher ground. Topography also exerts a control on glacier profile, partly by controlling the basal shear stress (Chapter 7), and partly by influencing the longitudinal stress in glaciers resting on an inclined surface.

5.3 GLACIER STRATIGRAPHY

It was a miracle of rare device,
A sunny pleasure-dome with caves of ice.
S.T. Coleridge: *Kubla Khan*

A glacier can be considered simply as a block of ice with an upper and a lower surface, bounded by the atmosphere above and by the ground, or water, below. Glaciers are commonly described in stratigraphic terms as a series of layers one on top of another. The top layer, comprising the glacier surface, is referred to as the supraglacial zone. The supraglacial zone is important as a source of material for the glacier, as an interface with the atmosphere and external energy sources, and for visible features that give clues to the glacier's behaviour. The body of the glacier between the upper and lower surfaces is referred to as the englacial zone. This comprises the bulk of most glaciers. The lowest part of the glacier, including the bottom surface or sole, is referred to as the basal layer. The basal layer includes ice and debris entrained from the bed. It is rarely more than a few metres thick, but plays an important role in glacier behaviour. The glacier bed and the underlying substrate comprise the subglacial zone. The materials within a glacier are sometimes described in terms of 'facies', originally a geological term referring to a body of material with a distinctive set of characteristics. Ice from different parts of a glacier might be referred to as supraglacial, englacial, or basal facies. The following sections describe the ice facies and other features that are characteristic of the different stratigraphic zones within a glacier.

5.4 SURFACE FEATURES

5.4.1 THE GLACIER SURFACE

The glacier surface is exciting because it is the part of the glacier that we most often see. Only the brave, the foolhardy, and the very unfortunate have ever seen much below the surface of a glacier. The glacier surface is a material and energy interface between the glacier and the atmosphere, and is very important in the context of glacier mass balance and heat budget. The supraglacial zone can be defined in a variety of ways. It is sometimes defined as the very surface of the glacier, with no vertical extent. In some definitions it extends downwards as far as the upper surface of ice, which is sometimes at the surface in the ablation zone but is usually at some depth beneath snow and firn in the accumulation zone. Sometimes it is considered to be the upper layer, about 10–15 m thick, that experiences seasonal temperature fluctuations.

5.4.2 FEATURES IN THE ACCUMULATION AREA

In the accumulation area, the glacier surface is generally dominated by snow in various stages of transformation to ice. Different zones within the accumulation area are characterised by different snow facies that can be distinguished in terms of their thermal and physical properties (e.g. Benson, 1962). Some of the many different snow facies classifications that have been used were summarised by Williams *et al.* (1991), who illustrated that surface facies were recognisable using satellite remote sensing. The main snow facies and zones that are commonly discussed are the following (Table 5.1 and Figure 5.2).

- **Dry-snow facies.** This comprises snow that is unaffected by melting at any time of year. Dry-snow compaction and recrystallisation therefore occur in a perennially dry environment that is characterised by continuously low temperatures. The dry-snow facies is therefore restricted to locations at high altitudes or latitudes. The dry-snow zone is the area above the dry-snow line, where negligible surface melting or percolation of meltwater into the snow occurs at any time of year, and the snow remains permanently dry. This zone is found extensively only in the central parts of the major ice sheets, on the Antarctic ice shelves, and on the highest mountains. In Greenland, the dry-snow facies occurs at present only in areas where the mean annual temperature is –25°C or colder, covering about 30% of the ice sheet's area.
- **Percolation facies.** This facies includes snow that has experienced surface melting, but in insufficient quantities to raise the pack temperature to 0°C. Meltwaters therefore percolate through the snow and refreeze entirely at depth, releasing latent heat and forming ice lenses or layers. In the percolation zone, meltwater is generated at the surface and penetrates downwards into the snow pack. The water may freeze at depth in the cold snow, or penetrate to the level of the first stratigraphic impediment. Summer melting is insufficient to wet the whole year's accumulated snow, but surface

Table 5.1 Mean density of different snow facies, and their relative extent on the Greenland ice sheet

Facies	*Mean density, upper 5 m*	*% Area of Greenland ice sheet*
Wet/soaked facies	> 0.5 g cm^{-3}	~10%
Percolation facies	0.43–0.39 g cm^{-3}	~45%
Dry-snow facies	< 0.375 g cm^{-3}	~30%

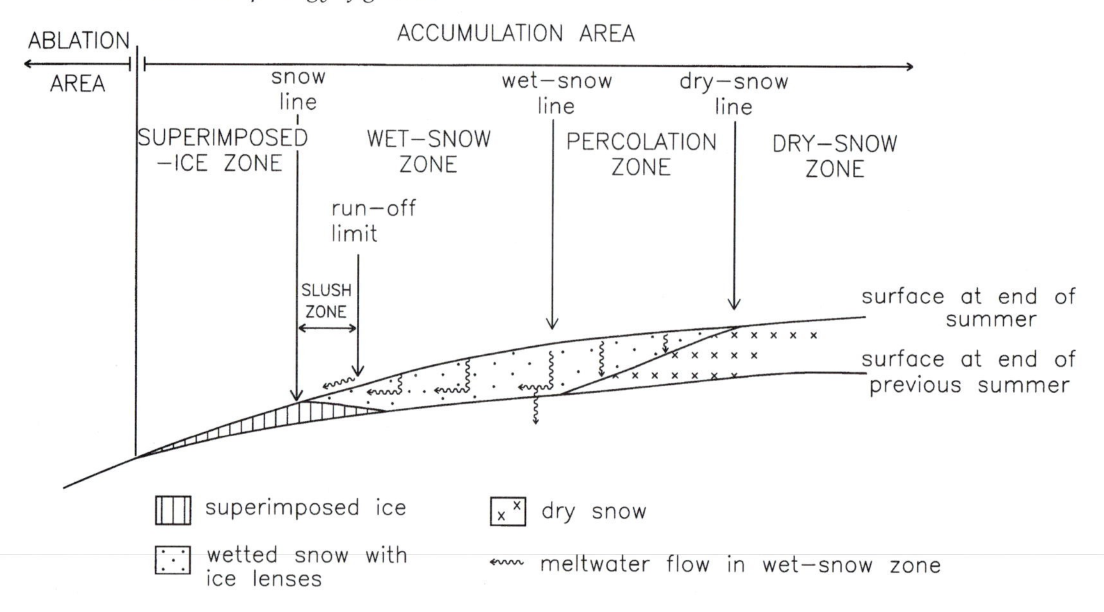

Figure 5.2 Snow facies zones on a glacier surface. Partly after Benson (1962).

meltwater percolates through pipe-like channels penetrating into the snowpack to form layers and lenses of ice in older snow. At Jakobshavn, Echelmeyer *et al.* (1992) identified a pervasive layer of refrozen ice 0.3–0.4 m below the surface in the wet-snow and lower percolation facies, but found that surface melt penetrated 3 m into cold firn by piping and caused significant warming by refreezing at that depth.

- **Wet-snow (soaked) facies.** This facies includes snow that has experienced sufficient melting to raise its temperature to 0°C for at least part of the year. The wet-snow facies is therefore generally isothermal (at 0°C) at the end of the melt-season, when any additional heat input exclusively fuels snow-melt. In the wet-snow zone, the surface is covered by wet or saturated snow at the melting point. Summer melting produces sufficient water to soak at least the whole snow accumulation of the past year. All the snow of the wet-snow facies has been wet at least once. At the lower limit of the wet-snow zone, snow cover becomes patchy, and where drainage is poor wet snow can form slush swamps or slush fields. In Greenland, slush fields are found up to 100 km from the ice edge east of Kangerlussuaq. The lower limit of the wet-snow zone is the snowline.
- **Superimposed facies.** Superimposed facies ice forms from the refreezing of meltwaters generated within the wet-snow zone. Refreezing may occur on the glacier (ice) surface, either beneath the snowpack or on the bare ice down-glacier of the snowpack.
- **Ice facies.** Below the snowline, bare ice is exposed at the glacier surface.

Temperate glaciers commonly have only two zones: wet-snow facies and ice facies. Dry snow and percolation zones are confined to only the coldest and largest glaciers and ice sheets. Not all snow zones are distinguishable by eye because they are defined by subsurface characteristics, but these can be recognised by other means. For example, snow facies were observed on Greenland ice-sheet surfaces by means of microwave radar backscatter (Long and Drinkwater, 1994), and passive microwave measurement was used to indicate the distribution of melting in the Greenland snowpack (Mote and Anderson, 1995). Snow zones can be observed over broad areas with relative ease by remote sensing.

The snow surface is commonly characterised by erosional and depositional features related to wind, and by features associated with weathering (ablation). Sastrugi, sometimes referred to as snow dunes, form in response to surface wind. They are usually directed parallel to the resultant wind vector. As they cast shadows that vary substantially with the angle of the sun, they can be an important influence on surface albedo and on temporal variations in albedo. The preferred orientation of sastrugi shows up in Seasat scatterometer data, and can be used to indicate wind regimes (Wendler and Kelley, 1988; Ledroit *et al.*, 1993). Snow surfaces are also characterised by a range of features related to weathering or ablation of the snow. These include sun-cups or ablation hollows, which form as networks of cuspate hollows with a relief of up to ~50 cm. In their extreme form, the cups can deepen to pits as deep as 3 m, and the intervening ridges stand up almost as pillars or spikes. This form, referred to as penitent snow or penitentes, is common only at low latitudes at high elevation. Sun-cups and penitentes arise from differential ablation of the snow surface and can result both from solar radiative heating and from turbulent heat transfer (Matthes, 1934; Lliboutry, 1954; Post and LaChapelle, 1971; Takahachi *et al.*, 1973). Under direct sunlight, slightly differently oriented parts of the snow surface receive slightly different amounts of insolation, so minor irregularities on the snow surface can gradually become accentuated. On a low-latitude glacier, for example, the flat bottom of a hollow will receive more insolation than the sides of the hollow, hence causing the hollow to deepen. Mature cups and penitentes tend to be oriented to the sun in such a way that upstanding snow faces tend to receive the minimum insolation, and maximum insolation is received in the troughs or cups. The presence of fine debris in the snow tends to accentuate sun-cup development when ablation is dominated by turbulent heat exchange, but to limit sun-cup development when ablation is dominated by radiation (Rhodes *et al.*, 1987).

5.4.3 FEATURES IN THE ABLATION AREA

The ablation zone extends from the edge of the glacier up to the lower limit of perennial snow. Winter snowfall melts away during each summer, and the glacier surface is characterised for part of the year by bare ice. The surface may be seasonally snow-covered, and for part of the year will be characterised by melting snow. Surface materials include ice, water and debris. Many of the characteristic features of the ablation zone are derived from the interaction of ice, water and debris in response to external energy input at the surface.

The breakdown of polycrystalline ice by ablation at the glacier surface is sometimes referred to as weathering of the surface. The surface of a glacier experiencing ablation frequently consists of a weathering crust up to ~50 cm thick, produced by differential melting and preferential water flow along grain boundaries. The result is enlargement of the veins, disaggregation of the polycrystalline mass, and the production of a surface layer comprising porous ice with loosely interlocking crystals (Müller and Keeler, 1969) (Figure 5.3).

Debris reaches the glacier surface from a variety of sources. Supraglacial sources include extraterrestrial influx (meteorites), airfall of tephra and other atmospherically borne particles, and rockfall or wash-on from valley walls and nunataks. Englacial material, entrained from the bed or from the surface in the accumulation zone, can be exposed at the surface by ablation in the lower part of the glacier. Sometimes thick and laterally extensive debris bands transverse to the flow direction emerge at the ice surface and can be traced as debris planes dipping downwards up-glacier. These can result from entrainment of sediment from

Figure 5.3 Weathering of ice. Ablation of the ice surface in the foreground has produced a rough 'weathered' surface by preferential melting along crystal boundaries. Russell Glacier, West Greenland.

the bed along thrust planes in the ice, or from the deposition and burial in the accumulation area of supraglacial debris that travels englacially before being exposed further down glacier. Many Icelandic glaciers, for example, exhibit prominent black debris layers in the ablation zone that represent tephra layers deposited on the ice surface by past volcanic eruptions. From a distance the glacier surface down-glacier from such bands can appear black because of debris dumped onto the ice as the tephra layer is exposed by ablation. The zones of dead ice that are typical of post-surge glacier snouts are often covered in subglacially derived debris that has been thrust into the ice during the surge and dumped onto the surface by subsequent ablation.

Debris dumped onto the glacier surface is commonly concentrated by streams, wind and gravity into supraglacial ponds, stream channels and crevasses, so the distribution of debris over the ice is rarely uniform. Differential ablation of ice with different amounts of debris cover can lead to a variety of phenomena, including perched boulders, dirt cones and cryoconite holes. Whether debris protects or melts into the ice depends on the thickness of the debris cover and its properties of heat absorption and transmission. Thin debris layers, or small individual particles, are more likely to transmit heat through to the ice than thick layers or large particles. Dark-coloured clasts heat up more than light ones, and are more likely to transmit heat into the ice rather than reflect it away. Patterson (1984) modelled the processes by which rocks on the glacier surface will either form pedestals or sink into the ice. If the solar energy supplied to the ice directly beneath the debris is greater than the energy supplied to the cleaner ice around it, the debris will sink by melting into the ice, forming a surface depression. This occurs when surface debris (which generally has a lower albedo than the ice) absorbs energy and transmits it to the ice beneath it. Surface pits formed by debris melting into the ice surface are sometimes referred to as cryoconite holes (Figure 5.4).

If the solar energy supplied to the ice directly beneath the debris is less than the energy supplied to the cleaner ice around it, the ice beneath the debris will be protected while the surrounding ice melts. In consequence, boulders and pebbles will become perched on ice pedestals (Figure 5.5), and patches of loose debris will form ice-cored dirt cones. Rock pedestals form at an angle determined by the elevation of the sun. In tropical locations, tall near-vertical columns of ice can develop beneath boulders. At high latitudes, pedestals tend to develop beside, rather than beneath, the boulders (Figure 5.6). Different types of dirt cones form in different circumstances, their form controlled largely by the manner in which the debris is concentrated. For example, debris concentrated in a small, shallow surface pool will produce a relatively small and short-lived cone, whereas debris concentrated in a deep in-filled moulin can generate a long-lived dirt cone that is resupplied with debris through a long period of surface ablation (Figure 5.7). Debris that accumulates in the bottom of a crevasse can lead to the formation of a line of dirt cones or a dirt-covered ice ridge as the crevasse bottom is exposed by ablation (Figure 5.8).

Where debris is supplied regularly to a particular point on the glacier surface (for example, downflow from a glacier confluence), major moraine ridges can be produced as the movement of ice carries the debris downglacier and as ablation exposes debris entrained at glacier bed against the valley walls of the two tributary glaciers (e.g. Østrem, 1959). As the debris cover becomes thicker, the ice below it is protected from melting and an ice ridge forms beneath the protective cover.

Figure 5.4 Cryoconite holes formed by melting of the ice by particles of surface debris that absorb heat from the sun. Russell Glacier, West Greenland.

Figure 5.5 At low latitude, where much of the insolation comes from directly overhead, protection of ice shaded below a boulder can lead to the formation of a tall ice pedestal. Cotopaxi, Ecuador.

When the debris cover increases to about 1.5–2 m, melting of the ice beneath virtually ceases (Dolgushin *et al.*, 1972). Bozhinskiy *et al.* (1986) suggested that diurnal temperature fluctuations are masked beneath a debris cover of about 0.16 m, seasonal oscillations beneath a cover of about 1. 54 m, and annual oscillations beneath a cover of about 3. 1 m. Where substantial areas of a retreating glacier terminus become buried beneath a thick debris cover, the ice can survive for thousands of years after regional deglaciation is complete (Chapters 8 and 9).

Surface debris can play an important role in the response of glaciers to climate fluctuations. Rogerson *et al.* (1986) showed that surface debris cover plays a major role in enabling small glaciers in Labrador to survive unfavourable climatic events. Average surface melting on portions of the glaciers covered by avalanche debris are approximately one-third of that on exposed ice. Sturm *et al.* (1986) described the impact of volcanic fallout from Mt Redoubt on the flow of Drift Glacier, Alaska. The lower portion of the glacier was blanketed with sand and ash, reducing ablation from about 5 m a^{-1} to virtually nothing. Chinn and Dillon (1987) suggested that accumulations of surface debris derived from basal debris bands might cause development of rock glaciers. They reported observations of Whiskey Glacier on James Ross Island where basal debris is transported to the surface along shear planes at a transition from warm to cold basal thermal regime. The debris blanket effectively inhibits ablation, and the perimeter of the glacier below the thermal transition resembles a rock glacier. The possibility exists of artificially altering surface melting. Introduction of a thin, low-albedo surface cover such as coal dust to the glacier would enhance melting, and increase run off. Material with a high albedo, or a thick covering of material, would have the opposite effect. This could be potentially important in controlling water supply to glacierised catchments.

Surface water plays an important role in supraglacial processes in the ablation zone of glaciers that are subject to melting. Melting glacier surfaces commonly feature stream networks, ponds and lakes, and drainage connections to the interior of the glacier. Arborescent drainage patterns similar to those of land-based rivers can develop, and many features of terrestrial fluvial geomorphology can be identified in supraglacial streams (e.g. Leopold and Wolman, 1960; Ferguson, 1973) (Figures 5.9 and 5.10). The glacier drainage

Figure 5.6 At high latitude, where the sun is low in the sky, ice protected by shading often forms a mound beside, rather than beneath, protecting rocks. West Greenland, ~70° N.

Figure 5.7 A dirt cone formed by the exposure of debris from a filled moulin. Solheimajökull, Iceland.

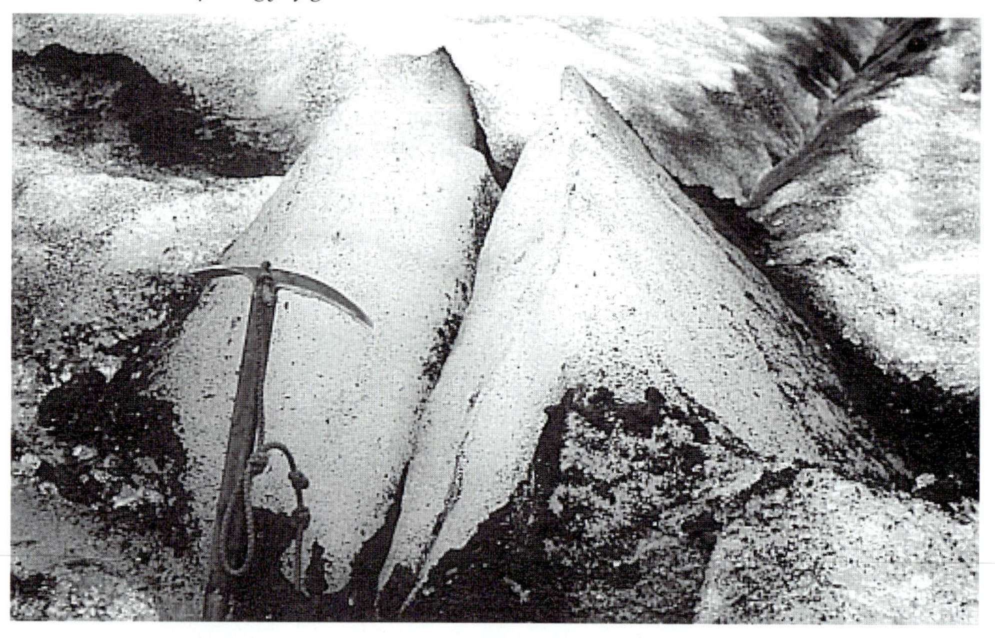

Figure 5.8 A dirt cone formed by the exposure of debris from a melting crevasse trace. Solheimajökull, Iceland.

Figure 5.9 Incised meanders in the ice surface caused by a supraglacial meltwater stream. Russell Glacier, West Greenland.

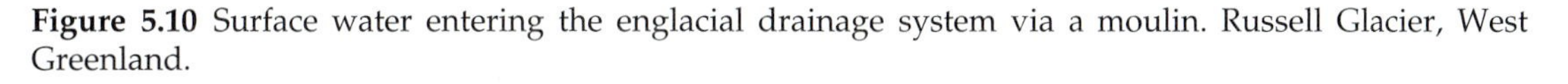

Figure 5.10 Surface water entering the englacial drainage system via a moulin. Russell Glacier, West Greenland.

system is discussed more fully in Chapter 6. Supraglacial lakes form in closed basins on the ice surface. Sometimes these are unstable features which are carried down-glacier as the ice moves, and these disappear as the ice topography changes. Sometimes they disappear abruptly if a crevasse or other drainage route opens in the basin and the water escapes. Sometimes the lakes are stable features as the ice surface topography can be controlled by fixed subglacial topography. For example, at Jakobshavn, some of the surface lakes seem tied to subglacial bedrock features: lakes form in flow-related depressions and are perpetuated by albedo effects (Echelmeyer *et al.*, 1991). These lakes are not advected down-glacier with the moving ice but remain stationary as quasi-permanent fixtures. Some of these lakes exhibit banding structure referred to as lake ogives. These form when open water forms and seasonally freezes over at the up-glacier end of the lake while the down-glacier end remains covered with ice being advected down-glacier: each year a distinct band of lake-surface ice forms at the up-glacier end of the lake. This then travels down-glacier and new open water appears at top end, freezing in turn to produce next band. The distribution of these annual bands down-glacier of the lake provides indication of ice speed. Supraglacial lakes are also associated with the formation of 'ice blisters'. These may be formed when frozen-over lakes or water-filled crevasses move into a zone of compressive flow zone and the ice lid over the water is buckled (Echelmeyer *et al.*, 1991).

Where the bulk of the glacier comprises cold ice, surface-derived meltwater does not penetrate below the surface and the drainage network is confined to the supraglacial zone. In temperate ice, however, the glacier surface is often characterised by conduits from the surface to the internal drainage network. Surface streams often fall into openings known as moulins. Rink (1877) described how 'a stream flowing between its blue walls with great velocity . . . suddenly came to a stop, and forming a most magnificent waterfall fell into a perpendicular hole in the ice. Another smaller stream likewise fell into a beautiful blue-tinted cleft, but a little further on it again reappeared, rising in the shape of a jet of water mingled with air and agitated by the wind.' Moulins originate from meltwater exploitation of surface openings such as incipient crevasses, and once formed they can become long-term features of the surface drainage of a glacier. Moulins are most commonly formed when water flows into a crevasse that connects with the drainage system. Even as the crevasse closes, the flow of water maintains an opening by melting the ice walls, so that a sink-hole, the moulin, remains after the crevasse is closed. The moulin thus formed will be carried down-glacier as the ice moves, but because crevasses tend to open repeatedly in the same location, a new drainage route will eventually develop, upstream from first moulin, at its original position. Water flowing at the surface will then drop into the new, upstream moulin, and first moulin will be abandoned. Individual moulins thus migrate short distances during their lifetime, but general positions of moulins or moulin fields on the glacier surface can remain constant for long periods. A large moulin on the Mer de Glace known as the Grand Moulin reforms every year at virtually the same location slightly upstream of an area of crevasses, and a series of old abandoned moulins from previous years occurs downstream (Reynaud, 1987). Kohler (1995) described moulin locations and surface drainage networks that had remained constant for nearly 25 years on Storglaciären.

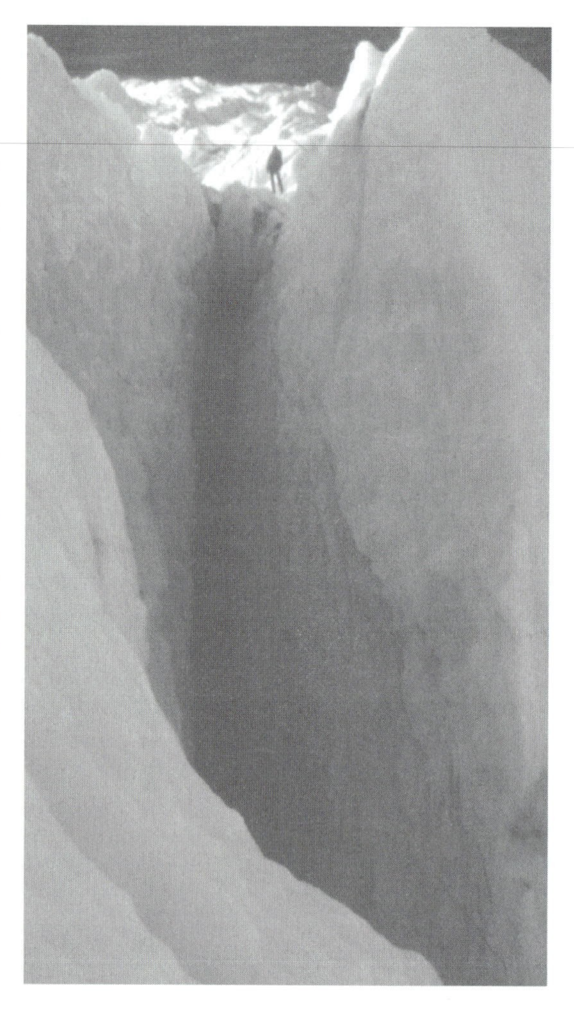

Figure 5.11 A crevasse close to the margin of the Greenland ice sheet. Russell Glacier, West Greenland.

5.4.4 FLOW-RELATED FEATURES

Many of the characteristic features of the ice surface, in both the accumulation area and the ablation area, are caused by ice movement.

Crevasses are tensional features that form at the surface when the principal extending strain rate exceeds some threshold value (Figure 5.11). The crevasse opens as a linear crack or chasm oriented in a direction perpendicular to the principal extending strain. As ice is deformed under increasing stress, it undergoes a transition from elastic, to ductile, to brittle behaviour. Maximum stress occurs at the ductile/brittle transition, which can be considered as a critical stress for failure (Vaughan, 1993b). Crevasse fields thus commonly mark positions of localised extending flow, such as ice falls, and crevasse orientation can be used to infer stress conditions in the ice (e.g. Whillans *et al.* 1993) (Figure 5.12). Caution must be exercised when interpreting snow-covered crevasses, however, as a crevasse might change orientation during flow before it widens sufficiently to cause sagging or collapse of the snow-bridge that masks it. Crevasses can occur in both accumulation and ablation zones, but can be observed most easily in the ablation zone, where they are not buried by accumulating snow. Crevasses may form at one location on the surface and then be carried passively down-flow. Snow accumulation above the crevasse can then be used to date the feature. Trains of buried crevasses may stretch downstream from a point of repeated crevasse formation. At ice streams, changes in activity have been dated on the basis of the timing of changes in rates of crevassing indicated by depth of burial of crevasse traces. However, not all crevasses initiate at the ice surface:

Figure 5.12 Crevasses at the margin of Solheimajökull, Iceland. The view is nearly vertical downwards. The ice flow is from right to left. At the right of the picture the ice flows over a subglacial tunnel and crevasses form perpendicular to the extending flow. At the left of the picture, ice flow is retarded by rising ground, and crevasses form parallel to the flow direction, perpendicular to the laterally extending strain.

crevasses can be initiated at a little depth below the surface and propagate upwards. This leads to difficulty in calculating crevasse ages from their depths of burial beneath the surface.

Large crevasses, and those with thin or zero snow-cover, are visible in satellite imagery. Merry and Whillans (1993) illustrated a fascinating range of flow-related features visible on the surface of Ice Stream B revealed by SPOT HRV (High Resolution Visible) images. Features that they identified include shear margins, flow traces, lumps and warps, lineations, crevasse bands and mottles, and features which they labelled horsetails, chromosomes, seagulls and ladders on the basis of their appearance. The forms of these different features reflect the strain regimes that are responsible for their development, and the features can thus be used as interpretative tools. Vornberger and Whillans (1990) argued that crevasse shapes can be used to make inferences about the velocity field of a glacier, because once formed they are subject to rotation and bending according to the velocity field through which they travel. Where a glacier is straight, the main crevasse systems are symmetrical around the centre line, but curved flow changes the pattern as the stress centre-line is shifted to the inside of the bend. This sort of asymmetrical pattern was described for the Blue Glacier, which follows a regularly curving route for much of its length, by Echelmeyer and Kamb (1987).

Particular types of crevasse sometimes referred to in the literature include randkluft and bergschrund. The term randkluft refers to a gap between ice and rock at the headwall of a glacier. Experimental observations by Mair and Kuhn (1994) demonstrated that these form by movement of the ice away from the wall, although melting of ice in contact with rock has also been postulated as a contributing factor. The term bergschrund refers to crevasses formed towards the back of a cirque basin. These have traditionally been attributed to tension between ice flowing down-glacier and ice adhering to the cirque back wall. Mair and Kuhn (1994) showed that they develop by rapid down-glacier increase in velocity. Where the ice is thin there is a clear influence of subglacial topography on surface strain and crevassing. Fine-resolution strain measurements on the ice surface within a few metres of the ice margin in West Greenland (Knight, 1992) show a link between ice flexure over subglacial obstacles, extension of the ice surface, and the formation of crevasses.

Other flow-related features of the ice surface include ogives (Figure 5.13). Ogives are among the most visually spectacular features of glacier surfaces. There are two main types of ogives: wave ogives and Forbes bands. Both types are transverse linear features that form in association with ice falls and travel down-glacier from the ice fall at the speed of the ice flow. Wave ogives are topographic waves on the glacier surface; Forbes bands are pairs of alternating light and dark ice zones. Waddington (1986) summarised the history of research on ogives. Wave ogives are attributed to differences in melting between ice that traverses an ice fall during the summer and ice that traverses the ice fall during the winter. Nye (1958) showed that a vertical column of ice passing through an ice fall is stretched horizontally by the local velocity gradient through the ice fall. This ice exposes a greater surface area than ice immediately above and below the ice fall, and so is subject to greater summer melting. Ice that passes through the ice fall in the winter is not so affected. Thus the summer ice forms a cross-glacier trough at the foot of the ice fall, while winter ice forms a cross-glacier ridge. The waves tend to degenerate down-glacier as snow accumulates preferentially in the troughs and the ice deforms back to a level surface. Forbes bands are attributed to summer deposition of dust and winter deposition of snow in crevasses in the upper part of an ice fall. Ice passing through the crevasse zone in summer forms a dark band; ice passing through the crevasse zone in winter subsequently forms a light band. Wave ogives and Forbes bands commonly occur together.

Figure 5.13 Aerial view of part of the western margin of the Greenland ice sheet about 63° N. Ice flows around a nunatak, and ogives are visible on the ice surface in the lower part of the picture.

Although they are formed by different processes, they are both formed in response to localised zones of rapid ice acceleration. As ogives form seasonally and move at the speed of the glacier, the wavelength is a measure of the annual velocity of the ice.

The surface of floating ice tongues and ice shelves can also show the effects of ice flow, especially related to underlying topography. For example, where floating ice runs aground on an island or submarine feature, the surface can be raised up to form an ice rise: a domed area which might display its own radial flow pattern. Smaller features, not generating independent radial flow, are referred to as ice rumples. Rumples or pressure rollers sometimes form upstream from ice rises in the form of wave-like sequences of ridges and troughs, formed by compressive creep buckling (Collins and McCrae, 1985). These phenomena illustrate how the surface of a glacier can give clear signals as to the processes and conditions below the surface.

5.5 THE ENGLACIAL ZONE

The englacial zone is the portion of the glacier between the supraglacial zone and the basal zone. It is not directly affected by surface or subglacial processes, and it usually constitutes the bulk of the glacier thickness. For example, of the 3053 m thickness of the Greenland ice sheet at Summit, about 3025 m (99%) is englacial facies ice. However, very close to the margin of some glaciers where ablation has removed most of the ice, the englacial zone may virtually disappear and the ice margin might be made up entirely of basal ice. The englacial zone can be observed directly by coring or via limited access through crevasses or drainage channels from the surface. It can be observed indirectly via remote-sensing methods such as surface-penetrating radar (e.g. Bamber, 1987; Hamran *et al.*, 1996) and radio-wave velocity (e.g. Macheret *et al.*, 1993).

Englacial ice is derived from surface accumulation, and is formed predominantly by the compression and recrystallisation of snow. It is usually characterised by low debris content, but the amount of debris included in the ice depends on the character of the glacier surface in the accumulation zone. In mountain environments where exposed rock surfaces are common, large amounts of surface-derived debris can be incorporated into the ice. In ice-sheet environments the only source of debris is atmospheric fallout, so debris concentration in the ice tends to be much lower. Lawson (1979) distinguished between the bulk of englacial ice, which he referred to as diffused facies, and distinct bands of dirt, which he referred to as banded facies. Dirt bands derived from surface accumulation tend to follow the sedimentary layering in the ice. Other bands that cross-cut the sedimentary stratification might be derived from the incorporation of debris into crevasses that subsequently close and rotate to dip up-glacier as the ice flows, or from thrusting upwards of material from the bed. Complex patterns of internal debris bands and other structural phenomena can arise when several tributary glaciers combine to form a compound trunk glacier.

Englacial ice is characterised by both primary and secondary structural features. Primary features are derived from the accumulation and transformation of snow. These include snow settlement structures and sedimentary stratification. Annual layering arises from differences between summer and winter snow surfaces. Additional stratification arises from the formation of ice layers by refreezing of meltwater in the firn. Secondary structures develop during the flow of ice after its initial formation. These include structures related to ductile and brittle deformation, such as crevasse traces, compression and flow structures, foliation, folds, thrusts and boudinage. These secondary structures within englacial ice reflect, and can therefore be used to interpret, the dynamic history of the ice. Foliation is a common feature of glacier ice, comprising a layered structure of bubble-rich and bubble-poor ice, on a scale of about 0.01–1.0 m, sometimes associated with differences in crystal size. Foliation can give exposures of glacier ice a striped white and blue, or light and dark, appearance (Figure 5.14). Foliation is usually explained in terms of rotation of pre-existing inhomogeneities, such as stratification or crevasse traces, during ice flow (e.g. Hambrey, 1975; Hooke and Hudleston, 1978; Hambrey *et al.*, 1980; Hudleston and Hooke, 1980). This would imply that foliation develops in response to total strain or deformation in the ice. However, Pfeffer (1992) showed that bubble foliation developed in the terminus lobe of Variegated Glacier during the 1982/83 surge at locations that experienced only low total strain. Pfeffer attributed the development of foliation to short-term very high compressive stress during the surge, and suggested that the formation of foliation might involve bubble migration through the ice in response to high stress, rather than deformation or rotation of pre-existing bubble inhomogeneity.

Layers of ice of different age and character in the englacial zone have different mechanical and rheological properties. For example, many borehole studies have demonstrated that pre-Holocene ice in the lower sections of many surviving ice sheets is characterised by higher dust content, smaller crystal size, and significantly lower resistance to deformation than the younger ice above it. Cunningham and Waddington (1990) demonstrated how discontinuities between the rheological characteristics of different ages of ice beneath Summit would be expected to cause boudinage during ice flow. This would involve fragmentation of individual layers, and would pose difficulties in the interpretation of ice cores. The thickness of ice

Figure 5.14 The contact between glacier ice and the bed. The ice is given a striped appearance by alternations of bubble-rich and bubble-poor layers (foliation). Note the crystal boundaries made visible by differential ablation of the surface. The patch of white material in the bottom left of the picture is recent snow that has been overridden and incorporated by the glacier. Athabasca Glacier, Canada.

layers associated with specific ages or climatic periods would be controlled by rheology rather than by age/depth relations. Ice of interglacial age beneath Summit might be expected to have thickness variations and possibly even layer-breaks generated by boudinage.

The englacial zone is frequently characterised by cavities and voids in the ice. For example, Pohjola (1994) identified that 1.3% of the observed ice column in four boreholes in Storglaciären was composed of englacial voids. These were interpreted as openings into englacial channels and cavities intersected during drilling. Where the drainage network is dense, a karst-like system of pipes and channels can exist, playing a major role in the transfer of water and sediment through the glacier. Such voids can be significant for density calculations. Cavities can also be important in the formation of regelation ice, both in the englacial zone and also at the bed. In temperate ice, where there is a pressure difference between ice under hydrostatic stress and a cavity at lower pressure, water will be squeezed out of the ice into the cavity and will refreeze in the lower pressure environment, forming ice crystals that grow inwards from the cavity walls (e.g. LaChapelle, 1968).

Some features of the englacial zone are discussed at greater length in the context of ice cores in Chapter 11.

5.6 THE BASAL LAYER

The basal layer is the part of the glacier in which the nature of the ice is directly affected by proximity to the glacier bed. Although there are substantial differences between descriptions of basal ice from different glaciers, two key characteristics seem to be typical:

1. Whereas the supraglacial and englacial zones comprise ice formed by compression and recrystallisation of snow that accumulated at the surface, basal ice is commonly formed either by the freezing of water at the bed of the glacier, or by metamorphism of surface-derived ice by thermal, strain and hydraulic conditions close to the bed.

Because of this, the chemistry and the physical structure of the basal ice are different from the ice above.
2. Whereas englacial ice usually contains only debris derived from the glacier surface, the basal layer can contain large amounts of debris derived from the bed of the glacier. This affects not only the chemistry and structure of the ice, but also its rheological properties and its geomorphic potential.

Where the glacier bed is melting, the basal layer may be thin or non-existent, but where melting is limited and basal accretion occurs, basal ice several tens of metres thick may develop. Table 5.2 lists the thickness of basal ice measured at a variety of sites, and Figures 5.15–5.17 illustrate some of the characteristic forms of the basal ice layer. The upper boundary of the basal zone is the upper limit of ice directly affected by basal processes, specifically: (i) formation of basal ice; (ii) entrainment of subglacial sediment; and (iii) flow through the vein network of basally derived water.

Basal ice forms a lower boundary layer within the glacier. As such it is critical to many aspects of glaciology, and deserves detailed attention. This is the business end of the glacier: the interface between the ice and its substrate; the interface between glaciology and geology. The basal ice acts as an agent of glaciers' impact on the landscape: the processes of debris entrainment, and the presence of debris in the

Figure 5.15 Basal ice is characteristically debris-rich and bubble-poor. Ice and debris structures reflect both entrainment and deformation processes. This photograph shows a layer of stratified basal ice about 2 m thick at the margin of the Greenland ice sheet. Russell Glacier, Greenland.

Table 5.2 Thickness of basal ice reported at various sites

Location	*Maximum thickness of basal ice (m)*	*Source*
Blue Glacier, Washington, USA	0.029	Kamb and LaChapelle (1964)
Myrdalsjökull, Iceland	0.05	Humlum (1981)
Breiðamerkurjökull, Iceland	0.2	Boulton (1975a)
Grubengletscher, Switzerland	1	Souchez and Lorrain (1978)
CAROLINE, Antarctica	6	Yao *et al.* (1990)
GISP2, Greenland	13.11	Grootes *et al.* (1993)
Camp Century, Greenland	15.7	Herron and Langway (1979)
Matanuska Glacier, Alaska, USA	23	Lawson (1979)
Thule, Greenland	>70	Swinzow (1962)

Figure 5.16 Debris in the basal layer can occur as distinct bands. These are often formed by freezing of material at the bed combined with compressive folding or thrusting in the basal ice. The ice close to the band appears dark because it is bubble-free. Russell Glacier, Greenland.

basal ice, are intimately linked with the geomorphic processes of erosion and sedimentation. Basal ice preserves a physical and chemical signature of the conditions of its formation, and thus serves as an indicator of processes and environments that exist in the inaccessible subglacial zone. The rheology of the basal ice layer, which accommodates the bulk of movement in many glaciers and forms part of the interface between the ice and its substrate, is critical to glacier dynamics. An understanding of the basal ice layer is thus central to realistically formulated models both of ice-sheet behaviour and of the development of glacial landscapes.

The basal layer is accessible either by drilling through the glacier to the bed, or by accessing subglacial cavities, or by making observations where the basal layer is exposed at the very margin of the glacier. Only a very small proportion of the basal ice in glaciers is open to inspection, but over a century of observations and analysis have permitted some detailed descriptions of basal ice from a wide range of glaciers. The history of research on the basal ice layer makes an interesting case study of scientific progress in glaciology, and is considered in more detail in Chapter 12.

Most descriptions of basal ice have revealed several distinct ice types in the basal layer at each site. Tables 5.3–5.6 illustrate some of the different classifications of basal ice that have been proposed to describe these different ice types or facies. A seminal contribution was made by Lawson (1979) on the basis of observations of a basal layer 23 m thick at Matanuska Glacier, Alaska (Table 5.3). Lawson identified two main basal facies: the Basal Stratified and the Basal Dispersed. At the bottom of the sequence was up to 15 m of Basal Stratified facies ice, which Lawson attributed on isotopic and sedimentological grounds to freezing of water at the bed of the glacier. This

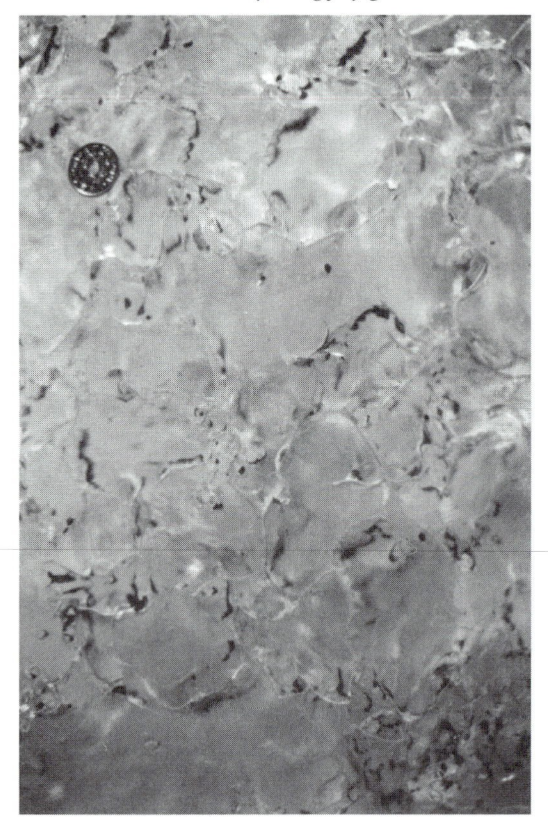

Figure 5.17 Debris can occur within the vein network in basal ice. This might be due to entrainment through the vein network or to 'sweeping' of particles to the grain boundaries during recrystallisation. This is 'clotted' ice (Knight, 1987) from the basal layer of Russell Glacier, Greenland.

facies was composed of interstratified layers, lenses and pods of debris-rich and clean ice of highly variable thickness (0.001–2.0 m), lateral extent (maximum 50 m), and debris-content (0.02–74.0%). Lawson defined three sub-facies within the stratified ice: the Basal Stratified Suspended sub-facies contained suspended debris particles and aggregates ranging from 0.02 to 55% by volume; the Basal Stratified Solid sub-facies was composed of well-defined layers of sediment-rich ice (> 50% by volume), often with preserved internal structures; and the Basal Stratified Discontinuous sub-facies contained irregular aggregates of predominantly fine-grained sediment. The stratified facies was overlain, above a sharp boundary, by up to 8 m of Basal Dispersed facies, the origin of which Lawson attributed inconclusively to a combination of subglacial and surface processes. This dispersed facies was more uniform and less debris-rich than the stratified facies, and comprised bubble-poor ice containing a uniform distribution of clay- to pebble-sized debris (0.04–8.4% by volume).

Lawson's classification has been widely adopted and built upon by subsequent researchers, and descriptions of glaciers from a range of other locations have been couched in Lawson's terminology. However, not all features observed at other glaciers fit comfortably into Lawson's scheme, necessitating a degree of adaptation and, in some cases, new definitions. Knight (1987) and Sugden *et al.* (1987a), for example, largely followed Lawson's scheme, but also recognised the importance of tectonic disturbance in emplacing debris bands from the bed into the basal sequence at the margin of the Greenland ice sheet (Table 5.4). Knight (1987), Sugden *et al.* (1987a) and Knight and Knight (1994) proposed that ice in Greenland that was similar to Lawson's stratified facies was formed by freezing of water at the glacier bed, but that a so-called 'clotted ice facies' that was similar to Lawson's

Table 5.3 Classification of basal ice facies at Matanuska Glacier, Alaska, according to Lawson (1979)

Zone	*Facies*	*Subfacies*	*Origin*
Basal	Dispersed		Combined firnification and basal regelation
	Stratified	Stratified *discontinuous*	Basal regelation
		Stratified *suspended*	Basal regelation
		Stratified *solid*	Basal regelation

Table 5.4 Knight's (1987) classification of basal ice facies at Russell Glacier, Greenland

Basal ice family	*Composition of family*	*Facies number and description*	*Origin*
Clotted	Clotted ice	6. Clotted ice	Basal entrainment by regelation-related processes in interior of ice sheet
Banded	Tectonic intercalation facies of clotted and solid families	5. Clotted ice between debris bands	Entrainment (facies 6), transport towards margin, and folding/thrusting
		4. Debris bands	Basal entrainment (solid family) and folding/thrusting
Solid	Basal entrainment	3. Stratified ice	Regelation
		2. Frozen till	Congelation/overriding of margin material
		1. Old snow	Overriding of margin material

dispersed facies was formed by a combination of small-scale regelation around bedrock obstructions and flow of debris-rich water through the inter-crystal vein network close to the bed. Sharp *et al.* (1994) also developed a facies scheme similar to Lawson's on the basis of research at surge-type Variegated Glacier, Alaska. They defined two basal facies: Basal Diffused and Basal Stratified. Basal Diffused was tectonically incorporated glacier ice (the basal equivalent of Lawson's Englacial Diffused facies). Basal Stratified was characterised by a higher debris content and more variable ice characteristics, and divided into three sub-facies: laminated, clear and solid (Table 5.5). Hubbard and Sharp (1995) extended the facies-based approach to the analysis of basal ice at warm-based glaciers. They identified seven basal ice facies and two sub-facies, some of which reflect distinctive processes operating close to the beds of predominantly temperate-based glaciers (Table 5.6). Even this brief review highlights a major problem that has beset recent research in basal ice: a proliferation of overlapping and conflicting terminology!

Two major and related issues in the description and interpretation of the basal layer are the processes by which the ice is formed and

Table 5.5 Ice facies of the Variegated Glacier, Alaska, according to Sharp *et al.* (1994)

Zone	*Facies*	*Subfacies*	*Origin*
Englacial	Englacial	Diffused	Firnification
	Englacial	Banded	Supraglacial debris
Basal	Basal Stratified	*Basal Stratified – laminated*	Incremental basal accretion
		Basal Stratified – clear	Metamorphism of englacial or basal diffused facies
		Basal Stratified – solid	En-masse basal accretion
	Basal Diffused		Tectonically incorporated englacial ice

Table 5.6 Hubbard and Sharp's (1995) classification of basal ice facies from glaciers in the Western Alps

Facies	*Subfacies*	*Origin*	*Implication*
Clear		Stress-induced metamorphism close to the bed; Lliboutry (1993) regelation	Basal interface at melting point. Bedrock roughness decimetres–metres
Laminated		Weertman (1957, 1964) Regelation	Base at melting point, cm–dm bed roughness, low basal melt rate, compressive strain
Interfacial	Interfacial layered	Freezing of flowing basal water	Sub-freezing base, metre-scale bed roughness, cavitation, linked cavity hydrology
	Interfacial continuous	Freezing of standing basal water	
Dispersed		Incorporation and transport of interfacial facies	Sub-freezing base, metre-scale bed roughness, cavitation, linked cavity hydrology, variable sliding rate
Solid		Adfreezing of saturated sediment	Unconsolidated sedimentary substrate
Stratified		Any basal facies tectonically thickened and interstratified with glacier ice	Marginal compression
Planar		Closure of ice-marginal fracture planes incorporating aeolian debris	Marginal fracturing/crevassing

the processes by which rock debris from the bed is entrained into the ice (Table 5.7). Entrainment of debris at the bed occurs by a variety of processes including regelation, congelation, flow of water through the vein system, shearing and folding. These processes of debris entrainment are discussed more fully in Chapter 9. Here we consider the formation of basal ice itself. As shown in Table 5.7, processes by which basal ice can be formed can be divided into:

- formation and accretion of new ice at the base of the glacier;
- metamorphic alteration of existing ice close to the bed.

New ice can be formed at the bed by a variety of processes of regelation and congelation. Regelation is the localised melting and refreezing of ice. This can occur in several ways. As ice flows around a bedrock obstacle, melting occurs in the high-pressure area up-flow of the obstacle, and meltwater flows past the obstacle either along the interface or through the ice to lower pressure areas where it refreezes (Weertman, 1957; Lliboutry, 1986). During this so-called Weertman regelation, water might be squeezed away from the obstacle. In this circumstance, absorption of latent heat on refreezing in the low-pressure zone downstream of the obstacle can, by a 'heat-pump' effect (Robin, 1976), produce a

Table 5.7 Summary list of mechanisms of basal ice formation (see text for details)

Event	*Mechanism*
Accretion of ice	
New ice	Regelation Congelation
Existing ice (subglacial, proglacial)	Adhesion Incorporation
Diagenesis of ice	Strain Hydrological processes Chemical processes
Entrainment of debris	Bump regelation, Substrate regelation/creep engulfment Congelation Vein-flow Cavity/crevasse infilling (squeezing) Structural deformation Enfolding/thrusting Traction/shearing
Thickening of sequence	Subjacent accretion Congelation Regelation Deformation Folding Thrusting

cold patch several metres in diameter in which water at the bed is frozen to the glacier sole. According to Shoemaker (1990) water can also be forced out of the ice to refreeze at the bed in low-pressure subglacial cavities. Where low-pressure water-filled subglacial cavities exist, melting can occur in boundary layer up to 10 cm thick in the ice above. A regelation cycle can operate whereby water drains from veins and refreezes at the ice/water interface. This can lead to the existence of a basal layer of soft ice with small crystals, and to the relative uplift of sediment in the boundary layer. Observations in cavities have revealed formation of regelation ice from water emerging into both air-and water-filled cavities (Kamb and LaChapelle, 1964; LaChapelle, 1968; Theakstone, 1979). Congelation is the freezing-on, at the base of the glacier, of water which may not be derived from local melting. Models of basal thermal regime (e.g. Boulton, 1972) reveal multiple transitions between thermal states at glacier beds that would be expected to lead to the alternate production and refreezing of basal water. Thermal zones can also be influenced by circulation of cold air from the glacier surface. Water flowing at the bed from warm to cold thermal zones, or from high to low pressure zones, can freeze to the glacier sole. Water penetrating to the bed from surface and internal melting, and also meteoric and extra-glacial water, can also contribute to congelation ice formed at the bed.

Ice already existing beneath or in front of the glacier can be attached to the glacier sole to form part of the basal layer. Proglacial ice sources include ground ice, lake ice, and glacier-margin accumulations of snow and ice. Souchez *et al.* (1994) have attributed the origin of the silty ice forming the basal layer of the GRIP ice core to non-glacial ice formed in front of the developing young Greenland ice sheet. They infer from the isotope characteristics of the silty ice that it probably formed as wind-drift or ground-surface ice outside the limits of the ice sheet and was incorporated into the basal ice as the ice sheet advanced over it. Hooke (1973a,b), Shaw (1977a,b), Evans (1989) and others have recognised the incorporation of ice and debris into the basal layer by the glacial overriding of proglacial accumulations of snow, superimposed ice and ice-cored debris. Evans suggested that the incorporation of alluvium-dominated proglacial debris fans or 'aprons' is a major source of basal debris where glaciers re-advance over extensive areas of dead ice. Sharp *et al.* (1994) recognised apron entrainment as an important process at Variegated Glacier. Where glaciers advance over frozen ground including ground ice or permafrost, the subsequent subglacial entrainment of material from the substrate would also involve attachment of pre-existing ice, which would have distinctive chemical, isotopic and structural characteristics.

Existing ice can also be entrained from subglacial sources. Tison and Lorrain (1987) consider in detail the processes of attachment and transport of ice formed as floor-coatings in subglacial cavities. Knight (1987) suggested that this was a major source of stratified basal ice, and Hubbard and Sharp (1995) proposed it as the source of their 'interfacial' and 'dispersed' facies.

Basal ice can be formed by the metamorphism of ice close to the bed, as well as by the entrainment of new ice. A range of processes operating in the basal zone have been suggested as possible sources of ice diagenesis, the most prominent of which relate to strain and hydrological conditions affecting ice crystallography and chemistry. Lliboutry (1986, 1993) has suggested the existence of a basal layer through which water is mobile in the vein network. This model has various implication for basal ice. For example, sediment as well as dissolved impurities could travel into and through the ice via the vein network. Knight and Knight (1994) have demonstrated this in the laboratory and suggested that it might account for the characteristics of the dispersed or clotted basal ice facies. Squeezing of water out of the ice through the vein network could result in a chemical flushing of the basal ice, as nearly all soluble impurities are concentrated in liquid in the vein network. This migrating water could also interact chemically with sediment in the ice. These processes can result in the development of a chemically depleted layer, and the formation of chemically enriched basal water which could refreeze to produce chemically distinctive new ice. The pressure conditions that are invoked to explain the driving of water through the vein network also indicate the possibility of squeezing gas. Hydrological flushing could also be associated with gas flushing, and with the bubble-poor nature of basal ice. Strain recrystallisation of ice in the rapidly deforming basal layer subject to variable stain from the impact of bed irregularities can be predicted to result in continuous reorganisation of the crystal fabric, involving migration of the vein network and expulsion of gases, solids and soluble impurities into the vein network. Partial melting at grain boundaries and the expulsion of vein water under pressure can be predicted to work in combination to produce a distinctive basal ice. The predicted characteristics of this ice match many of the characteristics of the dispersed (Lawson, 1979), clotted (Knight, 1987) and clear (Hubbard and Sharp, 1995) basal facies. It is not only non-basal ice that can be transformed by flow diagenesis near the bed; ice in the basal layer, also, is subject to modification during ice flow. For example, intricate structures imparted to laminated ice during entrainment by regelation may be entirely destroyed by subsequent recrystallisation of the ice, and dispersal of the debris (Hart, 1995a; Tison and Lorrain,

1987). Windblown snow entrained in marginal cavities can undergo dramatic changes over the course of one or two seasons.

The basal layer can be thickened by subjacent accretion or by compressive deformation. Weertman (1961) proposed thickening of the basal layer by sequential addition of layers of new ice at the bed by freezing-on. This process has been supported at several sites, and the origin of sequences of debris bands has been attributed to Weertman's process. However, not all debris band sequences can be attributed to this. For example, Knight (1989) demonstrated that the co-isotopic signature of the clean ice layers in a sequence of debris bands in Greenland was different from the refreezing signature displayed by ice in the debris bands, and was isotopically equivalent to the ice directly overlying the banded sequence. This is not consistent with an origin by freezing-on for the clean layers. The implication is similar to that of the non-basal ice intercalated with frozen-on material at Variegated Glacier (Sharp *et al.*, 1994): namely, that tectonic deformation is responsible for the development of the sequence. Souchez and Lorrain (1975) and Shoemaker (1990) described how water that was squeezed out of the basal ice could refreeze underneath a basal sediment layer attached to the glacier sole. By this refreezing, the debris layer is effectively raised above the sole and into the basal layer, often in the form of a debris band. If the process operated repeatedly or cyclically, it would facilitate the generation of a sequence of ice/debris laminations.

Several observers have reported folding and thrusting within the basal ice. For example, Knight (1994) illustrated the formation of debris-band sequences from deformation of the stratified facies, and Sharp *et al.* (1994) demonstrated compressive thickening of the basal layer of the Variegated Glacier during its surge. Folding of stratified and banded sequences is clearly visible in many marginal exposures (e.g. Fitzsimons, 1990) and could be invoked to explain patterns observed in the base of some ice cores. Folding can also incorporate non-basal ice from above into the basal sequence, causing intercalation of basal and non-basal ice. Boulton (1974) recognised that thickening of the basal layer could occur by flow divergence and concentration around subglacial obstructions, and Knight *et al.* (1994) demonstrated that such effects could operate at very large scales. Folding and thrusting within the basal layer is a major mechanism for the development of thick sequences of debris-bearing basal ice.

The basal layer of ice shelves is different from that of land-based glaciers, because the substrate is water and a different set of basal processes are likely. Ice-shelf basal ice can be accreted either by freezing-on of water from beneath the shelf, or by processes similar to land-based glaciers where the ice shelf is locally grounded. Eicken *et al.* (1994) recovered a core 215 m from the Ronne ice shelf at a point where the ice was approximately 240 m thick, and observed at a depth of 152.8 m a sharp discontinuity marking the transition between meteoric ice accumulated from above and marine ice accreted from below. Goodwin (1993) found that the basal layer in coastal ice cliffs of Law Dome, Antarctica, comprised a mixture of marine congelation ice formed by basal freezing of desalinated sea water, granular marine ice formed by episodic mixing of basal meltwater and seawater, and ice formed by basal regelation. The process of basal accretion beneath ice shelves was described in Chapter 3 (Figure 3.4).

5.7 THE GLACIER BED AND THE SUBGLACIAL ZONE

5.7.1 THE BOTTOM OF A GLACIER

The lower boundary, or bottom face, of a glacier is the glacier sole – like the lower boundary or bottom face of a foot or shoe. The upper boundary of the substrate is the glacier bed, upon which the glacier rests. This distinction is frequently blurred by glaciological writers. The term bed has sometimes been applied to the whole of the boundary or interface between the glacier and its substrate, including the bottom of the ice, the top of the substrate, and anything that lies between the two. The term sole has been used to refer to the whole of the basal layer of ice in which regelation processes occur (e.g. Barnes and Robin, 1966). The subglacial zone includes everything that lies beneath the glacier, including the substrate and anything at the ice/substrate interface. This term, also, has been subject to flexible usage in the literature. Sometimes it is used more narrowly, to include only the substrate, and sometimes more widely, to include everything in the bottom part of the glacier.

The glacier bed might be marked by ice/substrate contact, by a water layer, or by air-filled or water-filled cavities. The character of the glacier bed is critical to glacier behaviour. Basal friction, sliding and erosion are all affected by whether the glacier sole comprises clean ice or ice with incorporated debris; whether the bed is marked by dry contact between ice and substrate, by a water layer, or by cavities; whether the interface is frozen or warm, rough or smooth; and whether the substrate is rigid bedrock, deformable sediment, permeable or an aquitard. Until quite recently, most glaciological models assumed a rigid, impermeable, ice/rock contact. In fact, major parts of former ice sheets and of present-day glaciers rest not on bedrock but on sediment; often glacial till produced by the glacier itself, and often permeable and deformable. The effects of different bed conditions on glacier hydrology, glacier movement and geomorphic processes are discussed in Chapters 6, 7 and 9.

The glacier bed is very difficult to observe directly. Many studies of subglacial phenomena have been based on observations of former glacier beds exposed by retreating ice. Active glacier beds can be accessed via boreholes, through cavities at the margin, and via tunnels through the ice or through subglacial rock. Kamb and LaChapelle (1964) were among the first to make detailed observations at the glacier bed, making their observations of glacier sliding over bedrock by means of a 50 m tunnel excavated through the Blue Glacier, Washington. Other studies employed both natural (e.g. Theakstone, 1967) and anthropogenic (e.g. Wold and Østrem, 1979) tunnels and cavities. Pohjola (1993) made observations of basal sliding at the bed of Storglaciären, Sweden, by using a video camera lowered down a borehole. Nevertheless, direct observation of active contacts between ice and substrate remains understandably infrequent.

5.7.2 SUBGLACIAL CAVITIES

Substantial areas of the beds of many glaciers, especially glaciers resting on bedrock, are marked by subglacial cavities. Where cavities are accessible from the margin or the surface, and where they are filled with air rather than with water, they can afford valuable access to the bed. Subglacial cavities reflect particular dynamic and hydraulic conditions at bed, and result in particular geological and geomorphological phenomena. Natural cavities of several different types can exist. They can be formed either by the flow of ice over an irregular bed or by the erosion of ice by subglacial water.

Flow-related cavities form as a result of ice flow over irregular bed. They can be air-filled or water-filled, and occur when the speed of ice flow over bed irregularities exceeds the rate at which ice can deform to maintain contact with the bed. This condition arises when normal pressure fluctuation (ΔP) across the obstacle is greater than the mean effective normal pressure on horizontal surfaces in the vicinity of the obstacle (N_m). According to Lliboutry (1968):

$$\Delta P = [36\pi V_i(a/\lambda^2)]^{1/3}$$

where V_i = ice velocity; a = amplitude of bed relief; and λ = wavelength of bed relief.

Cavitation is therefore most likely in conditions of low normal pressure (thin ice),

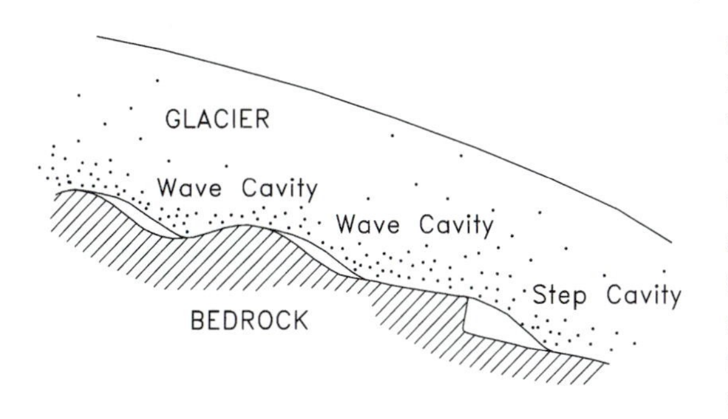

Figure 5.18 The form of step and wave cavities.

high ice velocity and high relief ratio a/λ (i.e. abrupt bed topography). Cavities of this type are referred to either as step cavities or as wave cavities, depending on the form of the topography causing their formation (Figure 5.18). Wave cavities form in the lee of bedrock bumps when tensile stress between ice and substrate exceeds the cryostatic 'closing' force, or where basal water pressure exceeds the overburden pressure and water is forced between the ice and the rock to create water-filled cavities. Step cavities form when the ice flows over an abrupt drop in the glacier bed. The processes of cavitation and the relationship between water pressure, ice pressure, cavitation and glacier sliding are discussed in more detail in Chapter 7. Flow-related cavities can be thought of as equilibrium features of ice flow for particular glaciological conditions. They persist for as long as the glaciological conditions that give rise to them. From a scientific perspective these cavities are especially valuable for the exceptional access they offer to the glacier sole for study *in situ* of subglacial processes such as sedimentation, regelation, and ice movement (Figures 5.19, 5.20 and 5.21). However, they do not generally offer good cross-sectional exposures through the basal layer, and may not be representative of conditions at the bed over a wide area.

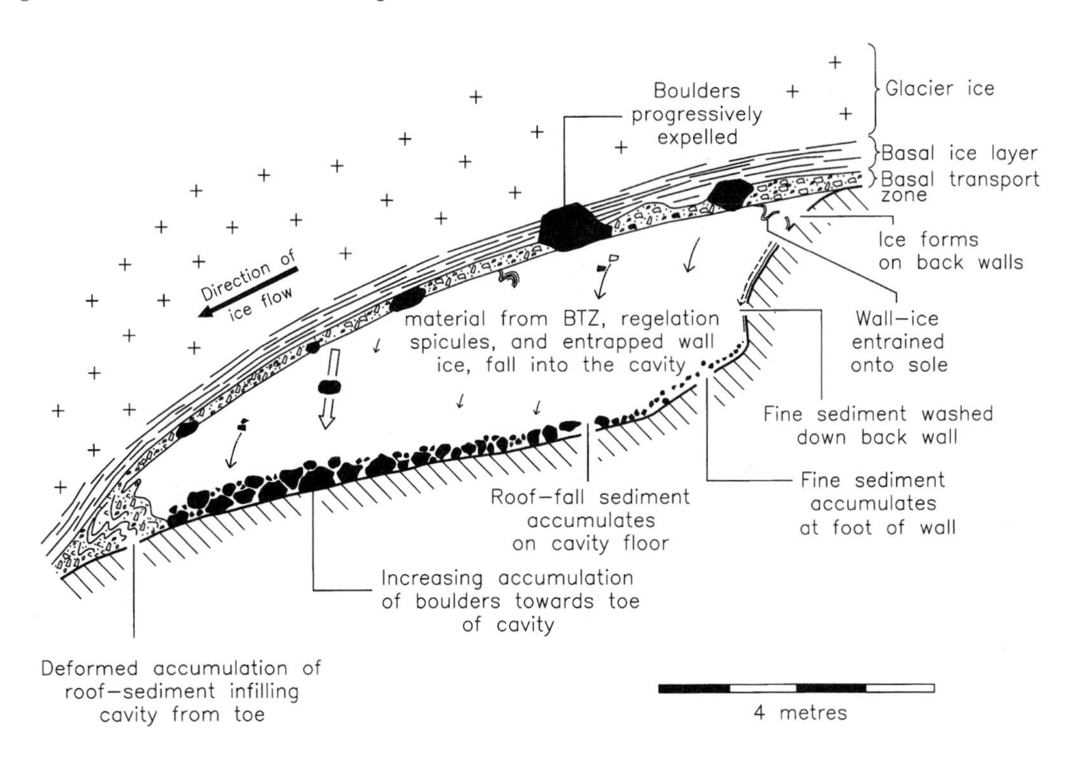

Figure 5.19 Some of the features typical of a subglacial cavity.

Figure 5.20 Debris-covered glacier sole and ice-coated back-wall of a subglacial cavity. The ice-coating from the rock is here being detached by the glacier sole passing above. Breiðamerkurjökull, Iceland.

Figure 5.21 The toe of a cavity. The view is down-glacier, with the sole of the glacier passing overhead. The lower left portion of the picture shows contorted ice and debris accumulating at the cavity toe. This derives partly from material transported across the cavity attached to the glacier sole, and may include an element of back-flow of ice from down-glacier into the cavity. Breiðamerkurjökull, Iceland.

Erosional cavities form as a result of natural or artificial tunnelling or quarrying into the ice. This can arise, for example, from erosion by water flowing beneath the glacier. The pattern of ice flow in these cavities tends towards the closure of the cavity, and the survival of the cavity depends on the continued through-flow of water. When formed by ephemeral flows such as periodic lake drainage or seasonally variable meltwater, these cavities can be short-lived. Formed by processes essentially external to the ice dynamic system, these can be thought of as exogenous phenomena in the subglacial ice-dynamic environment, and as such can occur in, and cross-cut, a range of glaciological settings. They are especially valuable for study of vertical sections through the basal ice layer, but are less useful for study of basal processes as the exposures of sole and substrate in these tunnels are likely to be atypical of local subglacial conditions generally.

Erosional and flow-related cavities thus display different features. Erosional cavities offer a cut-away view of the basal zone but do not allow direct observation of the sole of the glacier. Lee-side (step or wave) cavities offer a view of the glacier sole, but only in a particular subglacial context where the sole is not in contact with the substrate. Some of the differences between the two types of cavity are illustrated in Figure 5.22.

5.7.3 SUBGLACIAL LAKES

Subglacial lakes of various sizes are a feature of many glacier beds. The pressure gradient beneath a glacier normally serves to drive subglacial water outwards towards the margin, but water flow can be inhibited by basal

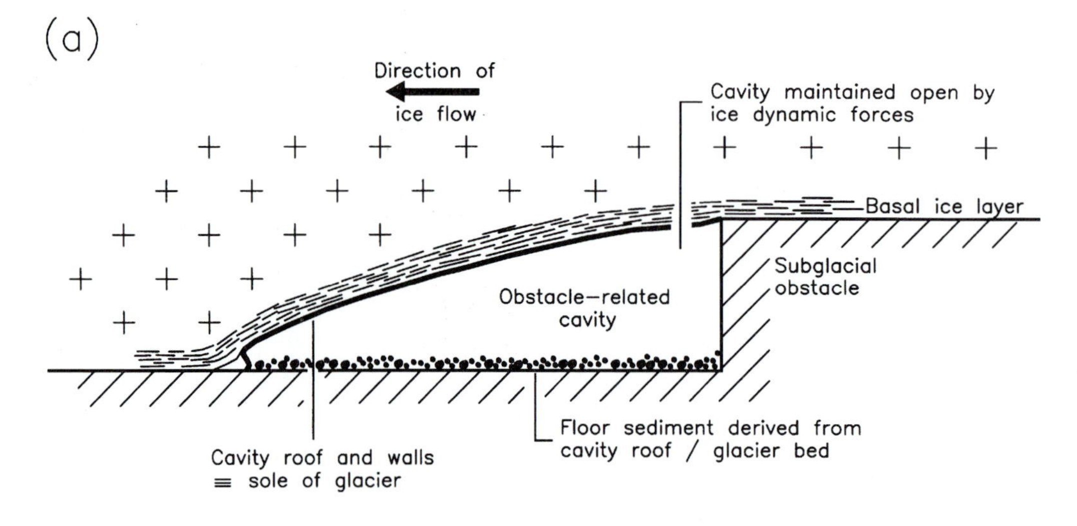

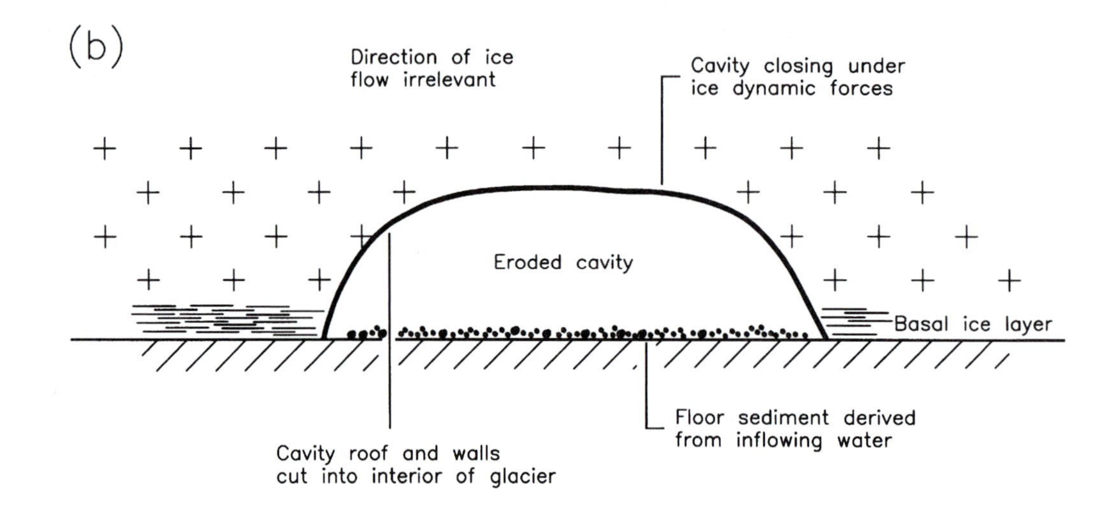

Figure 5.22 Cross-sections illustrating the different conditions of different cavity types.

topography if there is a sufficiently steep basal gradient opposing the ice-surface gradient. Oswald and Robin (1973) suggested that reverse bed gradients about 10 times the surface gradient would be sufficient to inhibit water flow and cause ponding and formation of subglacial lakes. Favourable conditions for lake formation would thus be expected where the ice surface gradient is low, such as under domes or saddles on the ice sheet. Ridley *et al.* (1993) described the distribution of 57 subglacial lakes identified by previous workers beneath the Antarctic ice sheet, typically of the order of 1 ice thickness in lateral extent, and clustered in a few locations suggesting areas where the basal ice is close to the pressure melting point over considerable distances. The ice sheet surface over the lakes is flat, owing to local decoupling of the ice from the bed. The largest subglacial lake yet discovered is Lake Vostok, covering 10,000 km^2 of the bed of the Antarctic ice sheet (Chapter 6).

5.8 THE ICE MARGIN

Glaciers are commonly divided areally into the subglacial zone that is covered by the ice and the proglacial zone that lies beyond the limits of the glacier. The boundary between these two zones – the ice margin – is of special interest for several reasons. Geomorphologists and sedimentologists tend to think of the ice margin in terms of the proximal proglacial environment: the proglacial area close to the ice. Glaciologists think of the ice margin as the area of the glacier close to its outer limit where particular glaciological conditions pertain. It is also especially exciting because it is the most easily accessible part of most glaciers: you can have a good look at it without having to get too intimately involved with the logistics of glacier surface explorations. The ice margin can provide a wealth of information for relatively low-cost expeditions.

The ice margin can take many forms. It might be a gentle ramp, or a vertical cliff. It might be floating on water or based on land. It might be fringed by meltwater streams or lakes, or it might be flanked by dry land. Glaciers with high englacial or basal debris content, especially when they are in retreat, may be covered in thick mantles of debris, and the precise position of the margin might be difficult to ascertain. Likewise, where glaciers with debris-bearing basal layers rest on ice-rich subglacial sediment, the base of the glacier might be hard to identify even in vertical cliff sections. Glaciers in retreat commonly leave fragments of themselves behind in ice-marginal moraines and proglacial sediments, preserved from ablation by the supraglacial debris cover, so the ice margin is often a broad and ill-defined zone rather than a distinct boundary. By contrast, advancing glaciers sometimes present the impressive spectacle of a clean and distinct ice-front bulldozing its way into virgin territory. The ice margin is a sensitive indicator of aspects of glacier health, and also contains information about processes and conditions in the inaccessible interior zones of the glacier. Ice derived from the highest parts of the accumulation zone is exposed closest to the ice margin, while ice derived from low in the accumulation zone is exposed to ablation higher up in the ablation area. A transect through a section of ice exposed at the margin, or along a surface traverse from the margin towards the interior of the glacier, is therefore a traverse upwards through the oldest, furthest-travelled ice that would appear at the bottom of an ice core. Furthermore, the ice exposed at the glacier margin includes the basal ice layer, which, as discussed previously, contains abundant information from the interior. Reeh *et al.* (1987) described the ice margin as a mine of ice for palaeoenvironmental study.

As ice flows towards the glacier margin, it suffers loss by ablation. The margin marks the position at which ablation finally succeeds in reducing the glacier to nothing. Glaciers typically become both thinner and steeper close to the margin, and thin ice has different dynamic characteristics from thick ice. Near the margin, the influence of bed topography on ice flow affects a greater proportion of the total ice thickness, and basal flow conditions are often reflected in structural features at the ice surface. For example, surface crevassing in response to ice flexure, caused by tidal forces on floating glaciers or by irregular bed topography beneath grounded glaciers, is more likely in thin ice than thick. Changing stress conditions under thinning ice close to the margin also affect subglacial drainage routes, and the confining pressures under which geomorphic processes operate.

The form of the margin can be affected not only by mass balance and ice-dynamic factors, but also by ice marginal and subglacial events such as jökulhlaups. The 1996 jökulhlaup from Skeiðarárjökull in Iceland was accompanied by removal of substantial amounts of ice from the margin by floodwaters that burst from beneath the ice. The escaping waters cut ravines into the ice margin, and ponded floodwaters formed a proglacial lake that caused flotation of sections of the margin and removal of marginal ice as icebergs that were subsequently deposited on the proglacial area when the flood subsided. Sugden *et al.* (1985) described the effects of a jökulhlaup on the Russell Glacier, part of the margin of the Greenland ice sheet. The ice margin was undercut by floodwaters, causing increased calving at the ice front, accentuation of the frontal ice cliff and accelerated ice flow towards the margin. Glacial and proglacial processes operating at the ice margin are also very important to glacial geomorphology, and are considered in more detail in Chapter 9.

5.9 CONCLUSION

Glaciers can be characterised in terms of their morphology and their structure. Gross glacier morphology is largely controlled by glacier dynamics and (except in the case of large ice sheets) topography. Glacier structure is a function of the way in which glaciers in general are formed and behave. Most glaciers therefore have broad structural characteristics in common, including their principal stratigraphic components: the supraglacial, englacial and basal layers. It is convenient to discuss glaciers in terms of these common features. Differences between glaciers, which are the crucial clues for understanding how glaciers relate to their different environmental and dynamic settings, can then be readily identified. The subdivision of glaciers into areas where different processes dominate and where different controlling variables play key roles is also a convenient device for organising descriptions and understanding of glaciers and glacial processes.

GLACIER HYDROLOGY 6

Out of a bellicose fore-time, thundering head-on collisions of cloud and rock in an up-thrust, crevasse-and-avalanche, troll country, deadly to breathers, it whelms into our picture below the melt-line . . .

W.H. Auden: *River profile*

6.1 WATER IN GLACIERS

All glaciers contain water. In some it exists only in tiny amounts between ice crystals; in others it forms huge lakes and river systems within the ice. The amount of water present in any glacier is determined primarily by the glacier's thermal regime and environmental setting, but can vary through time in response to both external conditions and internal dynamic processes. Cold or polar glaciers generally contain little water. Warm or temperate glaciers generally contain much more, and exhibit strong diurnal and seasonal variations in their hydrology. The presence of water is critical to glacier behaviour and to the impact of glaciers on the landscape. Water is central to most mechanisms of glacier movement, provides a mechanism for transferring heat through the glacier, and plays a major part in glacier sediment production and transport. The geomorphological and geological record of water activity associated with former glaciers provides valuable clues as to their characteristics. Hydrological activity associated with glaciers is one of the major sources of glacier-related hazards. Volumes of water in glacier systems can be prodigious. The spectacular discharges of meltwater associated with rapid deglaciation at the end of the last ice age were mentioned in Chapter 2. Equally impressive figures apply to present-day glacier hydrology. For example, subglacial Lake Vostok, a freshwater lake beneath the Antarctic ice sheet, has an area equivalent to Lake Ontario and a volume of about 1800 km^3. The glaciers of Alaska, in July and August alone, produce meltwater equivalent to 2.5 million m^3 of water for each km^2 of glacier surface (Meier and Post, 1991). By contrast, some cold glaciers accomplish all their ablation by calving and sublimation, producing no meltwater at all (Chapter 3).

This chapter considers sources of water in glaciers, water storage in glaciers, and routing of water flow through glaciers. In addition, specific issues such as water pressure, water chemistry and jökulhlaups are treated separately; not because they are conceptually isolated but because the research literature in glacier hydrology contains a great deal of work that focuses on topics such as these in isolation, and some readers may therefore find it helpful to have summaries of these topics drawn out from the main thread of the glacier hydrology story.

6.2 SOURCES OF WATER

Water in glaciers is derived from a variety of sources. The main source in most glaciers is the meltwater derived from melting of snow and ice. In some glaciers, rainfall can contribute significantly to the water content, and, in a few cases, input from external sources such as ice-marginal lakes and rivers can also be significant. Water can also be supplied to the glacier bed via groundwater flow.

Meltwater can be produced supraglacially, englacially and subglacially. Supraglacial melting of snow and ice is driven by insolation, turbulent heat transfer, precipitation, and other heat sources that were discussed in Chapters 3 and 4. Englacial melting occurs due to strain heating during flow, or heat transfer into the body of the glacier. This heat transfer can occur by percolating meltwater and by conduction from the surface or from the bed. Melting can occur at the base as a result of a variety of processes including geothermal heating, frictional heating, variations in melting point due to pressure fluctuations during flow across the bed, heat transfer by water flowing at the bed, flow of ice between zones of different basal thermal regime, and volcanic activity. In most glaciers, basal melting due to geothermal and frictional heating occurs at the rate of a few millimetres or centimetres of ice per year. Extraordinary events such as volcanic eruptions, however, locally produce much more spectacular values.

Water supplied from rainfall behaves much like water produced by melting at the surface, except that liquid precipitation often supplies a small amount of heat to the surface and can induce a small amount of additional melting. Individual precipitation events can supply relatively large amounts of water in a very short time compared with surface melting.

Rivers and lakes adjacent to a glacier can supply water to the glacier's drainage system. Water ponded up in topographic basins adjacent to the ice is in some cases periodically released into the glacier, and streams flowing along glacier margins can flow onto the ice surface, or into the interior via moulins or crevasses. Sometimes the position of a stream flowing along the glacier margin against a valley-side is controlled by water pressures within the ice. When the water pressure in the glacier is low, such streams might flow englacially, appearing at the surface only when the water pressure in the glacier rises. Lakes can be periodically 'perched' on the ice surface in the same way. Where glaciers flow across the mouths of tributary valleys, streams in those valleys might maintain a route through a tunnel beneath the ice, or might be ponded up to form lakes which may discharge over or beneath the ice.

The relative significance of these different sources varies between glaciers, but measurements from a variety of glaciers give some indication of their orders of magnitude. For temperate glaciers, basal melting is of the order of millimetres or centimetres per year, surface melting of the order of metres or tens of metres. Basally produced water thus makes a relatively insignificant contribution to total water flux. Measurements at Storglaciären, a small valley glacier in Sweden, indicated that of the water produced by the glacier 68% came from surface melting, 25% from direct precipitation, 6% from melting of snow on the valley walls above the glacier, and only negligible amounts from strain and geothermal heating (Östling and Hooke, 1986). On the basis of chemical analysis of the bulk meltwaters issuing from the snout of Gornergletscher, Switzerland, Tranter and Raiswell (1991) calculated that basally derived water accounted for as much as 50% of the discharge when total flow was at a minimum, but accounted for only a negligible proportion at maximum flow when surface-derived water dominated the water output from the glacier. In cold glaciers, conditions are somewhat different. Alley (1989a) pointed out that the majority of glacier surface area on the Earth is in dry snow and percolation zones, beneath which very little water, if any, reaches the bed from the upper portions of the glacier, and water flow at the bed is largely confined to basally derived water. In cold ice, vein diameters are smaller, and the vein network less permeable, than in warmer ice. However, liquid exists in the veins even at very low temperatures because the concentration of impurities at grain boundaries reduces the melting point. Thus even cold glaciers are not wholly without water.

6.3 WATER STORAGE IN GLACIERS

Between the uneven surface of the underlying land and the bottom of the ice, as well as throughout the ice itself, we may suppose large reservoirs to exist, containing almost the same quantity of uncongealed water even in the winter.

(Rink, 1877)

6.3.1 WATER STORAGE

Not all the water produced by a glacier, or introduced into the glacier from external sources, passes directly through the glacier. Some is delayed and held as stored water within the glacier. A delay in throughput of the water can last anything from hours to thousands of years. Part of this delay is accounted for by the time taken for the water simply to flow from its point of entry to its point of exit, but additional delays are caused by storage. Water can be held in storage supraglacially, englacially or subglacially. Volumes of water in storage, and the duration of storage, vary through time and between glaciers. An extreme example of water storage in the glacier system is subglacial Lake Vostok, which occupies an area of about 10,000 km^2 underneath ice 4 km thick in East Antarctica. The 1800 km^3 lake is fed by the melting of about 1 mm a^{-1} of ice from the glacier bed, and the residence time of water in the lake is estimated to be ~50,000 years. The mean age of the water in the lake, since it originally fell as snow at the surface of the ice sheet, is about one million years. (Kapitsa *et al.*, 1996).

Lake Vostok is by far the largest single water store that has been discovered, but temporary water storage is a major factor in the hydrology of most glaciers. A large element of storage involves seasonal changes in the capacity of the drainage routes through the glacier. Observations at several glaciers have revealed that, in the early part of the ablation season, water output from the glacier exceeds water input from melting and precipitation, and that the reverse is true later in the season, indicating that storage of water occurs within the glacier for parts of the year. For example, Östling and Hooke (1986) found that at Storglaciären between mid June and early August, about half a million cubic metres of water were delayed from runoff and held in storage within the glacier. That amount of water is equivalent to about 16 cm of water distributed across the whole area of the glacier. Stenborg (1970) found water storage equivalent to about 76 cm of water (~6 × 10^6 m^3) at nearby Mikkaglaciären, which is equivalent to about 25% of that glacier's total summer discharge. Tangborn *et al.* (1975) found that summer runoff from South Cascade glacier exceeded ablation by 38%, indicating the seasonal release of previously stored water. Similar storage can occur on a diurnal basis; Collins (1979) suggested that some of the meltwater at Gornergletscher was temporarily stored in basal cavities during high-pressure daytime flow, and then released at night as the water pressure decreased.

Storage areas in the glacier hydrological system include the following.

1. **Deep slush, snow and firn.** Slush can contain liquid water up to half its volume, and can be widespread over parts of glacier surfaces during the ablation season. Stenborg (1970) reported slush layers up to 2 m deep over areas of Mikkaglaciären. In West Greenland, slush fields are found on the ice-sheet surface up to 100 km from the ice margin.
2. **Crevasses and surface basins unconnected to drainage routes or temporarily dammed by snow.** Thomsen *et al.* (1988) reported that many seasonal lakes form on the surface of the Greenland ice sheet near Ilulissat (Jakobshavn) when moulins that drain closed basins are blocked by snow in the winter.
3. **Voids within the ice**, including blocked former drainage courses and cavities at the bed. Pohjola (1994) found that 1.3% of the length of a borehole through Storglaciären comprised voids.

4. **The vein network**, and inclusions within and between ice crystals (Lliboutry, 1971) (Figure 6.1).
   (a) Very flat liquid inclusions on the planes separating two crystals (two-grain boundaries).
   (b) Liquid in veins at three-grain intersections, and nodes where veins join at four-grain intersections (Nye, 1989).
   (c) Thin liquid films surrounding air bubbles within and between crystals.
   (d) Intracrystalline water inclusions, either dendritic or in the form of discs flattened in the basal plane of the crystal. Internal melting structures are sometimes referred to as 'Tyndall figures' after John Tyndall (1858), who first described them (Gagnon *et al.*, 1994).
5. **Porous substrate.** Many glaciers and ice sheets are underlain by permeable sediments with capacity to store water. Brown *et al.* (1987) suggested that 100 m of permeable sediments underlay the Puget Lobe of the Laurentide ice sheet. Boulton *et al.* (1995) argued that meltwater produced at the bed of an ice sheet can be forced into a permeable substrate by the overburden pressure of ice to produce very large groundwater flows. They suggested that subglacial aquifers in northern Europe could have been sufficient to drain all subglacial meltwater produced beneath the last ice sheet, although alternative models, discussed later in this chapter, reached different conclusions.
6. **Snow-dammed and ice-dammed basins on ground adjacent to a glacier margin.** These include, at the grandest scale, ice-dammed ice-marginal lakes such as Glacial Lake Missoula and Glacial Lake Agassiz, which abutted the retreating margin of the Laurentide ice sheet at the end of the last ice age (Chapter 2).

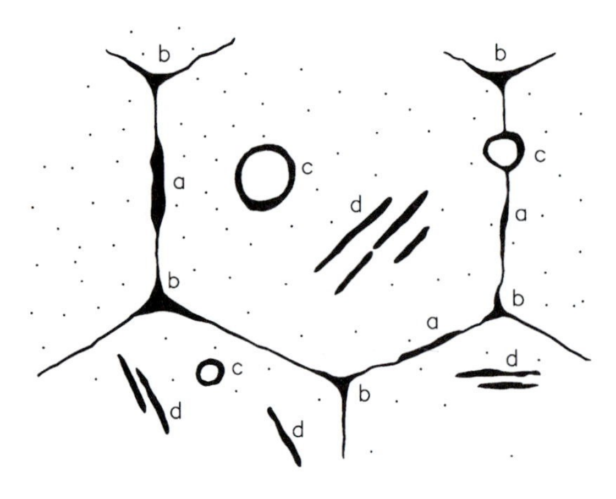

Figure 6.1 Water can exist at locations within and between crystals. See text for details.

Water storage exerts a major control on ice-bed interaction and glacier dynamics as well as on runoff from glacierised catchments. Periodic inputs of water that exceed the capacity of the drainage system can lead to high-pressure water storage at the bed, and when the pressure of the water is sufficient, water can escape abruptly. Such events have been described from many glaciers. The temporary storage of quantities of water at the bed has been seen to cause uplift of the glacier, followed by enhanced basal sliding and the release of trapped water. In many cases these events have been related to early-season meltwater production exceeding the capacity of the drainage network, and the term 'spring event' has sometimes been applied to them. Iken *et al.* (1983) describe early-season uplift of the Unteraargletscher in response to temporary storage of water at the bed, and Iken and Bindschadler (1986) describe increases in ice velocity accompanying increases in basal water pressure associated with increased water input derived from seasonal snow melting. At Variegated Glacier, waves of high water pressure propagating downglacier in association with ice-surface uplift and accelerated flow have been termed 'mini-surges' (Kamb and Engelhardt, 1987) (Chapter 7), and similar events have been observed on other glaciers. Holmlund and Hooke (1983) recorded short-term events akin to mini-surges at

Storglaciären, with elevated basal water pressures, high water levels in moulins, and enhanced sliding. It is clear that water pressure in glaciers is closely linked to a range of hydrological phenomena. As many previous authors have considered water pressure independently of its wider role, a brief introduction might be of value prior to discussion of specific elements of water flow through glaciers.

6.3.2 WATER PRESSURE

Water pressure in glaciers is central to the nature of water flow through the glacier system and controls a variety of glacial processes. Water sometimes flows in glaciers in what is termed 'free flow', at atmospheric pressure. This is equivalent to the open channel flow familiar to fluvial hydrologists. It generally occurs at and close to the ice surface, and close to the ice margin where the ice is thin. If all the water in a glacier passed directly through the glacier in open channels that were more than large enough to carry all the water, then all water flow in the glacier could occur at atmospheric pressure. However, in most glaciers water does not drain freely, and builds up within the drainage system. Water-filled passages exist at scales from intracrystalline veins to major tunnels, and the water is at a pressure higher than atmospheric pressure. The pressure is controlled by the freedom or otherwise of the water to drain directly through the ice and by the pressure of the ice itself. Controlling factors thus include the overburden pressure, the water supply, and the facility with which water can pass through the ice or the substrate. The ice overburden pressure (ρgh), which is effectively the weight of ice above any point in an ice column, defines the water equivalent line. This is the height to which a column of water would rise from a body of water at the glacier bed to balance the hydrostatic pressure of the ice, or in other words the surface level obtained if the ice column was replaced with an equal mass of water. At hydrostatic pressure, the water level in boreholes connected to the bed should be equivalent to the ice overburden pressure, so water levels should be about 90% of the ice thickness. If the water pressure at the bed exceeds the overburden pressure, such that a column of water from the bed would rise higher than the water equivalent line, then the basal water pressure is theoretically sufficient to lift the ice off the bed, effectively floating the glacier. Several observers have recorded water levels in boreholes and in natural moulins reaching levels higher than this, and observed water flowing or fountaining upwards out of the ice, indicating pressures much greater than could be accounted for by the simple standing of water in the glacier. The piezometric head or pressure head of water at any point in the glacier is defined by the elevation to which water would rise in open pipes or boreholes. This height is sometimes defined as a line or a surface through the glacier, referred to as the hydraulic or piezometric surface, and gives an indication of the distribution of pressure through the drainage system. In boreholes that penetrate to the bed, or moulins that connect with the basal drainage system, the pressure of water at the bed can readily be monitored as the water level in the hole rises and falls in response to changes in pressure (Hodge, 1979).

Water pressure varies across the glacier, and through time. Direct measurements have shown diurnal fluctuations of more than 0.5 MPa (e.g. Kamb *et al.*, 1985; Iken and Bindschadler, 1986). Major causes of change in water pressure are changes in ice pressure, or, more commonly, changes in the amount of water being input to the drainage system. Water input varies diurnally and seasonally, causing periodic variations in the water pressure environment and hence in water flow. For example, Hooke *et al.* (1990) reported water pressure measurements made in boreholes at Austdalsbreen, Norway. Water pressures tended to increase gradually during the winter, as the efficiency of glacial drainage routes gradually decreased, and to increase more rapidly in the early spring, as meltwater was added into the system. In late May or early June water pressures dropped abruptly as the drainage routes opened to full efficiency, and began to oscillate diurnally in response to fluctuations in water production. The relationships between water pressure and flow are discussed in the following sections.

6.4 WATER FLOW THROUGH GLACIERS

The facts here stated hint at the existence of extensive channels in the depth of the ice, which are subjected to continual changes on account of being closed and opened by its movement.

(Rink, 1877)

6.4.1 CONTROLLING FORCES

The direction of flow of water within a glacier is controlled by the ambient pressure in the ice. At the surface, or where the drainage system is connected to the surface at atmospheric pressure, flow will be driven simply by gravity. Thus supraglacial streams follow routes determined by the surface topography in the same way as terrestrial streams. In the upper sections of moulins, water cascades simply under the influence of gravity, and explorations of moulins have shown that they often drop more or less vertically for several tens of metres in their upper sections before connecting with an internal drainage system. A fundamental question in glacier hydrology is whether englacial and subglacial drainage flows at atmospheric pressure or under a higher pressure. Traditional modelling (e.g. Shreve, 1972; Röthlisberger, 1972) involves extensive pressurised flow, and field observations have in many cases supported this view (e.g. Kohler, 1995). However, Lliboutry (1983) and Hooke (1984) suggested that zones of gravity-driven open-channel flow should be more common in glaciers than previous studies had assumed. Lliboutry (1983) suggested that steady-state flow in a water-filled conduit with a balance between melting of conduit walls by water flow and contraction of the conduit by ice flow was relatively uncommon. He argued that conduits would commonly experience fluctuating discharges in which conduits would be periodically enlarged by high flows and then left unfilled by subsequent lower discharges, therefore experiencing atmospheric pressure for extended periods. Hooke (1984) argued that, except for very low flows on very low gradient beds, streams should be able to melt conduit walls much more quickly than the walls could close by plastic flow of ice, and that water flow was thus not likely to remain at high pressure. Even in the models of Hooke and Lliboutry, however, periods of high pressure flow are likely to occur when discharge is increasing after a period of low flow.

Where water is not draining freely at atmospheric pressure under the influence of gravity alone, the driving pressure will be controlled by a pressure gradient within the ice and within the drainage system. This pressure gradient is controlled by ice thickness, and hence by ice surface and bed topography. Most recent analyses draw on the work of Shreve (1972). In Shreve's analysis, a hydraulic potential can be defined for each point in a glacier, and the flow of water is driven by a potential gradient in the excess of water pressure over the hydrostatic pressure, which he denoted as Φ, such that:

$$\Phi = \Phi_0 + p_w + \rho_w g z$$

where Φ_0 is an arbitrary constant, p_w is water pressure (which can be approximated to the ice overburden pressure), ρ_w is water density, g is acceleration due to gravity, and z is the vertical position or elevation of the water. The hydraulic potential at any point can thus be thought of as the weight of ice above the point in question combined with the potential energy of the water itself (its weight and its vertical elevation). It is thus possible to identify planes or contours through the glacier of equal hydraulic potential (Figure 6.2). Water flow within the ice will be perpendicular to these contours or

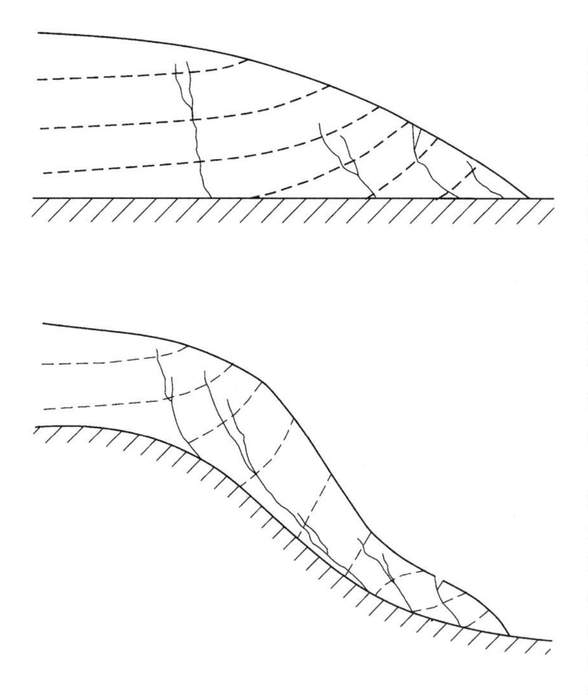

Figure 6.2 Equipotential planes (dashed lines) and drainage directions (solid lines) in different glaciological settings.

equipotentials of Φ, following the most direct line from higher to lower pressure locations. The equipotentials dip-up glacier from the ice surface at an average angle (α) controlled by the ice surface gradient (grad H) and the relative densities of ice (ρ_i) and water (ρ_w) and given (Shreve, 1972) by:

$$\alpha = \arctan [\rho_i |\text{grad } H| / (\rho_w - \rho_i)]$$

The gradient of the equipotential planes should theoretically be about 11 times the ice-surface gradient and in the opposite direction. For Midtdalsbreen, Norway, with a surface gradient of 8°, Willis *et al.* (1990) calculated the gradient of the equipotentials to be 58° (up-glacier) and hence the angle at which water should flow under pressure into the ice as 32° (down-glacier). On this basis it is possible to estimate the position where water from a point on the surface will intersect the bed. At the bed, subglacial drainage theoretically follows the line of steepest decrease of Φ on the bed. On a level bed this would follow the surface slope (and hence the flow direction) of the ice, but on a non-level bed it would tend towards topographic low points. The steeper the ice surface, the less is the influence of bed topography on water routing. Locally, flow at the bed will be affected by basally induced pressure fluctuations that distort the pressure distribution implied by Shreve's model. The size of passageways is determined by the gradient of Φ along the passage, the discharge Q, and the channel roughness.

This model fits field observations both of the distribution of features such as eskers that mark out the glacial drainage patterns of formerly glaciated areas, and of extant glacial drainage systems. On the basis of dye-injection experiments, Sharp *et al.* (1993) found that the major structural characteristics of the subglacial drainage system of the Haut Glacier d'Arolla could best be explained by Shreve's model of water flow perpendicular to contours of subglacial hydraulic potential.

6.4.2 NEAR-SURFACE WATER FLOW

For water near the glacier surface, drainage depends largely on surface conditions. Where the surface is permeable snow or firn, water percolates into the surface. The extent of its downwards penetration depends on the presence of impermeable layers in the snowpack, or the depth of the firn/ice transition where permeability decreases. Water can flow laterally through the firn and emerge onto the surface at the firn line, or penetrate downwards into the ice via crevasses or other drainage routes. Where the glacier surface is composed of ice, surface water flows or is stored at the surface. Substantial networks of supraglacial streams and lakes can develop, and have been shown to remain stable for periods of several years. Thomsen *et al.* (1988) mapped the surface hydrology of part of the Greenland ice sheet in the ablation zone close to the western margin near Jakobshavn and described streams up to

3 m deep. Where the ice surface is impenetrable, surface streams can discharge directly to the glacier margin. However, it is much more common for surface drainage to penetrate into the body of the glacier before reaching the margin. Water penetrates via open crevasses or by moulins, which are shaft-like openings in the surface (Chapter 5). At Jakobshavn, Thomsen *et al.* (1988) showed that drainage catchments on the ice surface very close to the edge of the glacier discharged their waters directly across the ice surface to the margin, but that all the more up-glacier catchment areas drained via moulins into the englacial zone (Figure 6.3). Seaberg *et al.* (1988) reported that 15% of surface water at Storglaciären runs off directly to the margin, while the remainder penetrates into the englacial and basal drainage system. The proportion reaching the margin directly tends to be inversely proportional to the size of the glacier.

6.4.3 ENGLACIAL WATER FLOW

Water which penetrates into the glacier from the surface, and water produced in the englacial zone, flows through an englacial drainage network. The englacial network can be very complex, and involves routeways at several different scales. A piezometric distinction can be drawn between englacial routes connected in open flow to the surface, and routes where water pressure is independent of atmospheric pressure. Routes with free connection to the surface, in which water flows at atmospheric pressure, can penetrate well below the surface. For example, channels draining from the base of moulin shafts even 50–60 m below the ice surface have been observed not to follow the equipotential pressure field predicted by Shreve (1972), and channels approaching the ice margin attain atmospheric pressure some distance up-glacier from it. For englacial routes where water pressure is independent of atmospheric pressure a thermomechanical distinction can be drawn between two scales of flow: (i) flow in inter-crystalline veins where flow dynamics are a function of the physics of small spaces; (ii) flow in larger conduits where more conventional hydraulic principles apply. The boundary between the two scales of flow is around a discharge of $Q = \sim 10^{-3}–10^{-4}$ m^3 s^{-1} (Röthlisberger and Lang, 1987).

The smallest scale of englacial water flow operates within the ice vein network. As discussed in Chapter 4, the vein network consists of interconnected passages at the boundaries between ice grains. Vein geometry has been

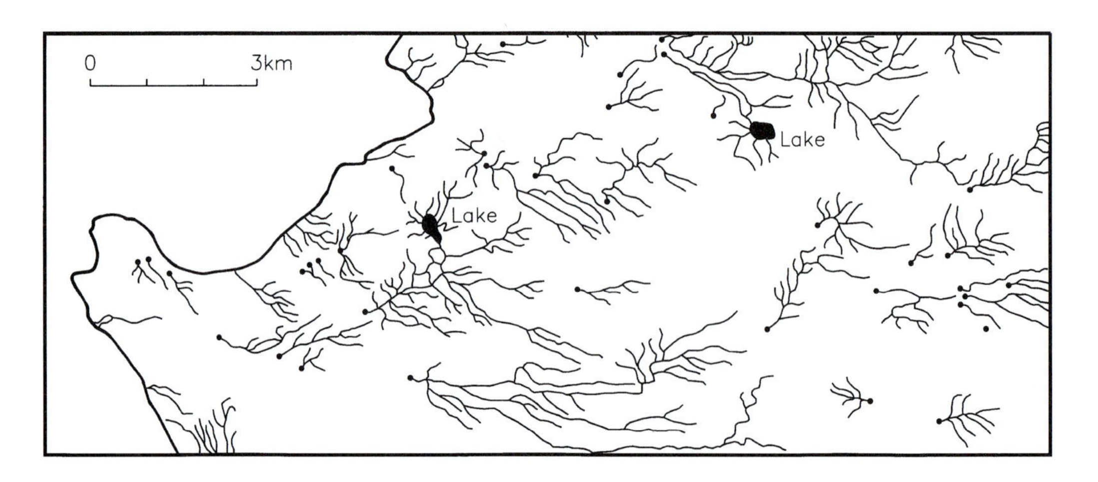

Figure 6.3 Map of surface drainage near to the margin of the Greenland ice sheet near Jakobshavn. The dots represent moulins. Based on Thomsen *et al.* (1988).

described by Nye (1989). Where three ice grains meet, the boundary is marked by a vein, which is effectively a tube with a curvilinear triangular cross-section. At four-grain intersections, veins meet and join to form nodes, which have the form of tetrahedra with open corners from which the veins extend (Figure 6.4). It has been suggested (Shreve, 1972; Nye and Frank, 1973) that these interconnecting passages form a route through which water can drain, making glacier ice permeable. Vein diameter is largely temperature dependent, but theoretical and measured vein widths of around 25–100 μm support the possibility of water flow through the network, given a suitable pressure gradient. However, reorganisation of the vein network upon recrystallisation (which might be continuous in active ice) and obstructions such as air bubbles or dirt particles in the veins might easily prevent drainage. In the light of such constraints, Raymond and Harrison (1975) concluded from observations of ice from Blue Glacier, Washington, USA, that drainage through the vein network was possible, but likely to be negligible, with a water flux of around 0.1 m a^{-1}. The reality of extensive water flow through the vein network can also be questioned on theoretical grounds. Lliboutry (1971) pointed out that if the vein network were fully permeable, then if there were an unlimited supply of water, melting caused by water flow through the network could completely melt a glacier in a period of a few years. On the other hand, Lliboutry (1986, 1993) also pointed out the possibility that water produced at the bed could be mobile through the vein network in ice close to the glacier bed.

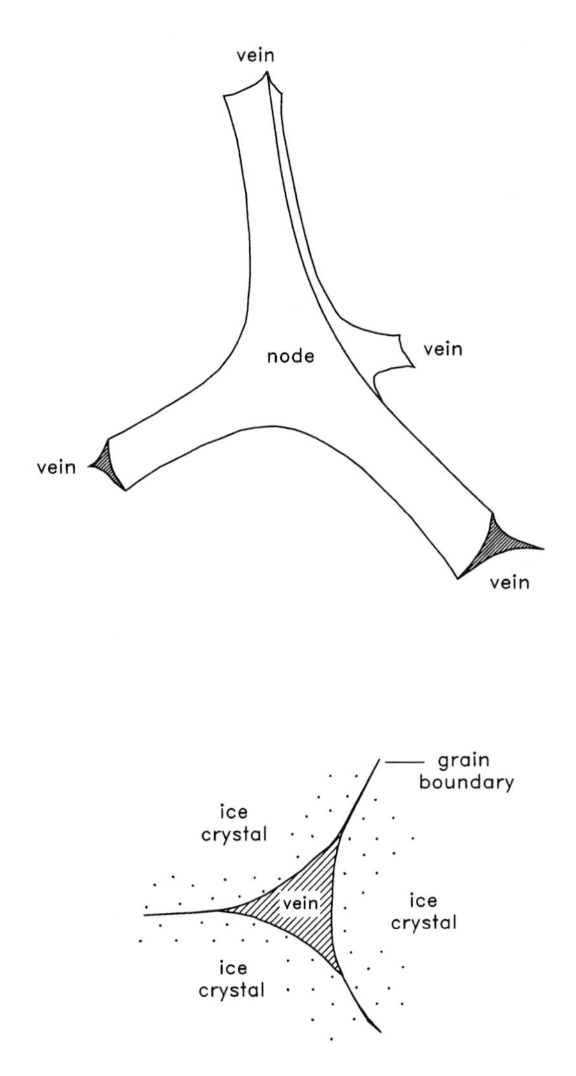

Figure 6.4 Three-dimensional and cross-sectional views of the structure of an intergranular vein in polycrystalline glacier ice. After Nye (1989).

Much more efficient drainage can be achieved through larger conduits. Pohjola (1994) found the interior of Storglaciären to comprise 1.3% englacial voids, which seemed to form part of an extensive englacial drainage network. Englacial passageways ranging in size from millimetres to metres seem to provide a rapid and effective drainage system, which in many cases connects directly to both the surface and the bed of the glacier. Raymond and Harrison (1975) identified tubular conduits several millimetres in diameter, exhibiting branching connections, that would be sufficient to drain meltwater through the glacier. By contrast with the intergranular vein network that could accommodate a flux capacity of only about 0.1 m a^{-1}, they found that the system of millimetre-scale tubules could accommodate the annual water production at their site of ~9 m a^{-1}. At an even larger scale, Näslund and Hassinen (1996) described englacial tunnels up to 15 m in diameter at Myrdalsjökull, Iceland. Many moulins connect directly into an englacial drainage network. Commonly, vertical moulin shafts connected by horizontal or sloping sections penetrate all the way to the glacier bed. In temperate glaciers it thus seems that the ice is riddled with drainage conduits of various sorts. Larger passages tend to grow at the expense of smaller ones, partly because pressure is lower in conduits with a higher discharge, and partly because water supplies more heat per unit wall area in larger conduits. Arborescent or dendritic networks of passageways thus develop, with small passages connecting with progressively larger ones as flow is concentrated into major drainage arteries (Shreve, 1972). This model is supported by the predominance of large outlet rivers emerging from glacier margins, rather than large numbers of minor streams.

Conduits through the ice respond to a variety of controlling forces (Röthlisberger, 1972). The flow of water through the conduit serves to keep the conduit open by mechanical and thermal erosion. Deformation of ice in response to overburden pressure and glacier movement tends to close the conduit. The formation and survival of the conduit is thus a function of the interaction between closing and opening forces. Röthlisberger and Lang (1987) provided a brief review of some of the relationships involved. Heat transfer between the water flowing in the conduit and the ice forming the conduit walls is governed primarily by friction, which is related to discharge. Also important, albeit to a lesser extent, is variation in the melting point with pressure at different depths in the ice, which is a function of the route taken by the conduit. The rate of closure of the conduit by ice deformation is controlled by the flow parameters of the ice (A and n from the ice-flow power law, Chapter 4) and by the difference between the ice overburden pressure and the water pressure in the conduit. The size and shape of the conduit is related to the water discharge and to the hydraulic roughness. Modellers generally assume a circular cross-section, and some observations of englacial tunnels support this (e.g. Näslund and Hassinen, 1996). This is most likely if the conduit is full, and experiencing pressurised flow, so that radial forces of tunnel closure by ice deformation are opposed by radial forces of tunnel opening by water pressure and melting. If the conduit is not full, and flow is at atmospheric pressure approximating open channel flow, then asymmetrical tunnel development can occur, for example, by vertical erosion of the ice by the water. Sugden and John (1976) speculated by analogy with karst that channels in the upper, vadose (unpressurised flow) zone would be narrow and deep since the water occupies and affects only the bed of the channel, while channels in the lower, phreatic (pressurised) zone would approximate a circular cross-section.

6.4.4 BASAL WATER FLOW

The flow of water at the glacier bed has been a major area of recent research interest in glaciology. Basal hydrology is intimately linked to mechanisms of flow, and a full understanding of glacier dynamics hinges on an understanding of basal water flow. The basal water system is difficult to observe directly, and our understanding of it has evolved gradually by a combination of theoretical work and indirect observation. Willis (1995) and Hubbard and Nienow (1997) have provided useful reviews of approaches to describing subglacial hydrology.

Water can reach the base of a glacier from the supraglacial, englacial and subglacial (groundwater) zones, or it can be produced locally at the bed by melting. Water can flow at the base of a glacier in a variety of different ways. It can flow as a thin film at the ice/bed interface, through a system of interconnected subglacial cavities, or through channels cut into the ice or into the substrate. Water can also flow through the vein network in the ice very close to the bed, and through the material underlying the glacier if the substrate is permeable. Different styles of subglacial drainage are associated with different glacier bed conditions and different styles of glacier flow. A distinction is sometimes made between 'channelised systems' involving pipes or tunnels and 'distributed systems' involving linked cavities, films or porous flow. These are sometimes referred to as 'fast-flow' and 'slow flow' components. A further distinction is commonly made between hard-bed (or rigid-bed) hydrology and soft-bed hydrology. With hard-bed hydrology, where the bed is rigid and impermeable, flow is dominantly by water films, channels and cavities. With soft-bed hydrology, where the bed is permeable and deformable, porous flow and flow in soft-walled canals cut into the substrate are common. These key aspects of subglacial drainage are considered in turn in the following paragraphs.

(a) Rigid bed, basal water film

The idea of water flowing in a thin lubricating film at the ice/bed interface has traditionally been part of the explanation of a glacier sliding over a rough bed (e.g. Weertman, 1957, 1972). A water film is also an integral part of the theory of basal regelation (Chapter 7) in which water produced by melting at high pressure zones flows as a film between the ice and the bed to areas of lower pressure. The idea of thin, localised water films associated with transport of water between high- and low-pressure areas as part of regelation sliding is well supported by observation. For example, Hallet (1979) inferred the existence of a film, usually only micrometres thick, from the morphology of subglacial carbonate precipitates. The implication of Hallet's observations was that the bulk of basal water transport occurred in channels and cavities, but that localised regelation flow was accommodated by the water film. Walder and Hallet (1979) found that 80% of the base of Blackfoot Glacier was separated from the bed by only a thin water film.

The idea of an extensive, thick water film transporting substantial amounts of water and facilitating sliding of the glacier over bedrock irregularities has been questioned on the grounds that such a film would be unstable and liable to break down into a series of channels. Any unevenness in the film would be amplified as more water flowing in thicker areas would lead to melting of the ice and further thickening of the film until a channel was effectively cut into the ice (Walder, 1982).

Weertman (1986) described a theoretical situation in which a basal water film could constitute a major part of the basal drainage network (Figure 6.5). Weertman considered a steady-state environment where all of the water at the bed of a glacier was produced by basal melting, and no surface water reached the bed. Water flux was low, and the water pressure was equal to the ice overburden pressure. Cavities that formed behind bedrock protuberances remained unconnected, as the pressure on the stoss sides of the obstacles was greater than the water pressure. Water flow past these high-pressure zones was thus accomplished through a film, rather than through conduits between cavities. However, if additional water was supplied (for example, from the surface), then higher water pressures would permit orifices between cavities to be enlarged, and the water film system would replaced by a linked cavity system. Weertman and Birchfield (1983) suggested that a

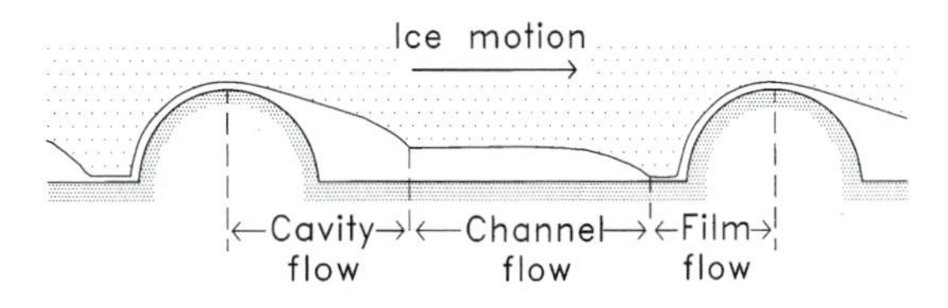

Figure 6.5 Basal water flow through cavities, channels and films relative to bumps on the glacier bed. After Weertman (1986).

channelised or linked cavity system could not be supported by basally derived water, and that as soon as a channel system exhausted the supply of water from the bed it would collapse, returning flow to a film system. They envisaged the possibility of cyclic transitions between film and channel flow. They suggested that surface-derived water might flow through channels, but that basally derived water might often remain in the basal water film. They also suggested that channels could draw in water from the film from only a relatively small distance, so that extensive areas of the bed would retain a film that was not captured by channels.

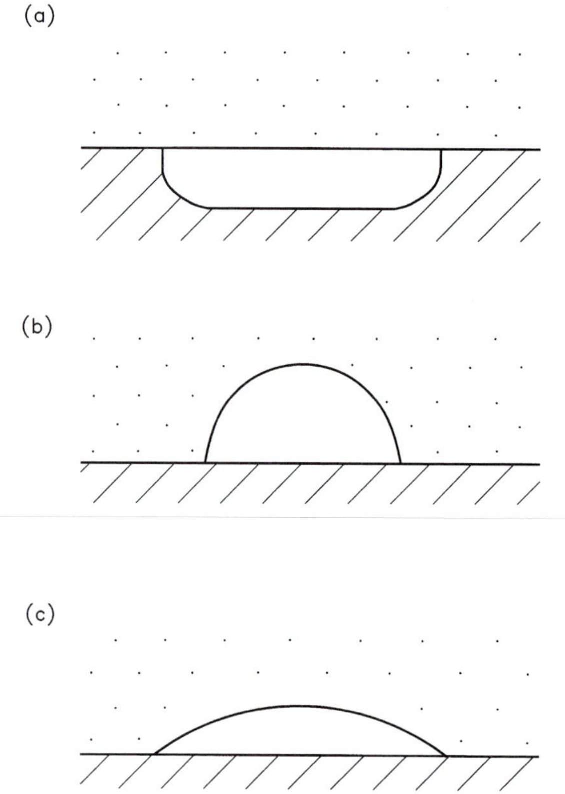

Figure 6.6 Different types of subglacial tunnels: (a) Nye (N) channel; (b) Röthlisberger (R) channel; (c) modified version of R-channel.

(b) Rigid bed, conduit system

While implications of water film for glacier dynamics are substantial, the efficiency of such a drainage system is relatively low. To accommodate the volume of water discharge produced by many glaciers, and in view of observations of major meltwater streams issuing from glacier margins, meltwater flow through large conduits at the bed must be considered. Several types of subglacial conduit have been considered in the literature (Figure 6.6). Channels or tunnels flowing across the bed with their roof cut upwards into the ice are sometimes referred to as R-channels, or Röthlisberger channels, after Hans Röthlisberger, who described them. Channels incised downwards into bedrock beneath the ice are sometimes referred to as N-channels, or Nye channels, after the descriptions given by John Nye. Channels similar to N-channels but cut into deformable sediments beneath a glacier can also exist.

Röthlisberger (1972) postulated semi-circular, water-filled channels incised upwards into the base of a glacier. In a steady state, these channels would be maintained by a balance between melting of the walls (channel opening) by water flow and channel closure by ice flow. However, Röthlisberger's model required very deformable ice around the channel to maintain this balance. With more realistic assumptions of ice hardness, Röthlisberger's model would predict water pressures rather lower than those that are actually measured in many field situations. Basal ice is indeed often more deformable than the upper layers of a glacier for various reasons, but not sufficiently so to make enough difference to satisfy the requirements of the model. Hooke *et al.* (1990) pointed out that this problem could be solved by assuming a different shape for the channels. They assumed that channels were not circular or semi-circular, but rather resembled the space between the arc and the chord of a circle (Figure 6.6c). With that assumption they found good agreement between modelled and measured water pressures in channels.

Nye (1973) postulated channels incised downwards into bedrock beneath the ice. Channels of this type have been observed frequently in exposed glacier forelands. Sharp *et al.* (1989) described channels 0.1–0.2 m wide and *c.*0.1 m deep, spaced at intervals of about 1.5 m across the bed at Glacier de Tsanfleuron. In this case the channels constituted links or orifices between cavities. Walder and Hallet (1979) described channels 0.1–0.3 m in width, and 0.05– 0.25 m in depth, on the former bed of Blackfoot Glacier, Montana. N-channels necessarily take longer to form than R-channels, and therefore reflect either a very stable glacier configuration, or a drainage system that is insensitive to changes in glacier conditions. By contrast, R-channels are relatively mobile and able to adjust or relocate rapidly in response to changing glaciological or hydrological conditions. As they are cut into the ice, R-channels can migrate down-glacier as the ice moves. Channels cut into soft subglacial sediments are also relatively responsive and flexible. The position and morphology of subglacial channels is controlled by the hydrodynamic forces of water pressure in competition with ice movement and deformation.

An important characteristic of flow through a subglacial tunnel network is that, when water flow is in a steady state, the water pressure in a channel is inversely related to the water discharge through the channel. In other words, as the discharge increases, the pressure in the channel decreases. A consequence of this is that, where two channels intersect, the water pressure will be highest in the smaller channel, so water will tend to flow from the smaller to the larger. This explains why larger channels tend to grow at the expense of smaller ones, why subglacial channel networks tend to develop into dendritic patterns, and why meltwater becomes progressively concentrated into a small number of major arteries.

(c) Rigid bed, linked cavity system

A linked cavity drainage system was proposed by Lliboutry (1969), and has been discussed in detail by Iken and Bindschadler (1986), Kamb (1987) and Walder (1986). Cavities can exist in the lee of bedrock protuberances, and are kept open when the rate of sliding exceeds the rate of plastic closure. These cavities can be water-filled, in which case heat exchange between water and ice contributes to cavity survival, and the water pressure acts in opposition to the ice closure pressure. In a steady state, the plastic closure of the cavity is matched by the sliding of the ice combined with melting of the roof by the water and the pressure of the water in the cavity. It was mentioned above that the isolated and unconnected cavities that existed with a water film at low water discharges could become connected via linking orifices as water discharge increased. Walder and Hallet (1979) found a practically continuous network of interconnecting cavities and channels on the exposed forefield of Blackfoot Glacier, with the connections between cavities formed by N-channels. Kamb (1987) offered a model of such a linked cavity drainage system with basal cavities ~1 m high and ~10 m across linked by narrow connections or orifices with a bed separation of ≤ 0.1 m. The orifices linking cavities will tend to open in areas of relatively low ice pressure, and hence avoid the high-pressure zones at the stoss face of bed protuberances. The linked cavity network would thus be expected to develop with a strong lateral component to the flow between cavities. Water would thus follow a much more tortuous route when draining through a linked cavity system than when draining through a tunnel system. In fact, Sharp *et al.* (1989) found that N-channels exposed on the deglaciated forefield of Glacier de Tsanfleuron were much less tortuous than Kamb's model suggested they should be, and followed a path roughly parallel to the ice flow direction.

It is possible that linking channels cut as R-channels might behave differently from linking channels cut as N-channels. Observers of deglaciated forefields observe only the record of former N-channels. In Kamb's (1987) model, the positions of the orifices are controlled by bedrock configuration, not necessarily following channels incised into the bed like conventional N-channels, but following ice/bed separations determined by bedrock roughness features of diverse kinds. In such a situation the linked cavity network can become a stable feature. If the orifices were cut into the ice (R-channels) the system may be more sensitive to fluctuations in water pressure and hence less stable. In this linked cavity system, the water flow is controlled primarily by the geometry of the orifices, which effectively throttle the flow. The system is relatively insensitive to changes in cavity geometry, but very sensitive to changes in orifice geometry. Orifice size is controlled by the ratio between sliding rate and overburden pressure (controlling ice/bed separation) and by melting of the orifice roof by water flowing through the orifice. The orifice geometry is potentially very sensitive to catastrophic growth, as increasing discharge will cause increasing frictional heating of the orifice roof, and hence orifice enlargement, thus allowing further increase in discharge.

Kamb identified a melting stability parameter Ξ that provides a measure of the influence of viscous heating on the size of orifices. Where $\Xi \geq 1$, the system is 'viscous-heating dominated'; in other words, melting by water flow is the main control on orifice size, and the orifices are unstable against rapid growth in response to variations in water pressure or small changes in orifice size. In systems where $\Xi \leq 1$, viscous heating is a less critical parameter in controlling orifice size, so orifices remain stable against rapid growth even if water pressure or orifice size does change slightly. Where ice velocity is higher, Ξ is lower and the orifices are less sensitive to meltback. Systems that are more sensitive to meltback are more likely to undergo transition from linked cavity to a channel system in face of rising water pressure, as unstable orifice growth leads to the development of tunnel or channel segments.

A significant difference between linked cavity flow and channel flow is that, in the cavity network, water discharge and water pressure are directly related such that an increase in water flux leads to an increase in water pressure in the system. One consequence of this is that larger passages within the cavity network will not necessarily capture water from smaller passages as was found to be the case for a channel system. The linked cavity system can therefore maintain a stable configuration with a large number of small water routes, extending over broad areas of the bed, in contrast to the large-artery preference of the channel system. At sufficiently high water pressures the linked cavity system can break down, as high water pressure either outweighs the closure forces within cavities, or leads to uplift of the ice and accelerated sliding. The consequent release of water from the confinement of the system could lead to rapid development of R-channels at the expense of the linked cavity network.

Walder and Driedger (1995) attributed outburst floods from South Tahoma Glacier to breakdown of linked cavity system in response to high water input, and similar instability could be associated with the so-called 'spring events' in which glaciers experience periods of high water pressure, uplift, enhanced sliding and increased meltwater discharge (e.g. Iken and Bindschadler, 1986). The mechanism of glacier surging (Chapter 7) might also be linked to basal water pressure and the transition between channelised and linked cavity flow systems (Kamb, 1987). Sharp *et al.* (1989), on the basis of their observations at Tsanfleuron, suggested that a threshold water pressure controlling sliding rate can be identified independent of major reorganisation of the drainage system. Below a certain discharge, water flow through a linked cavity system will occur at atmospheric pressure and the sliding velocity will be independent of water flux. Above the threshold, the cavity system fills and water pressure exceeds atmospheric pressure; effective basal pressure (ice overburden minus water pressure) then varies with water discharge, and increased discharge can lead to enhanced sliding.

(d) Soft-bed hydrology

Water at the bed of a soft-bedded glacier can be discharged either by flow along the ice/bed interface or by porous flow into subglacial aquifers. In most situations, certainly those usually considered by modellers of glacial hydrology, subglacial aquifers will have a finite thickness, and be underlain by an aquitard or aquiclude that limits the downward flow of water. Such cases would include, for example, till underlain by rock, sand underlain by clay, or gravel underlain by sand. In such a situation, for a low-gradient bed the pressure gradient generated by the overlying ice would serve to drive water laterally through the aquifer in the general direction of glacier flow. Discharge through the aquifer can be calculated on the basis of Darcy's law for water flow through a porous medium, whereby discharge varies with water pressure and material permeability. Such a route for the evacuation of meltwater within a glacier is relatively inefficient compared with the volumes of water typically produced. For example, to accommodate the meltwater produced by a theoretical ice sheet 1000 km long with a sliding velocity of 25 ma^{-1} under a shear stress of 10^5 Pa (1 bar) and a hydraulic gradient governed by an ice-surface gradient of 0.01, a continuous aquifer of clean, unconsolidated sand about 250 m thick would be required (Alley, 1989a). Boulton *et al.* (1974) calculated that a 1.5 m layer of till could accommodate a flux of 0.0004–0.0011 $cm^3\ s^{-1}$, and a 2 m layer of gravel a flux of about 6.4 $cm^3\ s^{-1}$. If the porous flow system cannot evacuate all the water supplied to it, then the aquifer effectively fills up, and pore water pressure within the aquifer rises. As well as enhancing glacier motion by weakening the bed and permitting greater deformation of sediment, increasing water pressure will also alter the style of water flow both within the till and at the ice/till interface. One possibility is that flow will concentrate into discrete channels and pipes within the till, potentially developing a conduit drainage system within the till. Clarke *et al.* (1984) envisaged the creation of preferred flow paths by the removal of fines from the substrate. These paths might take the form of channels cut into the sediment at the ice interface, or pipes within the sediment. More frequently considered is the case where water flows above the aquifer at the interface between the substrate and the ice. In that situation, flow of a water film or, at higher discharges, water in channels would be expected.

However, with a soft bed, erosion of the substrate is likely to play a major role in channel formation. Several issues need to be taken into account considering the form such a tunnel might take. Firstly, sediment from the deformable substrate would be under pressure to flow into the channel in the same way that ice would be under pressure to close the channel. For the channel to remain stable, not only must the rate of plastic closure of ice into the channel be matched by the water pressure and melting of the channel's ice roof, but in addition the squeezing of sediment into the channel must be matched by erosion of sediment by the flowing water. Channels cut into a mixed substrate such as a till might quickly acquire an armoured bed as fines are preferentially removed and coarser particles remain. Otherwise, as erosion progresses, the potential development of a deep subglacial gorge will be offset by collapse of the walls, and lateral inflow of sediment. As sediment flowing in from the sides is removed by erosion, the whole of the glacier bed in the area of the channel will be lowered, leading to the creation of what is known as a 'tunnel valley' (Figure 6.7).

Boulton and Hindmarsh (1987) considered the situation where the pore water pressure increased as high as the overburden pressure, so that the effective pressure

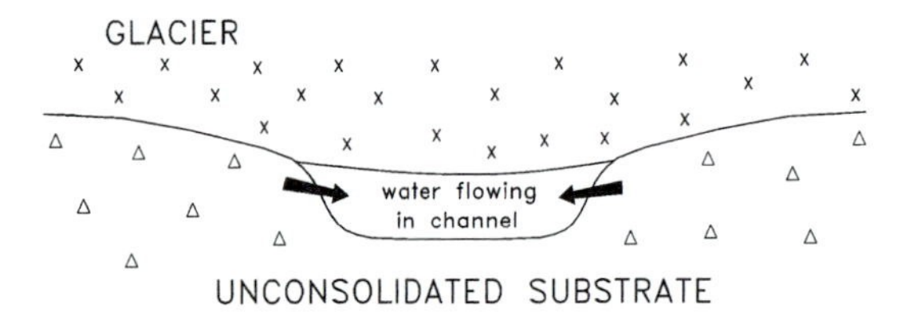

Figure 6.7 Theory of tunnel-valley formation by flow and erosion of subglacial sediments.

reached zero or less. Stable deformation would then be replaced by unstable deformation, but piping and channelised flow in the sediment could lead to more efficient drainage and reduction of pressure. The frequency of tunnel valleys would then be just sufficient to accommodate enough discharge to reduce the pore water pressure below the critical level. Brown *et al.* (1987) suggested that the 100 m of permeable sediments underlying the Puget Lobe of the Laurentide ice sheet was insufficient to drain more than a fraction of the water supplied to the bed, and that excess water was drained through tunnels 1–8 m wide. They calculated that tunnels would need to be spaced at intervals of 2–5 km to accommodate the water flux necessary to drain the ice. Alley (1992b) considered the apparent paradox of low-pressure channels existing within a deforming till layer, and concluded that non-steady creep of till into the channels would only occur within a very short distance of the channel, not more than about 10 times the thickness of the till layer.

Piotrowski (1997) proposed a model of subglacial drainage cycles in which groundwater flow through the substrate is followed, as water supply exceeds drainage capacity, by ponding of water at the ice/substrate interface, and eventually by catastrophic release of excess water through tunnel valleys. He calculated that for the marginal portion of the Scandinavian ice sheet in north-west Germany less than 25% of the meltwater could be accommodated by groundwater flow, and the remainder was evacuated periodically via tunnel valleys. This model differs from that of Boulton *et al.* (1995), who suggested that the overburden pressure of a large ice sheet can force basal meltwater into subglacial aquifers to produce huge groundwater fluxes. Modelling groundwater flow in Northern Europe through glacial cycles, they found that all basal meltwater could be accommodated by subglacial aquifers. In their model, regional groundwater systems are supposed to be completely reorganised by the impact of subglacial flow, and a link between hydraulic pressure and 'hydro-fracturing' within the substrate drives a self-organisation of the substrate hydraulic conductivity to accommodate changing water fluxes.

Walder and Fowler (1994) explored further the problem of water flowing at the interface between the ice and a saturated deformable substrate. Where water cannot escape into an underlying aquifer, the situation is in some ways analogous to rigid bed conditions, and the instability of film or sheet water flow. Walder and Fowler proposed two types of drainage conduits in such a situation. They suggested that sediment-floored R-channels could exist at high effective pressures, and wide, shallow, ice-roofed canals cut into the sediment (equivalent to N-channels) at low effective pressures. They found that for low gradients typical of ice sheets only canals can form, but for higher gradients typical of valley glaciers both channels and canals are possible. However, where both types of conduit exist, the water pressure will be higher in the canals than in the R-channels, and so the canals are unstable to the formation of channels. Their model thus predicts that channelised meltwater flow at the interface between an ice sheet and a saturated till bed should take the form of shallow canals at low effective pressure. More recent modelling by Engelhardt and Kamb (1997) on the basis of borehole measurements of water pressure beneath Ice Stream B suggested that canals of this type are 'the only currently viable candidate' for basal water conduits at that site.

6.4.5 OBSERVATIONS OF BASAL WATER SYSTEMS

The basal water system is difficult to observe directly, but several indirect observations have revealed details of the system at a number of glaciers. An important point that has emerged from such studies is that the drainage systems of individual glaciers can involve several different types of flow; channelised and distributed drainage systems can coexist, and the nature of the drainage system can change through time. This is of critical importance to our understanding of glacier dynamics.

Several different approaches have been taken to the indirect study of basal water systems. The most widely applied methods have been: observations of the exposed beds of retreating glaciers; use of dye or other tracers injected into the glacier and retrieved or monitored lower in the drainage system; analysis of the chemical composition of meltwater emerging from the subglacial zone; studies via boreholes drilled into the subglacial drainage system from the glacier surface; observation of the discharge of streams issuing from the margins of glaciers.

One approach has been to study relics of the drainage network preserved on ground exposed by retreating glaciers (e.g. Walder and Hallet, 1979; Hallet and Anderson, 1980; Sharp *et al.*, 1989). Sharp *et al.* (1989) reconstructed the former drainage system beneath Glacier de Tsanfleuron, Switzerland, on the basis of the micromorphology of rock surfaces exposed by the retreating glacier. They found that the system comprised two elements: a system of large step-cavities connected together by channels cut into bedrock; and a system of isolated cavities that would, when active, have been connected only by a regelation water film. The linked cavity / channel network covered 51% of the area of the glacier bed, and was deemed to have drained the bulk of the surface-generated meltwater that reached the bed. The water-film system with unconnected cavities was deemed responsible for draining water produced by melting at the bed. The observed linked cavity system corresponded in many respects with the configuration predicted by earlier theory. However, a difference was that the *N*-channels forming the orifices between cavities were oriented subparallel to the ice-flow direction, rather than transverse to it as predicted. This might be because the geometry of the bedrock roughnesses causing the cavities was different from the modelled case. Bedrock steps would not be expected to generate such high upstream pressures as bedrock bumps. In the description of the drainage system at Tsanfleuron, step-cavities seem to form the linked system, while bump-related cavities form the isolated cavity system. In addition, basal drainage systems on carbonate terrains such as the limestone bedrock beneath Tsanfleuron might generate special conditions. For example, limestone might be especially susceptible to the formation of N-channels by flowing meltwater. The relationships between discharge, pressure and system-type might therefore be different for glaciers with different bedrock geometry and lithology.

A second approach to the observation of subglacial drainage systems has been to observe the behaviour of dye injected into the system (e.g. Stenborg, 1969; Burkimsher, 1983; Hooke and Hock, 1993). Usually, dye is injected into the drainage system via moulins or boreholes from the glacier surface, and dye concentration is measured in water emerging in meltwater streams draining the glacier snout. Fluorescent dyes such as rhodamine and fluoresceine can be detected at concentrations of less than one part per billion. As the pulse or cloud of dye passes through the recording point, measurement of dye concentration through time produces a dye return curve or breakthrough curve. The shape of this curve is controlled by the dispersive processes that are controlled by flow hydraulics, so the shape of the curve can be interpreted to reflect the structure of the drainage route. More efficient routes will produce more peaked breakthrough curves. Key parameters include the time delay between injection and recovery, the proportion of the injected dye that actually emerges from the glacier, and the nature of the dispersion of the dye. The time from injection to recovery gives an indication of water velocity (v) if the sinuosity of the route is known or estimated. The dye recovery rate gives an indication of the amount of storage in the hydrological system. The dispersion coefficient (D, $m^2\ s^{-1}$) is the rate at which the dye is spread out in the water during its passage, and is controlled largely by turbulence. The dispersivity ($d = D/v$) is the rate of dispersion relative to the rate of advection of water, or the rate at which the dispersion coefficient increases with velocity. Dispersion is generally proportional to velocity for channels of a given roughness. For braided or interconnecting systems, higher dispersivity values reflect more intense braiding with shorter individual channel lengths. Values of dispersivity measured at Storglaciären, Sweden (Seaberg *et al.*, 1988) and at Midtdalsbreen, Norway (Willis *et al.*, 1990) gradually decreased from an early season high point, indicating progressive simplification of an initially complex distributed system into a system with fewer, larger conduits. Willis *et al.* (1990) were able to infer the characteristics of the drainage system of Midtdalsbreen from the way in which dye injected at different locations and at different times into the drainage system through surface crevasses travelled through the glacier to detection sites in outlet streams in front of the glacier. Variation in throughflow velocity, and the dispersivity of the dye in the proglacial streams, indicated two distinctive drainage systems in different parts of the glacier. The eastern side of the glacier was characterised by rapid throughput times and a very peaked, concentrated dye signal in the outlet stream. This was interpreted to indicate a conduit system within the ice. The western side of the glacier was characterised by slow throughput, and a very dispersed, gradual dye signal. This was interpreted to indicate water transit through a linked cavity system beneath the glacier. Variations through the summer in the behaviour of injected dye suggested a seasonal evolution of the drainage system, which is discussed in more detail below.

A third approach to investigating subglacial drainage has involved analysis of the chemistry of meltwaters draining the glacier (e.g. Collins, 1978; Tranter and Raiswell, 1991; Tranter *et al.*, 1993; Sharp *et al.*, 1995). Glacier meltwater is enriched in some ionic species relative to precipitation, and traditional analyses have been based on the fact that some of these species are derived mainly from interaction between meltwater and rock or rock debris at the bed. Meltwater with high ionic concentrations relative to precipitation might therefore be associated with a flow route involving slow basal transport (distributed flow system), while ionically dilute waters might be associated with non-basal flow routes, or with rapid water transit through the basal system (conduit flow system). From this distinction between delayed-flow and quick-flow components arose the use of chemical mixing models (e.g. Collins, 1978), which assumed that the bulk composition of meltwater emerging from a glacier could be interpreted as a simple combination of waters from these two routeways. The base signature of quick-flow waters could be inferred from the chemistry of precipitation, and the signature of delayed-flow waters from the chemistry of water emerging from the glacier during the seasonal discharge minimum when all the emerging water must have followed a delayed-flow route. The relative contribution of each flow route to the final composition of emerging meltwaters at different times could then be calculated, and the characteristics of the internal drainage system inferred. However, recent work (e.g. Sharp *et al.*, 1995) has cast doubt on the reliability of the mixing model approach, questioning the validity of the assumption that bulk meltwaters are composed of two discrete components that do not evolve or become further enriched in solutes after they mix. Mixing of water from different routes can occur at substantial distances in-glacier from the margin, and the mixed water can experience continuing chemical changes. Interaction between water and particles in suspension can contribute significantly to the solute in streams. Brown *et al.* (1996) found that as much as 70% of the Ca^{2+} load transported by bulk meltwaters draining from Haut Glacier d'Arolla may have been acquired in post-mixing reactions. In light of these problems, some recent work has taken a different approach, focusing on the analysis of individual ionic and gaseous components in meltwaters from a variety of settings (see Hubbard and Nienow, 1997, for a brief review). Chemical composition of glacier meltwaters is discussed separately later in this chapter.

A fourth approach to the observation of subglacial drainage systems has been through analysis of water level fluctuations and other parameters in boreholes and in natural moulins connecting with the basal drainage system (e.g. Hodge, 1976, 1979; Iken and Bindschadler, 1986; Waddington and Clarke, 1995). For example, Hubbard *et al.* (1995) described observations of water levels in a dense array of boreholes drilled through the Haut Glacier d'Arolla, Switzerland. Their observations suggested that the drainage system was composed of two interacting components: a conduit system and a distributed system, operating simultaneously. The studied area of the glacier was characterised by a strip or axis about 70 m wide, aligned roughly parallel to the flow direction, in which the minimum diurnal water pressures were low, but in which the diurnal range of water pressure was high. Outside this axis, water pressures were higher and more stable. The central area of variable pressure was interpreted as being dominated by flow through a subglacial channel, while the remainder of the bed was drained by a distributed system. The two elements seemed to be connected, in that water was forced out of the conduit into the cavity network during the afternoon when water pressures in the conduit were high, and flowed back from the cavity system into the conduit overnight when conduit pressures fell below cavity pressures. The water flow between the two systems occurred through a permeable sediment layer at the bed that could have served as a reservoir for the water forced out of the conduit. On basis of borehole measurements, Fountain (1994) concluded similarly that water could flow through subglacial sediment into a conduit system. Murray and Clarke (1995) measured water pressure in boreholes at Trapridge Glacier and found, like Hubbard, that two distinct drainage systems could be identified. Some of the boreholes accessed portions of the bed that were directly connected to an efficient subglacial water system. Other holes reached sections of the bed with no direct communication to the connected basal water system. In these holes pore-water migration is the main flow process. Murray and Clarke found that the basal water system was heterogeneous over very short distances, and that sections of the bed could fluctuate between periods of connection with the basal water system and periods of disconnection.

Inferences about the character of the basal drainage network can be made on the basis of observations of meltwater discharge in streams emerging from beneath the glacier margin. For example, Raymond *et al.* (1995) measured a range of hydrological characteristics of the discharge streams of Fels and Black Rapids Glaciers in Alaska, including stage, electrical conductivity and suspended sediment. They found that there was a very small-amplitude diurnal fluctuation in discharge but with only a very short delay between peak water input (ablation) and peak discharge output. They interpreted this as representing a transfer of up to 90% of the water through a slow system that did not show any significant diurnal pattern, and only as little as 10% through a fast system that generated a (small) discharge peak shortly after the peak water input. The rapidity of the response in the fast-flow system indicated also that the fast-flow system only extended up-glacier for about 30–40% of the glacier's length, beyond which all flow was in a slow system. They also found that dilution of the solute in the water from the slow system by periodic inputs of dilute water from the fast system caused a cycle in solute concentration equivalent but inverse to the discharge fluctuation.

Meltwater discharge at the glacier snout is affected not only by the basal water system, but also by hydrological processes in the supraglacial and englacial zones. The nature of water discharge from glaciers and how it can be used to infer glacier characteristics are discussed further in the following section.

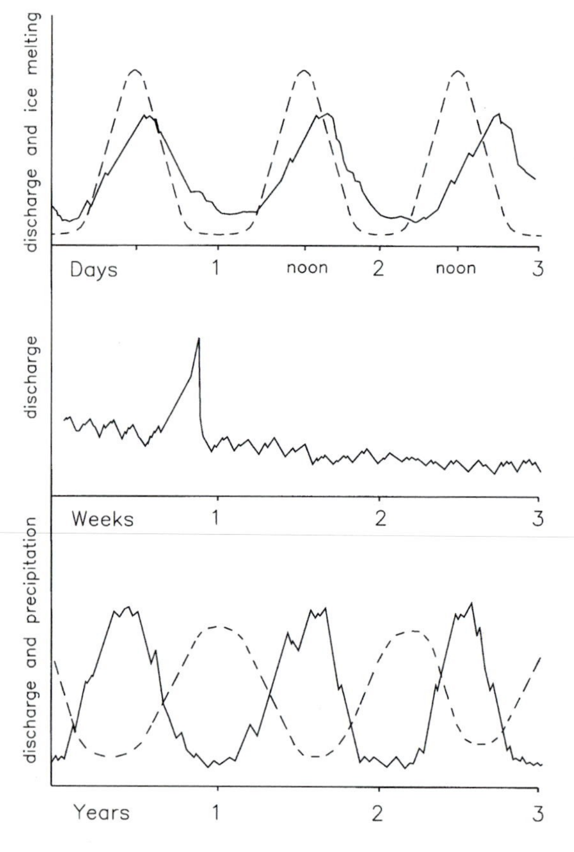

Figure 6.8 Typical meltwater discharge hydrographs. See text for details.

6.5 WATER DISCHARGE FROM GLACIERS

Water discharge from glaciers is controlled primarily by the production, storage and release of meltwater, and the storage and throughput of water from external sources such as precipitation. Discharge is characterised by variability at a range of time scales controlled by variations in water production and variation in the efficiency of the drainage network. In addition, aperiodic fluctuations occur in response to extraordinary events such as jökulhlaups. Discharge hydrographs are thus typically characterised by a seasonally varying curve with superimposed diurnal fluctuations and occasional isolated excursions (Figure 6.8). A baseflow component is supplied by groundwater runoff and runoff from storage or slow-release areas within the glacier such as firn, water-filled crevasses and cavities, and ice-dammed lakes from which seepage occurs. A small component of the baseflow can also come from the production of meltwater by stable geothermal heating and the slow transmission of surface water through the intergranular vein network.

A major control on discharge and discharge variability in meltwater streams is the production of meltwater. This is directly related to ablation, and hence to climate. This correlation between meltwater flow and surface ablation can be recognised at both seasonal and diurnal scales. Melting is at a minimum in the winter, and rises to a peak during the summer. The discharges of many glaciers reduce to zero for periods in the winter. At the diurnal scale meltwater production is generally at a minimum overnight and in the early morning, and increases to a peak in the middle part of the day. The meltwater discharge follows a similar pattern, but with the cycle delayed by a certain amount of time depending on the throughput time for water passing through the glacier from the various areas of melting to the discharge points at the snout. The shorter or more efficient the drainage route, the shorter is the time lag. Water stores such as snow on the glacier surface can be a major source of delay. The diurnal peak tends to occur earlier and earlier in the day as the ablation season progresses and the throughput of meltwater becomes progressively more efficient. The peakedness of the diurnal cycle can be affected by the size of the ablation area supplying the snout. Where there is a sufficiently long throughput time for water produced in the most up-glacier distant melting areas, water from the previous melt cycle might still be arriving at the snout when the first (closest derived) water from the next cycle begins to arrive. Measurements at a number of glaciers have indicated a range of throughput times. Mean values are unhelpful as some water runs off more or less instantaneously and some is held in storage for long periods. For example, at Hintereisferner (Ambach *et al.*, 1974; Behrens *et al.*, 1976), dye introduced into the upper part of the accumulation area did not arrive at the snout for 10 days, and the dye return curve peaked only after 17 days, indicating slow water transit through the snow in the accumulation area. Dye introduced into the lower part of the accumulation area reached the snout in 20 hours, and dye introduced into the ablation area took only 0.5–3 hours.

Seasonal-scale variations in discharge also reveal another major control on glacier hydrology: namely, seasonal evolution of drainage network. As was discussed above, the drainage system of most glaciers becomes more hydraulically efficient as the ablation season progresses. First, the storage potential offered by snow-covered areas of glacier surface decreases as the area of snow cover diminishes. Second, routeways within the glacier increase their capacity as the season progresses. As a result of this, throughput times for water decrease, and the peak discharge associated with maximum daily meltwater production therefore arrives at the glacier snout progressively earlier in the day. Because the throughflow times are decreased, the discharge curves become more peaked as the season progresses, as the flattening effect of the slow-travelled component is reduced. Early season storage of meltwater in snow and in cavities at the glacier bed leads to delayed runoff, and a subsequent late-spring discharge peak as the stored water is released. Summer runoff can exceed summer ablation as part of the runoff is accounted for by delayed release of winter and early spring meltwater.

Rainfall also affects discharge, but not always directly. Increase in water throughput due to precipitation can be offset by reduction in water production resulting from reduced insolation. Peak discharge from glacial meltwater streams often occurs at times when non-glacial streams are at their lowest flow, since glacial meltwater discharge correlates more with insolation than with precipitation. Catchments with both glacierised and glacier-free areas tend to exhibit less variability in river discharge than wholly glacierised or glacier-free catchments, as the glacierised and glacier-free areas respond in opposite ways to climate input (Braithwaite and Olesen, 1988). Catchments with about 30–60% of their area glacierised show the lowest amount of discharge variation (Röthlisberger and Lang, 1987).

The discharge, and even the location, of individual meltwater streams can vary through time due to changes in routing of subglacial meltwater flow. For example, at Solheimajökull in southern Iceland, for several years prior to 1996 the major outlet for water flowing through the primary subglacial drainage network of the glacier coincided with the outlet of a subglacial tunnel that carried water beneath the snout of the glacier from a tributary valley blocked by the advancing glacier snout. The two distinct flows met at the portal, and could be distinguished easily by their different sediment loads and hence their colour. The tributary stream that made its passage through the tunnel was distinctly brown, and the glacially derived meltwater distinctly grey. In 1996 the tributary stream still maintained its route through the snout, but the principal glacier drainage stream emerged ~150 m away at a different section of the margin, reducing the flow in the older channel to a fraction of its former value.

Water emerging from glaciers is often characterised by a high suspended sediment load. Seasonal and periodic variations occur in the suspended sediment load of meltwater streams, but sediment-discharge variability is not synchronous with water-discharge variability. The main controls on sediment discharge are: fluctuations in subglacial water velocity; sediment disturbance by basal sliding or channel migration; fluctuations in basal water pressure; and the seasonal evolution of the subglacial drainage network. Typically, alpine glaciers are characterised by high sediment concentrations early in the season as spring meltwater flushes sediment from areas of the bed that were not washed during the winter low-discharge period. During the ablation season, variability in sediment discharge is related to the evolution of the drainage network releasing stored sediment and flushing new areas of the bed. These effects play a greater role in sediment variability than do water discharge variations or progressive production of sediment by erosion. Sediment output from glaciers is considered further in Chapter 9.

A major source of variability in water discharge from glaciers is the periodic discharge of water that has been stored in lakes. This phenomenon is discussed in the following section.

6.6 ICE-DAMMED LAKES AND JÖKULHLAUPS

6.6.1 JÖKULHLAUPS

The discharge regimes of rivers draining from some glaciers are characterised by occasional huge, short-duration discharges that are orders of magnitude higher than the usual flow. A flood of this type is sometimes referred to as a jökulhlaup, an Icelandic term derived from the words *jökull* (glacier) and *hlaup* (river-rise or flood). Jökulhlaups can result from the sudden release of water stored within the glacier, from the sudden drainage of ice-dammed ice-marginal lakes, or from the sudden production of extraordinary amounts of meltwater as the result of volcanic activity. Jökulhlaups occur with a regular periodicity in many glaciers, and although they are short-lived, their magnitude is such that they can play a dominant role in proglacial geomorphology and sedimentology, and can affect the form and behaviour of the glacier itself. Among the largest freshwater flood discharges ever to have occurred were those associated with repeated jökulhlaups from glacial Lake Missoula at the margin of the Laurentide ice sheet around 13,000 BP, which had almost fantastic discharges up to 21×10^6 m^3 s^{-1}, about 20 times the present average global runoff (Baker, 1983; Dawson, 1992). Even the impact of much smaller floods than these on sedimentology and geomorphology makes them prominent in the geological record of former glaciations. Because of their repeating nature and impact on both physical and human landscapes, it is important to be able to predict the timing, peak flow, frequency-magnitude relationships and flood hydrograph form of jökulhlaups (Maizels and Russell, 1992). The impact of jökulhlaups on human activity is discussed further in Chapter 10. Major controls on the nature of jökulhlaups relate both to the hydrology and the glaciology of the water/ice system, and so the nature of jökulhlaup events can be used to interpret aspects of the hydrological and dynamic status of a glacier. Walder and Costa (1996) provided a substantial list of recorded jökulhlaups and their characteristics.

6.6.2 SOURCES OF JÖKULHLAUP WATER

There are two main sources for jökulhlaup water: the sudden production of large amounts of meltwater as a result of volcanic activity; and the sudden release of water which has previously been stored within, or adjacent to, the glacier.

Volcanic jökulhlaups appeal spectacularly to the popular imagination, and to the news media. For example, meltwater produced by a subglacial volcanic eruption that began beneath the Icelandic ice cap Vatnajökull on 30 September 1996 emerged as a 45,000 $m^3 s^{-1}$ jökulhlaup flood from the snout of the outlet glacier Skeiðarárjökull on 5 November to be greeted by the cameras of the world's press. At least 80 subglacial volcanic eruptions have been reported in Iceland since about AD 900 (Bjornsson, 1992; Gudmundsson *et al.*, 1997). However, volcanic jökulhlaups are not the most common form, nor, in the nature of their activity, do they offer to the glaciologist as much information on the operation of glacier processes as do jökulhlaups derived from the release of stored water.

The formation of temporary water stores upon, within and beneath glaciers was discussed earlier in this chapter. The release of water from temporary storage as the drainage system evolves through the ablation season is part of the seasonal cycle of glacier hydrological activity. Walder and Driedger (1995) showed how increases in water input to the bed caused by rainfall or warm weather could destabilise the basal cavity system and generate sudden outbursts. The seasonal variation in glacial water pressures and fluctuations in the level of what is effectively a water table within the glacier can lead to the seasonal formation and disappearance of surface and marginal lakes. Sudden drainage of surface lakes can be caused by water pressure fluctuations within the glacier or by the formation of connections between isolated bodies of water and the main glacier drainage system (Fisher, 1973). Large floods outside the seasonal cycle of glacier hydrology can occur when bodies of water are released from long-term storage in the glacier. Russell (1990, 1993a) described the winter-time drainage of ice-covered supraglacial lakes 20–30 km from the margin of the Greenland ice sheet that had previously been thought of as permanent features. Around 35–90 million m^3 of water drained from the lakes englacially or subglacially and emerged at the margin where the ambient temperature was –30°C. The ice surface where the water was removed collapsed to form craters up to 2 km in diameter, their margins marked by concentric fractures. Similar supraglacial collapse features in the Antarctic have been called ice dolines (Mellor, 1960). Glacial water stores supplying jökulhlaups may be supraglacial lakes, pockets of water held englacially, water in subglacial cavities or lakes, or water in ice marginal lakes. Haeberli (1983) reported that 60–70% of glacier floods in the Swiss Alps were derived from ice-marginal lakes, and 30–40% from water pockets within glaciers.

Marginal water storage can occur at a variety of topographic settings:

- where a hollow is formed between a valley glacier and its valley wall;
- where a glacier margin is enclosed by higher ground such as a fringing moraine;
- where a glacier blocks the mouth of a tributary valley through which a stream flows;
- where a glacier flows out of a tributary to block a major river valley.

For water to pond up behind an ice dam in any of these situations, a seal must be formed between the ice and the bed such that the rate of input of water to the basin exceeds the rate of output. The seal may be watertight, or may allow a certain amount of seepage. If no effective seal forms between the ice and the bed, water will not accumulate behind the ice, but will drain away as it is produced. For example, at the time of writing this book the glacier Solheimajökull in southern Iceland is blocking the mouth of a tributary valley but has not formed a watertight seal. Water flowing down the tributary valley maintains a tunnel beneath the snout of the glacier. As Solheimajökull advances further, the thickness of ice over the position of the tunnel will increase, increasing the overburden pressure and hence the rate of tunnel closure. A critical threshold must exist where the tunnel will close completely during the winter low-discharge period and a seal will form, allowing a lake to form in the tributary valley. This situation has occurred in the past, and historical maps show the lake present or absent at different points in recent history as the mouth of the tributary valley is blocked or exposed by Solheimajökull's advance or retreat (Grove, 1988). Lakes of the same type include Lake Donjek, in the Yukon Territory, Canada, which forms when the surge-type Donjek Glacier advances to block the path of the Donjek River, forming a lake of about 234×10^6 m^3 (Clarke and Mathews, 1981).

6.6.3 JÖKULHLAUP INITIATION AND DRAINAGE

Glaciologists' interest in jökulhlaups has focused on three main issues: the triggering mechanisms by which the release of water is initiated; the processes by which the flow of water is maintained and curtailed; and the controls on the periodicity of the events. The conventional model of jökulhlaup drainage involves several stages. Initially, water is stored in some form of reservoir, held behind a sealed ice dam. Next, the capacity of the lake exceeds the retaining power of the ice dam (either because the lake grows or because the dam weakens) and water somehow begins to be released from the reservoir. Next, even though the amount of water in the reservoir may have been reduced below the level that was securely held prior to the start of drainage, the outpouring of water increases – often very rapidly. Finally, either the water supply is exhausted or the drainage route closes and the flood ends. In many cases the dam then reseals and the reservoir refills until the cycle is repeated.

For non-volcanic jökulhlaups, several different triggering mechanisms by which drainage might be initiated have been suggested. Different mechanisms probably operate in different cases. Mechanisms that have been discussed most frequently include the following.

1. Flotation of the damming ice by pressure of water in the reservoir (e.g. Thorarinsson, 1953). Where the water pressure at the damming or sealing point matches or exceeds the ice overburden pressure, the water can escape by flotation of the dam as the ice becomes buoyant. For ice-marginal lakes this occurs when the lake depth is about 90% of the thickness of the ice, although the precise depth will depend on the bulk density of the glacier as determined by debris content and surface crevassing. Flotation can be brought about either by water input to the lake increasing the water depth, or by surface melting of the glacier causing the thickness of the ice dam to decrease. The water level required for flotation might be less than predicted if a floating ice tongue raised by increasing water level acts as a buoyant cantilever to reduce the overburden pressure at the seal (Nye, 1976).
2. Reservoir overspilling the dam (e.g. Miller, 1952; Liestøl, 1955). If the geometry of the ice and the lake are such that flotation cannot be achieved, the lake might drain by water overtopping the dam and spilling out of the basin. This is likely if the glacier is frozen to its bed, if the glacier has a high bulk density owing to high debris content, or if low points the dam are caused by the presence of crevasses. Breaching of the dam by crevasse propagation might be considered as a separate cause of drainage initiation.
3. Pressure of deep water forcing a route through ice by deformation. According to the theory of Glen (1954), if the water depth reaches as much as 150–200 m, then the water pressure at the bottom of the lake will be sufficient to induce plastic deformation in the bottom section of the ice dam, facilitating the opening of drainage routes for the water.
4. Undercutting of the dam by water flow through permeable subglacial sediments (e.g. Fowler and Ng, 1996). Seasonal variations have been invoked to explain periodic drainage of lakes where seepage occurs

through permeable sediments beneath an ice dam. In winter, seepage continues in a stable fashion, but in summer, heat from the seeping water melts ice at the base of the dam to open a more efficient drainage conduit (Röthlisberger and Lang, 1987). Piping of water through the sediment, or deformation of the sediment, could also contribute to destabilisation of the subglacial seal.

5. Rupturing of the seal by movement of the glacier. Lake drainage can be initiated by variations in glacier movement, even when the water level is constant. If the movement of a valley glacier relative to the valley sides is slow in winter, lakes may form in closed basins along the lateral margin of the glacier. In spring, as the glacier accelerates, drainage routes down the valley walls in the lee of valley-side irregularities are reopened and the lakes drain (Knight and Tweed, 1991).
6. Weakening of the dam by earthquakes, by landslides onto the dam or into the reservoir, or by iceberg calving causing waves in the lake. A variety of extraordinary phenomena could be responsible for mechanically weakening the dam or for temporarily inducing conditions such as high (wave-related) water levels that could initiate drainage. With lake drainage, as with most aspects of the physical environment, nature can always supply additional examples to any list of circumstances!

Drainage can thus be initiated either by changes in the water level in the lake or by changes in the thickness, strength or motion of the restraining ice.

The triggering mechanisms described above can account for the initial breaking of the seal and initiation of lake drainage, but most of them cannot explain the catastrophic release of the bulk of the stored water that is commonly observed. For example, in the flotation theory, when a small amount of water has been released the water pressure at the base of the dam will be reduced, and the overburden ice pressure would be expected to reseal the ice dam. The lake would then remain close to the flotation depth, periodically lifting the seal to spill small amounts of excess water and relieve the flotation pressure. Likewise with the Glen pressure mechanism and the oversell mechanism, the triggering mechanism accounts only for the initiation of drainage, not its continuation. In all of these cases the continued drainage is due to melting of ice along the drainage route by the water flowing out. By the time the initial triggering effect is exhausted, the escaping water has enlarged the drainage route sufficiently that water continues to flow. As more water flows, the drainage route enlarges progressively by melting. Heat is supplied by friction of water flow and by advection of heat from the reservoir. The heating and melting will be proportional to the discharge, so a positive feedback is likely to occur, leading to catastrophic drainage of the reservoir. The drainage ends either when the water supply is exhausted or when the escape route is blocked. Many measured jökulhlaups have been observed to stop draining before the reservoir is empty. The usual explanation of this is that, as the reservoir shrinks, the water pressure in the drainage tunnel falls, and is outweighed by the overburden pressure tending to close the tunnel, so the tunnel closes and the drainage route is pinched off (e.g. Nye, 1976).

The style of the release is reflected in the hydrographs typical of jökulhlaup releases (Figure 6.9). It follows from the triggering mechanisms listed above that drainage can be divided into two broad classes: those where a tunnel is somehow created through the base of the glacier, and those where drainage occurs supraglacially by circumvention or mechanical failure of the dam. Where an initial triggering is followed by rapid tunnel enlargement, the discharge is usually characterised by an exponentially steepening rising limb reflecting tunnel enlargement, then an abrupt falling limb as water supply is exhausted. Sometimes the hydrograph is more complicated because of temporary blockages in the drainage route,

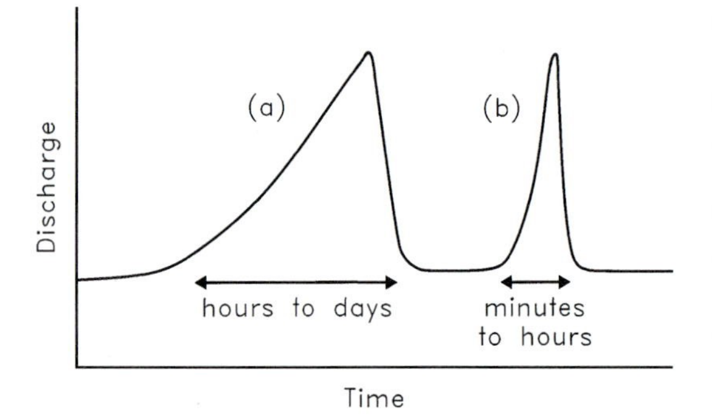

Figure 6.9 Hydrographs typical of different types of jökulhlaup: (a) tunnel drainage; (b) subaerial breach or emptying of subglacial water pocket. After Walder and Costa (1996).

caused, for example, by ice fragments. Where water drains into a complicated englacial drainage system the peak can be delayed and the rising limb of the hydrograph made more gentle. The exponential rise in discharge has been described by the function $Q_t = K(Vt)^b$ where K and b are constants, Q_t is the discharge at time t, and V_t is the volume of water drained up to time t (Clague and Mathews, 1973). The peak discharge can be predicted according to the Clague–Mathews relationship $Q_{max} = 75(V_{max})^{0.67}$, where Q_{max} ($m^3\ s^{-1}$) and V_{max} ($10^6\ m^3$) are the maximum discharge and the total volume of water involved in the event. This relationship is based on empirical observation, rather than on physically based theory, but seems to fit well with observed data although it lacks theoretical basis (Röthlisberger and Lang, 1987). For volcanic jökulhlaups, the hydrograph can be more peaked, reaching maximum discharge very quickly and abating equally abruptly (Thorarinsson, 1957). Walder and Costa (1996) discussed the influence of different types of dam-breaching on the characteristics of the discharge event, and suggested revisions to the Clague–Mathews relation. Reviewing published accounts of jökulhlaups, Walder and Costa suggested that different discharge relations are valid for different styles of drainage. For a given lake volume, they found that floods draining through subglacial tunnels tend to have lower peak discharges than floods draining supraglacially or along the ice margin as a result of the mechanical failure of an ice dam. They suggested $Q_{max} = 46(V_{max})^{0.66}$ for tunnel floods, a relation of the same order as that suggested by Clague and Mathews, but $Q_{max} = 1100(V_{max})^{0.44}$ for non-tunnel (breach) floods.

A situation that is sometimes considered under the broad heading of jökulhlaups but which has not been mentioned here is release of water by failure of a moraine dam. Unless the moraine is ice-cored, controls on dam failure are not really glacial in nature, though geomorphological and hydrological impacts are similar. Indeed, the discharge relations found by Walder and Costa (1996) for the mechanical failure of ice dams were deemed to be close to those for the failure of constructed earthen dams. The conditions for the formation of moraine-dammed lakes are also glacial in nature. For example, Lliboutry *et al.* (1977) described how disastrous floods afflicted the Cordillera Blanca in Peru in the 1930s after glaciers retreated behind the position of Little Ice Age moraines to form unstable moraine-dammed lakes (Chapter 10).

6.6.4 JÖKULHLAUP PERIODICITY AND LIFE CYCLE

It has been observed that many jökulhlaups are periodic in nature, recurring sometimes at regular intervals. This periodicity results from the fact that, in most cases, once drainage has occurred the drainage route closes as a result of ice flow. The reservoir then gradually refills to the point where the triggering mechanism again becomes effective. In some glaciers jökulhlaups occur annually, but in other cases the cycle is much longer, depending on the length of time taken for the lake to refill to the critical level.

The periodic cycle can be interrupted by major changes in glaciological or hydrological conditions. Many lakes only drain catastrophically in a very specific set of conditions, particularly related to advance and retreat of the ice margin. For example, Clarke (1982) attributed the transition from stable drainage to cyclic drainage at Hazard Lake, Steele Glacier, Yukon territory, to a reduction in ice thickness caused by a glacier surge. Before the surge, ice pressure at the seal always exceeded water pressure because a marginal spillway prevented the lake from attaining sufficient depth. Following the surge, ice ablation lowered the dam height to a level where flotation was possible when the lake was full. A reverse situation occurred at a draining lake at Russell Glacier, West Greenland, where sudden drainage occurred subglacially on a regular periodic basis for many years up until 1987 (Sugden *et al.*, 1985; Russell, 1989), but ceased thereafter following glacier advance and thickening of the ice dam. The lake has now filled to a deeper level than that at which drainage used to be initiated, and drains steadily through a bedrock overspill channel in hills at the distal end of the lake. The maximum lake level is thus limited, and the possibility of subglacial drainage now depends on a reduction in the thickness of the glacier. Major subglacial outbursts will not happen again until the ice retreats to its former position and the ice is thin enough for the seal to be broken by the depth of water that the lake basin can accommodate.

Jökulhlaups are thus tied to a narrow range of glacier positions. If the glacier is too far forward, drainage is impossible; if the glacier is too far in retreat, the lake will not form at all. A similar example of the same effect has been described by Knight and Russell (1993) who showed that, as a glacier margin advanced to dam a marginal lake, there was a window of jökulhlaup opportunity associated only with a certain range of glacier positions (Figure 6.10). Such windows of opportunity can be ended in a variety of ways, not only by glacier advance. For example, a small periodically draining lake in Iceland that was described by Knight and Tweed (1991) ceased to play a role in the glacier hydrological system by 1996 because the whole 13,300 m^3 lake basin had been filled by sediment, and the lake no longer existed.

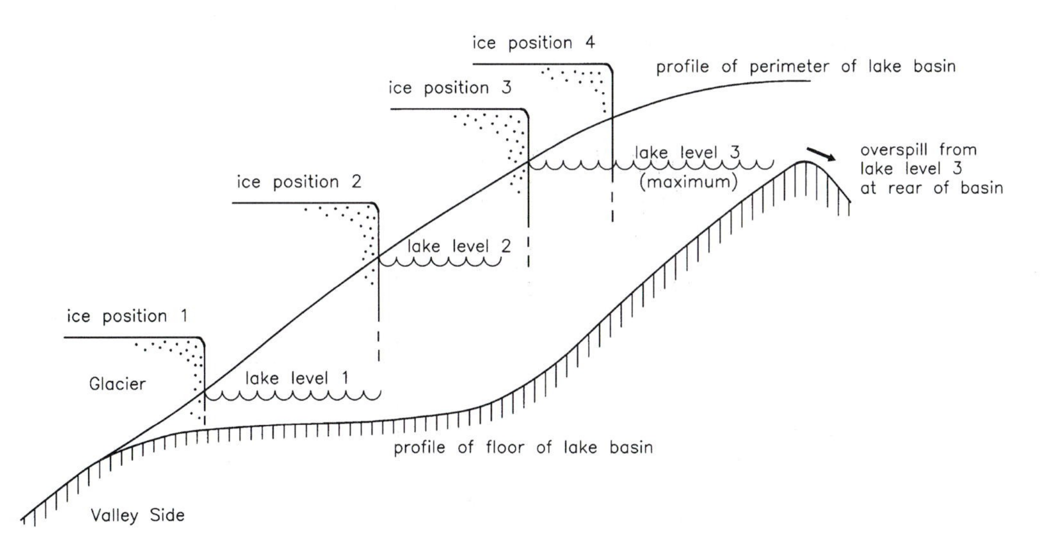

Figure 6.10 The propensity of a lake to drain depends in part on the position of the ice margin relative to surrounding topography. For a lake basin in Greenland, Knight and Russell (1993) showed that the maximum possible lake depth was controlled by the changing position of the margin, and that periodic drainage could occur only when the margin was within a limited range of positions.

For the jökulhlaup cycle to be restarted, the glacier would have to retreat, allowing the basin to be flushed clean, then re-advance to reform the lake.

The life cycle of jökulhlaups clearly has a bearing on glacier behaviour and human activity, and jökulhlaups are discussed further in Chapters 9 and 10.

6.7 CHEMICAL COMPOSITION OF GLACIER WATERS

The chemical composition of glacier ice was discussed in Chapter 3. Here we are concerned with the chemical composition of meltwater produced by glaciers. The chemistry of water in glaciers is important to glaciologists as the chemical signature contains clues to the history of the water and its interaction with different parts of glacier system. Knowledge of glacier hydrochemistry is therefore a valuable tool in the reconstruction of intraglacial drainage networks, and in identifying water sources, routes, and rates of flow through different sections of the glacier. The chemistry of water emerging from the glacier also gives an indication of the hydrochemical processes affecting the glacier bed, with implications for glacial geomorphology and sediment transfer.

Meltwater draining from glaciers is often rich in a range of solutes, including ions, cations and dissolved gases. Some ionic species, such as Cl^- and NO_3^-, are derived in precipitation or by dry deposition from atmospheric sources. Others, such as Ca^{2+}, Mg^{2+} and SO_4^{2-}, are derived principally from interaction between water and material within the glacier. The chemical composition of meltwater emerging from glaciers is controlled by: the chemical composition of the source water (which may be precipitation, melting snow, or melting glacier ice); sorting or flushing of chemicals within the ice and snow; and interaction of water with the atmosphere and with soluble materials within the glacier, especially at the bed. Several studies in Alpine glaciers have found that surface sources are of relatively low importance, and most of the solute load in waters draining from glaciers in Alpine catchments has been found to be subglacial in origin. For example, Souchez and Lemmens (1987) described how proglacial streams can contain several tens of times more dissolved cations than supraglacial streams, indicating that most of the dissolved content of the emerging water is acquired within the glacier. An exception to this situation can occur early in the melt season when chemical impurities from the winter snowpack are quickly leached into the drainage system in the first stages of snow-melt. Water from the later stages of the snow-melt is much more pure. The concentration of solutes in surface water is also affected by the total amount of precipitation (lower precipitation gives higher concentration of solute) and proximity to industrial sources. A vertical profile through a glacier, ignoring the effect of horizontal advection, would show increasing concentration of solute in water with depth owing to flushing and downwards migration of solute through vein network. Percolating water can also interact with englacial sediment, which can contribute to the solute content of water reaching the bed and exiting the glacier. Some chemical effects of this sort were considered in the context of debris-rich basal ice in Chapter 5. Water flowing at the bed interacts chemically with bedrock, with mineral particles constituting subglacial till, and with mineral grains produced by abrasion and comminution. In former ice sheets that had advanced over areas of boreal forest, there may also have been huge amounts of organic matter at the bed of the glacier which would have had an impact on subglacial water chemistry. Most modern glaciers are either in retreat, or in re-advance over relatively poorly vegetated areas, so this issue has not been well studied. Water in the subglacial environment can be an especially powerful agent of chemical weathering; low temperatures facilitate enhanced solubility of CO_2 in water, and an abundance of freshly ground rock flour provides a constantly renewing source of susceptible minerals. Interaction with rock particles in the basal zone is the major influence on the chemistry of meltwaters discharged from most glaciers.

A range of chemical processes are involved in the acquisition of solute by meltwater, including dissolution, hydrolysis, carbonation and ion exchange processes. Souchez and Lemmens (1987) and Tranter *et al.* (1993) provided useful reviews of some of the main processes involved. Key controls on the chemical interaction between water and material in the glacier are: the transit time of water; the availability of reactants; and the rates of the chemical reactions that occur during transit. Solute acquisition is greatest when water spends a long time in an environment where rapid chemical weathering reactions are possible. According to Tranter *et al.* (1993), mechanisms of solute acquisition or chemical weathering by water in glaciers generally involve two coupled reactions. One part of the weathering process is the supply of aqueous H^+ (protons) and the other is the subsequent weathering reaction in which those protons are employed. Different chemical histories will apply to waters that follow different routes through the glacier drainage system.

Quick-flow waters that transit rapidly through ice-walled conduits and open channels are likely to be relatively dilute. The weathering processes typical of this environment are relatively quick, but are limited because protons are supplied mainly by the relatively slow diffusion of CO_2 gas into solution. For water in quick-flow routes, a typical reaction can be illustrated by the weathering of calcite, which can be relatively rapid:

$$CaCO_3 \text{ (solid)} + H^+ \text{ (aqueous)} \rightarrow Ca^{2+} \text{ (aqueous)} + HCO_3^- \text{ (aqueous)},$$

fuelled by, relying upon and limited by the slow diffusion of CO_2 gas to provide the aqueous protons:

$$CO_2 \text{ (aqueous)} + H_2O \text{ (aqueous)} \rightarrow H^+ \text{ (aqueous)} + HCO_3^- \text{ (aqueous)}.$$

By contrast, delayed-flow waters, which transit slowly through a distributed basal drainage system, are generally solute-rich because they spend extended periods in an environment where relatively rapid reactions predominate in the proton supply. Greater contact between water and rock facilitates much greater provision of H^+. Paired reactions in this environment can be illustrated by sulphide oxidation (providing protons) and carbonate dissolution. An example of the production of H^+ by sulphide oxidation is the conversion of pyrite (FeS_2) to ferric oxyhydroxides:

$$4FeS_2 \text{ (solid)} + 15O_2 \text{ (aqueous)} + 14H_2O \rightarrow 16H^+ \text{ (aqueous)} + 8SO_4^{2-} \text{ (aqueous)} + 4Fe(OH)_3 \text{ (solid)}.$$

The H^+ thus produced can lead to further dissolution of calcite as described above, or of more slowly reacting phases such as anorthite (calcium feldspar):

$$CaAl_2Si_2O_8 \text{ (solid)} + 2H^+ \text{ (aqueous)} \rightarrow Ca^{2+} \text{ (aqueous)} + H_2Al_2Si_2O_8 \text{ (solid)}.$$

Solute acquisition is thus controlled partly by the rates of processes supplying H^+ protons. For example, water routed quickly though water-filled tunnels will not be able to acquire H^+ protons either from reactions at the bed or from gaseous diffusion, so weathering rates, even for minerals in suspension in the flow, are limited. Since gaseous diffusion of CO_2 into solution is a relatively slow process, even englacial water in open channel flow will be less well supplied with H^+ protons than will water able to interact with the bed. Along with the supply of rock flour at the bed, and the length of time the water resides in the weathering environment, this explains why englacially routed waters are relatively dilute in solutes compared with basally routed waters. An exception to this general rule might apply early in the melt season when proton-rich solute is leached out of the snowpack and into the englacial drainage system so that the englacial waters can become more aggressive: For example, H_2SO_4 (solid) $\rightarrow$ $2H^+$ (aqueous) + SO_4^{2-} (aqueous).

The chemical composition of bulk meltwaters sampled at glacier margins can vary through time. This is due partly to the fact that the composition is controlled not by equilibrium conditions, but by kinetic factors, or the rates at which different parts of the reactions involved in the weathering process occur in different circumstances within the seasonally changing glacier hydrological system. Variability results also from variation in the composition of the source water, which can be caused by variability either in precipitation or in the output from ablation of the snowpack and ice. Some further variability stems from variation in water discharge. For example, low winter discharge is often marked by very high solute concentration, but because the total discharge is so low, the solute flux is correspondingly low. In summer, solute is generally much more dilute as the total water discharge is higher, but the total solute flux is much greater in summer than in winter. This is largely because water accesses and flushes a higher proportion of the bed in summer, releasing water and solute from winter storage, and effecting solution over a broader area of the bed.

The chemical processes described above imply removal or transfer of material, so we will return to chemical weathering and solute removal in the context of glacial geomorphic processes (Chapter 9).

6.8 CONCLUSION

An understanding of glacier hydrology is central to an understanding of glacier behaviour. Glacier hydrology controls many of the major glacier dynamic and glacial geological processes. The behaviour of water in glaciers also reveals the structure of the ice and of the glacier at a variety of scales, and indicates how this structure changes through time in response to seasonal and longer-term changes in the glacier and its environment. Major issues in modern glaciology, including surges, ice streams and deforming beds, hinge on the role of water at the glacier bed. The development of our understanding of traditional glaciological problems such as sliding has relied on a growing knowledge of the effect of water within the glacier. However, substantial portions of our understanding of subglacial drainage are based on theoretical modelling in the absence of access to direct observation. Observations of englacial and subglacial hydrology rely on remote sensing techniques such as dye tracing, water-pressure monitoring and chemical analysis of meltwater which, in turn, lack a reliable theoretical basis for their interpretation. New developments in chemical analysis of meltwater have been put forward as promising avenues for progress in understanding glacier hydrology, but chemical analysis of meltwater is still limited by the problems that beset all forms of remote sensing of the englacial zone. The problems of studying parts of glaciers that we cannot practically access are discussed further in Chapter 12.

MOVEMENT OF GLACIERS 7

7.1 GLACIER MOTION

The fact that glaciers move is critical to the hydrological cycle, to the functioning of the ocean/atmosphere system, and to all the geological and geomorphological phenomena that arise from glaciation. In all glaciers ice moves from the accumulation area to the ablation area, the motion supplied by input of fresh ice in the accumulation area and drained ultimately by loss of ice in the ablation area. This movement of ice through the glacier is achieved by a variety of mechanisms, which combine to different degrees in different circumstances, so glaciers can behave in a range of different ways. This chapter focuses on the nature of glacier motion and the mechanisms by which driving forces are converted to movement. We are concerned here primarily with horizontal motion. Aspects of the vertical movement of ice are considered in Chapter 5 (regarding basal ice) and Chapter 11 (regarding ice cores).

Glaciers move at a wide range of speeds. In surging glaciers, speeds of up to 65 m d^{-1} have been recorded over short periods. The fastest moving non-surge glaciers include Jakobshavn Isbrae, which maintains continuous speeds of up to 7 km a^{-1} ($\approx$19 m d^{-1}) near its snout (Echelmeyer and Harrison, 1990). More typical speeds for glaciers are of the order of centimetres per day, or metres per year. Table 7.1 illustrates some of the range of glacier velocities that have been measured. Most glaciers also exhibit substantial temporal variations in movement. Surging glaciers alternate periodically between slow and fast flow, and many non-surging glaciers do the same, to a lesser extent, on a seasonal basis. Individual glaciers also show different rates of movement in different parts of their area, in response to local mass flux and dynamic controls.

Table 7.1 Measured velocities from a selection of glaciers

Glacier and location	*Characteristics of glacier and sources of data*	*Measured velocity ($m\ d^{-1}$)*
Variegated Glacier, Alaska	Warm-based surge-type valley glacier Kamb *et al.* (1985), Kamb and Engelhardt (1987)	65 (brief surge peak) 0.55 (quiescent phase)
Jakobshavn Isbrae, Greenland	Tidewater outlet from ice sheet Echelmeyer and Harrison (1990)	19 (continuous)
Ice Stream B, West Antarctica	Ice stream draining ice sheet McDonald and Whillans (1992)	1.22 (2-year mean)
Lewis Glacier,	Tropical mountain glacier, in retreat Kenya Kruss and Hastenrath (1983)	0.0067 (2-year mean 1981/2)

7.2 MECHANISMS OF MOVEMENT

7.2.1 TYPES OF MOTION AND DRIVING FORCES

Movement occurs by different mechanisms in different locations within a glacier. Movement can occur within the ice, or at the boundary between the ice and the substrate, or within the substrate itself. Internal movement is accomplished by creep and by thrusting. Movement at the bed is accomplished by sliding, by regelation and by enhanced basal creep. Movement within the substrate is accomplished by deformation of the subglacial material, and this material can itself slide over a rigid underlayer. These mechanisms might not all operate at a particular glacier, and operate in different proportions in different situations, but can be illustrated schematically as in Figure 7.1. On the whole, basal processes dominate in warm-based glaciers, and internal deformation in cold-based ones.

The surface velocity represents a combination of basal and internal flow, and is not the same as either the basal flow or the average flow for the whole ice column. Hamley *et al.* (1985) calculated the ratio of average column velocity to surface velocity and found that a typical value for an isothermal ice mass with a power-flow law exponent of 3.21 would be 0.81; and for cold ice sheets with a temperature increase towards the bed, between 0.85 and 0.92. Surface measurements will thus overestimate the overall ice flux. Furthermore, the rheology of ice is such that not all basal velocity perturbations will be transmitted to the surface. Modelling by Balise and Raymond (1985) suggested that basal velocity anomalies shorter in length than the thickness of the ice may not be transferred to the surface at all.

Table 7.2 illustrates the relative importance of different styles of movement in different glaciers. In many glaciers, dominant styles of movement change through the year and as glaciological conditions alter. This is illustrated by the data for Variegated Glacier in Table 7.2. In winter, nearly all the movement of Variegated Glacier is accomplished by internal deformation. In summer, when more water is present at the bed, more than 30% of the total movement is accomplished by basal processes. During surges, nearly all the movement of the glacier is achieved by basal processes. Even in

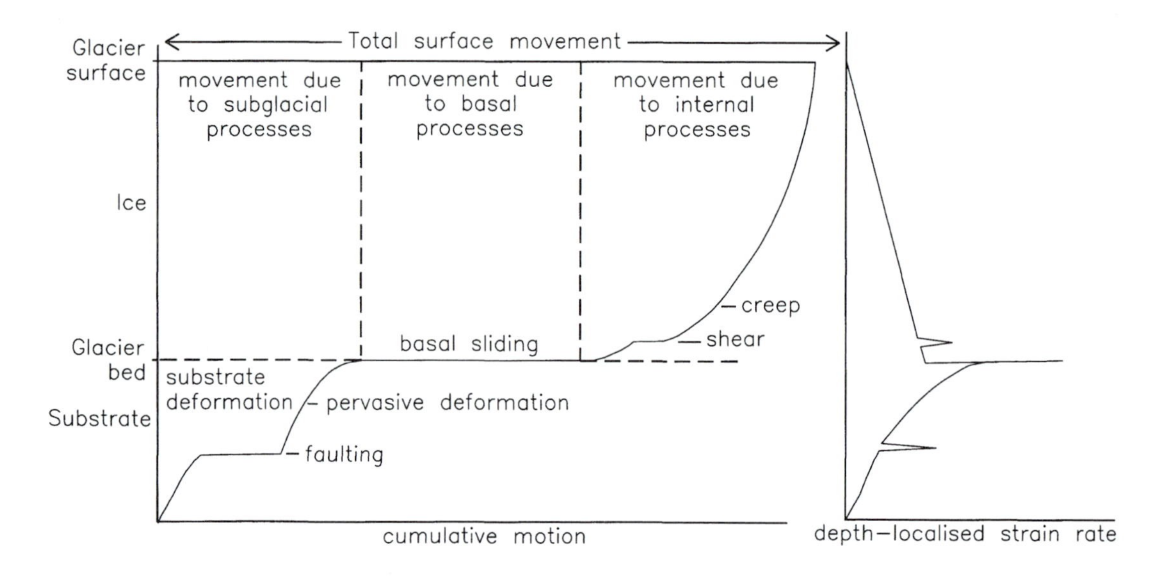

Figure 7.1 Schematic illustration of the principal mechanisms of glacier movement.

Table 7.2 Relative importance of different styles of movement in different glaciers (Data from Blake *et al.*, 1994; Boulton and Jones, 1979; Echelmeyer and Zhongxiang, 1987; Hooke *et al.*, 1992; Kamb *et al.*, 1985; Robinson, 1984)

Glacier	*Internal deformation (%)*	*Basal processes (%)*	*Subglacial deformation (%)*
Taylor (Antarctica)	40	60	0
Blue (Washington, USA)	10	90	0
Breiðamerkurjökull (Iceland)	0	12	88
Trapridge (non-surge) (Alaska, USA)	12	50	38
Urumqui No. 1 (China)	37	3	60
Storglaciären (Sweden)	23–43	57–77	0
Variegated (Alaska, USA) (winter)	100	0	0
Variegated (Alaska, USA) (summer)	≤70	≥30	0
Variegated (Alaska, USA) (surge)	5	95	0

non-surging glaciers, seasonal variations in basal water pressure beneath glaciers are likely to influence both the ice/bed contact and the stress field in the ice, resulting in seasonal variations in vertical deformation profiles, sliding rates, substrate deformation and the relative significance of different movement processes. Different styles of movement can be dominant in different parts of a glacier. Blake *et al.* (1994) measured movement at seven sites on Trapridge Glacier, Alaska, and found that sliding accounted for 50–70% of movement at six of them, but 90% of movement at the seventh.

The major energy input to drive glacier movement is gravity. Under the influence of gravity, glaciers move downhill and deform under their own weight. The energy or driving force imparted is determined by the weight of the glacier and its gradient, so the key parameters are ice density, ice thickness, acceleration due to gravity (constant) and gradient. For ice sheets, which overwhelm local topography, it is the gradient of the ice surface, not the gradient of the bed, that is usually most important. A useful analogy is the floor of a fish tank: a sloping or irregular floor to the tank does not generate water movement, but a sloping or uneven water surface does. It is because the driving forces are independent of bed topography that ice at the base of ice sheets can move ice uphill. For mountain glaciers, which are morphologically controlled by their bed topography, the bed gradient is important both because it can control the ice surface gradient and because the glacier can act in the same way as any free object on an open slope, liable to downwards motion. The total force acting on the ice can thus be divided into two parts: a lithostatic stress and a resistive stress (van der Veen and Whillans, 1989). The lithostatic stress is the gravitational driving stress, and the resistive stress involves both longitudinal stresses generated as one section of a glacier pushes or pulls adjacent sections, and drag from movement of the ice against the bed or valley walls. The following paragraphs consider the various ways in which these driving forces are manifested in glacier movement.

7.2.2 INTERNAL MOVEMENT

(a) Internal deformation (creep)

We must then regard the ice not as a mass entirely rigid and immobile but as a heap of coagulated matter, or as softened wax flexible and ductile to a certain point.

(Bordier, 1773)

It was shown in Chapter 4 that both individual ice crystals and polycrystalline ice masses deform by processes of creep. These processes can contribute substantially to the movement of glaciers. A variety of processes that allow deformation to occur operate at the scale of the crystal structure of the ice. These processes include: the movement of individual crystals relative to one another (intergranular adjustment); 'glide' or slip on the atomic basal plane of the crystals; movement along intra-crystal shear planes; and recrystallisation. Recrystallisation can involve the migration of boundaries of existing crystals, growth of some crystals at the expense of others, and the formation of new crystals. During these processes the crystal fabric is commonly reoriented so that the basal planes of individual crystals become aligned parallel to the direction of the driving stress, favourable to deformation by basal glide. The resistance of ice to deformation thus changes as the deformation proceeds. Glen's (1955, 1958) experiments on the relationship between stress and strain in ice, and the so-called 'flow law', were described in Chapter 4. Here we discuss deformation in more general terms in relation to glacier movement.

Creep deformation of polycrystalline glacier ice is driven by stress as described by Glen's flow relation, commonly expressed as:

$$\varepsilon = A\tau^n$$

where ε = strain, or the amount of deformation; τ = stress, or the driving force; A = a value to represent hardness, which depends largely on temperature; and n = an exponent, the value of which is commonly taken as 3.

The driving stress (τ) is a function of the weight of the ice and the gradient of the surface driving the ice movement. Weight is given by the density and thickness of the ice under gravity, and the driving gradient is the ice-surface gradient, so we can simplify and define the driving shear stress as:

$$\tau = \rho g h \sin \alpha$$

where τ = stress; ρ = ice density; g = acceleration due to gravity; h = ice thickness; and α = gradient.

This stress is the force available to drive movement in the glacier. For a stationary rigid slab of material resting on a surface this would define the shear stress across the bed (τ_b) (Orowan, 1949). If we ignore for a moment other styles of movement (basal sliding, deforming substrate, etc.), then we can consider this stress as the driving force for internal deformation of the ice. The distribution of the stress through the thickness of the ice (varying with h) can then be inferred, and from that the strain or deformation at different depths can be calculated from the flow law, to produce a vertical strain profile. Because (according to the flow law) strain is proportional to the cube of stress, and stress is proportional to depth, most of the strain is accomplished in the lower section of the profile. Early theories predicted that the velocity of ice flow would increase with depth, a phenomenon referred to as extrusion flow (Demorest, 1941). However, except in exceptional local circumstances (Hooke *et al.*, 1987; Gudmundsson, 1997a,b), motion in the lower parts of the ice column effectively carries ice higher in the column, so that the velocity at depth is transmitted to the surface and the velocity is greatest at the surface where ice motion is a cumulative function of the total strain through the depth of the ice column. The enhanced deformability of ice lower in the column is amplified further by the effect of temperature in the flow law; warmer ice deforms more easily, and ice temperature generally increases towards the bed of a glacier. We can therefore incorporate into this simple model a temperature gradient through the ice, and recalculate the theoretical strain profile taking into account the hardness parameter. However, there are a number of complicating factors which need to be taken into account to predict more realistically the strain profile of real glaciers. Some of these factors are considered in the following paragraphs.

One major complication is that ice hardness is affected not only by temperature but also by a range of other factors including particulate and dissolved impurities, liquid water content, crystal fabric, and hydrostatic or confining pressure (Hooke, 1981; Budd and Jacka, 1989). For example, Hooke *et al.* (1972) found that, except for very low debris concentrations, the creep rate of ice in laboratory experiments decreased exponentially with increasing concentration of fine sand in the ice, suggesting that debris strengthens or stiffens the ice. This can be attributed to the existence of dislocations in the ice crystal structure around each sand grain that disrupt the glide-plane dislocations. Lawson (1996) found that the influence of debris on ice strength was temperature dependent; she observed debris-rich ice to be stronger than clean ice close to the melting point but weaker than clean ice at temperatures below –5°C. The relationship between debris content and ice deformation is not yet fully understood, but is an important aspect of glacier rheology, especially in the context of the basal ice layer. The ice content of the deforming substrate/basal ice continuum probably exerts a strong control on movement. Nickling and Bennett (1984) demonstrated experimentally that the shear strength of a frozen coarse granular debris layer increased as the ice content was increased from 0 to 25%, then decreased progressively as the ice content was increased beyond 25%.

The distribution of the debris, as well as its concentration, may play a vital role in determining its rheological effect. Discrete layers of debris might be expected to have a different effect from dispersed particles or aggregates. Swinzow (1962) found that deformation seemed to be enhanced by scattered debris in the ice but impaired where many of the debris clasts were in contact, as in a debris band. Dissolved impurities in ice can also have a variety of effects, some leading to softening and some to strengthening of the ice (e.g. Jones and Glen, 1969). Liquid water content tends to soften the ice, as water in the vein network facilitates intergranular adjustment (Duval, 1977). Crystal fabric is also important: in ice with an initially random fabric, strain rate will increase as crystals become reoriented into position favourable to glide. Crystal size is also relevant. Small crystals are often associated with high strain rates, but this is partly because high strain inhibits crystal growth and encourages nucleation of small crystals. These factors tend to amplify the strain concentration at base as dirt and solute concentrations are higher near to the bed.

Furthermore, Pleistocene ice tends to be softer than Holocene ice, partly because of its higher impurity content and generally strong single maximum crystal fabric, and Pleistocene ice is confined to the bottom of ice-sheet profiles. Fisher and Koerner (1986) discussed the rheological differences that exist between Holocene and pre-Holocene ice in ice cores. They showed that flow enhancement in the pre-Holocene ice was related to particle content and crystal size (which were directly related to each other) but not to crystal c-axis orientation. They suggested that modellers of pre-Holocene ice sheets should use ice three times softer than modern ice in their models. Dahl-Jensen (1985) used measurements of bore-hole tilting at Dye 3 to identify variations in deformation with depth. The deformability of the ice at different depths in the 2037 m borehole could be considered as a function of temperature, ice crystal size, ice fabric and impurity content in the ice. Removing the effects of temperature, a flow enhancement factor was defined to compare deformability of ice from different depths. This flow enhancement factor was three times stronger in the Pleistocene ice making up the lower section of the core than in the Holocene ice above (see Figure 7.8). The greater deformability of the Pleistocene ice was attributed to the higher concentration of dust and other impurities, and to the smaller crystal size. Dahl-Jensen reported that the greatest flow enhancement occurred in the silty ice making up the lowest 25 m of the core. Hooke (1973b) found that bubbly Pleistocene ice near the base of the Barnes ice cap deformed about 2.5 times faster than bubble-poor Holocene ice above.

The existence of a relatively soft layer of ice in the lower portion of many ice sheets is critical to rheological models of ice sheets, and the change in thickness of this layer through time is an important variable in ice sheet rheology. The Pleistocene layer is expected to have gradually thinned since the end of the Pleistocene, and the ice sheet as a whole thus to have become stiffer. However, this change has yet to affect the lowest sections of the thick ice sheets, and the bulk of movement is accommodated in these lowest sections. Also, it will take a certain amount of time for ice accumulating at the surface to reach the base of the ice sheet, so a period of enhanced ice-sheet flow might be expected only after an interval of some time after the period at which the ice was originally formed. The zone of flow enhancement in the ice profile will migrate downwards through the thickness of the ice like a wave, over a period of as much as several hundred thousand years.

A second major complication in predicting internal deformation is the fact that basal shear stress is not the only stress driving movement in glaciers, although it is frequently treated as such. It is also necessary to consider longitudinal stress. Longitudinal stress can be envisaged as the force of one section of a glacier pushing or pulling another section. For example, in a glacier surge one section of a glacier might accelerate markedly due to changed basal conditions, and thus impart a 'push' to ice down-glacier and a 'pull' to ice up-glacier. Several authors have described both tensional and compressive structures imparted to different sections of surging glaciers, and Hughes (1992a) referred to the 'pulling power' of ice streams. Ice falls, where the glacier flows down a steep slope between two less steep sections, can generate strong longitudinal stress gradients, with visible surface manifestations such as crevasses and compression structures. Longitudinal stress can in some situations equal or exceed the shear stress (McMeeking and Johnson, 1985). The observed flow in a glacier often relates better to longitudinally averaged slope and thickness values than to highly local slopes and thickness, indicating a longitudinal coupling of flow (Kamb and Echelmeyer, 1986). It is also important to bear in mind that changes in surface gradient will not be transferred directly to changes in basal shear stress. Nye (1969b) demonstrated that τ_b did not fluctuate with changes in α over distances of about four times the ice thickness, but did so over distances about 20 times the ice thickness. This can be predicted to generate internal longitudinal stress.

A range of other complicating factors can also be taken into account. For example, strain heating in locations such as the edges of ice streams can lead to softening of the ice (Hughes, 1975). Echelmeyer *et al.* (1992) observed that ice at the margin of the Jakobshavn ice stream was a few degrees warmer than ice at the centre-line, and suggested that strain heating and softening could result in a positive feedback loop of softening, straining and heating that could foster the development of ice streams. Obstacles to flow such as protuberances from the bed can also affect deformation, and these are considered further in the following section on basal movement.

(b) Shear

In some cases, creep deformation seems to be concentrated in narrow bands or shear zones. These can originate from stress concentrations in the ice, or from minor heterogeneities in ice fabric or impurity content that become amplified by positive feedback as straining occurs. These bands do not really constitute shearing; rather, enhanced flow in thin layers of ice.

The existence of discrete shear planes where one body of ice is offset against another has been controversial in the past. Some investigators have suggested that shearing is a major mechanism for entrainment of debris into the basal layer of glaciers, and others that shearing is mechanically unlikely. Certainly coincident shear thrusts and debris bands have been observed in the terminal zones of glaciers, as discussed in Chapter 5 in the context of basal ice, but the contribution of shearing to glacier movement under thick ice remains unclear.

7.2.3 BASAL MOTION

(a) Hard-bed/soft-bed

Modern analyses of basal motion are generally divided into theories for hard-bed conditions and theories for soft-bed conditions. Traditional theory concentrated on hard-bed conditions, where glaciers were assumed to rest on bedrock. More recently, the recognition that many present-day glaciers, and large parts of former ice sheets, have been underlain by deformable sediment has fostered increasing interest in the dynamics of glacier motion on deforming beds.

(b) Basal sliding

In some cases perhaps the subglacial streams may, even in a slight degree, be considered as lifting the ice so as to facilitate its gliding towards the sea.

(Rink, 1877)

The term 'sliding' is commonly used to refer to a range of processes by which ice moves across its bed. In fact, several distinct processes can be identified. As we saw in the previous section, driving stress derived from the weight and gradient of the ice can be regarded as a shear stress across the ice/bed interface, $\tau_b = \rho g h \sin \alpha$. If the glacier were a rigid block of ice resting on a rigid smooth surface, this basal shear stress would be translated directly into a sliding motion across the bed. In what is sometimes referred to as 'classical' sliding theory, sliding velocity is a direct function of basal shear stress and bed roughness (e.g. Weertman, 1957; Nye, 1969a,b; Kamb, 1970). In fact, several additional factors need to be taken into account in calculating the relationship between the driving stress and the sliding velocity, and the history of attempts to model glacier sliding is, to a large extent, a history of attempts to include increasingly realistic parameters into the models. Major issues have been the mechanisms by which ice overcomes irregular rock surfaces, and the relationship between basal water pressure and friction at the bed. Unless there is complete hydraulic uncoupling, then even where a layer of water lubricates the interface, protruberances or 'obstacles' on the bed get in the way of the sliding. Only in extreme cases, therefore, can glacier movement be explained in terms of 'simple' sliding. Other basal movement mechanisms must be invoked to overcome obstacles. These are discussed in the following section.

(c) Regelation slip and enhanced basal creep

Weertman (1957) modelled sliding for a simple case where the bed consists of regularly spaced cubic obstacles on an otherwise plane surface (the tombstone model). He identified regelation (pressure melting) and enhanced basal creep (creep rate enhancement through stress concentration) as the two primary mechanisms by which glaciers can overcome obstacles to sliding on the glacier bed.

Regelation operates where ice close to the melting point encounters an obstruction to flow at the bed. There exists an area of relatively high stress at the up-flow side of the obstacle and an area of relatively low stress on the downstream side. This gives rise to a difference in pressure melting point between one side of the obstacle and the other. The decrease in the melting temperature on the upstream side leads to melting of the ice. The water thus produced will follow the stress gradient around the obstacle to the downstream side, where the pressure is lower and the melting point higher, and the water will refreeze on the downstream side having negotiated the obstacle. For this mechanism to operate more than momentarily, there needs to be a supply of energy to the upstream side of the bump to facilitate the melting. If that energy is drawn from the surrounding ice or substrate, the local temperature will quickly fall to the point where the pressure enhancement will be insufficient to lower the pressure melting point to the new temperature, and melting will cease. Weertman proposed that where the obstacle is small, the heat released by refreezing the meltwater on the downstream side of the bump is transferred through the bump to supply heat for melting on the upstream side. If the ice is at the pressure melting point throughout, then the ice in the low-pressure zone down-flow of the bump will be warmer than the ice up-flow, and heat will flow up-glacier through the ice and the rock. Weertman predicted that the maximum obstacle size through which a heat transfer could effectively be maintained was about 1 m, and for obstacles larger than that the regelation mechanism would not work. The sliding velocity due to regelation is a function of the amount of melting achieved, which in turn is a function of heat production and transfer such that:

$$U_{sr} = (kK_r/L\rho_i l)\tau_b(\lambda_b l)^2$$

where U_{sr} = regelation sliding velocity; k = constant; K_r = thermal conductivity of bedrock; L = latent heat of fusion of ice; l = obstacle length; ρ_i = ice density; τ_b = shear stress against obstacle; λ_b = obstacle spacing.

This relation shows that regelation sliding velocity is inversely proportional to the size of obstacles. Robin (1976) pointed out that if the water produced by melting escaped into the ice through the vein network or into the substrate, rather than flowing around the obstacle, the heat supply for melting would have to be drawn from the ice locally rather than from the downstream side of the obstacle, and a cold patch would form. The cold patch would inhibit regelation sliding until the pressure increased to the point where the pressure melting point fell to the ambient temperature. Robin predicted that this could lead to a jerky, intermittent, 'stick–slip' motion of the ice. Goodman *et al.* (1979) confirmed this theory both in the field and in the laboratory, and a variety of other field measurements have revealed irregular 'jerky' sliding movement (e.g. Theakstone, 1967; Vivian and Bocquet, 1973). However, Lliboutry (1986, 1993) also considered regelation where water flows away from the interface into the vein network, and pointed out that because melting occurs within the ice as well as at the interface, and because water flow is continuous through the vein network close to the bed, water should therefore always be able to reach the lee side of a bump. Weertman-regelation involving flow through a water film, and Lliboutry-regelation involving flow through the vein network, might operate as distinct processes and are discussed in Chapter 5 in the context of the basal ice layer.

Enhanced basal creep occurs where obstacles at the bed generate high stresses in the ice in the up-flow side of the obstacle, which in turn generate higher strain rates and hence faster flow. Obstacle length is an important control of flow enhancement both because larger obstructions generate larger stress concentrations, and because velocity is a function of strain rate over distance. The strain enhancement is a function of the flow law parameters (stress, hardness and the flow law exponent) and the size of the bedrock obstacles, such that the sliding rate past obstacles can be given by:

$$U_{sc} = lA(\tau_b\lambda_b^2/2l^2)^n$$

where U_{sc} = basal creep velocity; l = obstacle length; A = ice hardness; τ_b = basal shear stress; λ_b = obstacle spacing; and n = flow-law exponent.

This shows that creep enhancement is greater for larger obstacles, and negligible for small obstacles. Weertman's model thus indicates that increasing obstacle size inhibits regelation sliding but enhances creep. Large obstacles are overcome by enhanced creep, and small obstacles by regelation. Weertman identified a 'controlling obstacle size' at which $U_{sc} = U_{sr}$, where the efficiency of both regelation and creep is low and the combined sliding rate is at a minimum. Obstacles of this size are too large for regelation to operate effectively and too small for enhanced basal creep to be efficient. The speed of flow over obstacles of this critical size will control the rate at which the glacier as a whole can slide over the bed. Controlling obstacle size varies with the ratio of obstacle size to obstacle spacing but is of the order of somewhat less than 1 m. For obstacles of this critical size, assuming $n = 3$, the total (combined) sliding velocity can be given by:

$$U_s = D\tau_b^2 R^{-4}$$

where D is a constant involving thermal properties of ice and rock and the hardness parameter from the flow law, and R is bed roughness given by l/λ_b. Thus, the combined sliding rate for regelation and enhanced basal creep is proportional to the square of the basal shear stress, and inversely proportional to the fourth power of the roughness.

Later models by Nye (1969a, 1970) and Kamb (1970) incorporated more complex bed-roughness parameters than Weertman's 'tombstone' model, but reached broadly equivalent conclusions. In these models the bed was represented by either 'white roughness', where there is a continuous range of obstacle sizes, or 'truncated white roughness' where the range excludes the smallest scale of roughness. Truncated white roughness might equate to a bed smoothed by abrasion. In either case, the models showed that ice overcomes smaller obstacles by regelation and larger obstacles by enhanced creep; the boundary between the two is defined by a 'transition wavelength' (the equivalent of Weertman's controlling obstacle size) of the order of 0.5 m.

In examining the relationship between stress, bed-roughness and basal sliding, the Weertman and Nye–Kamb models neglected many of the realities of the subglacial environment. The models reproduced only partially the real behaviour of glaciers, failing to explain such phenomena as short-term velocity variations. Later models incorporated additional elements to approach reality more closely. Major issues in more recent models have been the incorporation of the effects of subglacial water pressure and the existence of cavities at the glacier bed that can grow and shrink as the glacier moves. These are discussed next.

(d) Sliding with cavitation, and the effect of water at the bed

Most glacier beds impart significant resistance to sliding. There is friction between the ice and the substrate; the ice may contain debris which causes friction in travelling across the bed; and protruberances or 'obstacles' on the bed get in the way of the sliding. This friction is partly a function of the weight of the glacier or the overburden pressure at the bed. However, the overburden pressure, and hence the friction, can be offset to some extent by basal water pressure. Forbes correlated glacier velocity with precipitation events and high temperatures (melting) as early as 1842, but early sliding theories ignored the fact that variations in basal water pressure, caused either by changes in the efficiency of drainage through the glacier or by variations in the input of water to the system, can significantly affect glacier sliding. High basal water pressure can effectively lift the glacier from its bed, swamping small bed roughnesses, increasing shear stress on bed irregularities that are not submerged beneath the basal water, and hence increasing sliding. Hanson and Hooke (1994) recognised that diurnal variations in flow in the north cirque of Storglaciären correlated with the diurnal melt cycle and were disturbed by rainfall events: local ice movement was controlled by water input in the cirque, not by longitudinal coupling with lower sections of glacier. Bindschadler (1983) suggested that, for time scales of days to weeks, variations in basal water pressure (and hence in bed-separation) can dominate the sliding process. Friction between the ice and the bed can be simplified to an equation of the form:

$$F = s(\rho g h - p_w)$$

where p_w = water pressure, and s = a factor for the shape and roughness of the bed.

This indicates that basal friction increases with the weight of the ice and the roughness of the ice/bed interface, but decreases with increasing water pressure. An 'effective basal pressure' ($N = \rho gh - p_w$) can then be considered, which is the difference between the overburden pressure and the basal water pressure. Friction at the bed is then proportional to a function of bed roughness and effective pressure. Lliboutry (1958, 1968, 1975, 1978, 1979, 1987) developed a series of models of basal sliding that incorporate not only sophisticated approximations of bed topography but also variations in effective pressure. According to Lliboutry, different things happen at different rates of sliding. At slow sliding speeds, sliding is related directly to shear stress and inversely to bed roughness, as it was in the Weertman and Nye–Kamb models. At higher sliding speeds, however, Lliboutry predicts that cavities are likely to open between the ice and the bed and a that a quite different sliding relation comes into effect. Cavitation occurs when the normal pressure or contact force at the bed falls below some minimum value. Cavitation can occur in the lee of bedrock bumps when tensile stress between ice and substrate exceeds cryostatic 'closing' force, or where basal water pressure exceeds the overburden pressure and water is forced between the ice and the rock to create water-filled cavities. Cavitation reduces contact area between ice and rock, reducing friction locally and increasing basal shear stress over areas of the bed that remain in contact. The overall effect is thus to enhance basal sliding. For sliding with cavitation, the sliding rate will be controlled by the cavities. Lliboutry predicts that sliding will be directly related to shear stress, and inversely related to both bed roughness and effective pressure. A similar conclusion has been reached by several other models (e.g. Fowler, 1987; Budd, *et al.*, 1979). Iken and Truffer (1997) suggested that the effect of water pressure on basal sliding at Findelengletscher was controlled by the interconnectedness of cavities. They found that isolated cavities served to reduce the variations in sliding that were associated with pressure changes, because the pressure in the cavities decreased as sliding speed increased. In contrast, interconnected cavities amplified the effect of pressure variations by transmitting the fluctuations to larger areas of the bed. They suggested that long-term changes in glacier velocity could be attributed to changes in the interconnectedness of the subglacial cavity system.

As lee-side cavities are commonly inclined down-glacier, increasing water pressure causing the growth of a cavity can impart a down-glacier motion. This hydraulic cavitation, or hydraulic jacking, effect was considered by Lliboutry (1964, 1968) and Iken (1981). The increase in forward motion imparted by the effect is greatest when the cavity first forms, and diminishes as the cavity reaches a steady state. Even a small drop in water pressure after cavities have formed, by as little as 1% of the overburden pressure, could lead to a temporary backwards motion in the ice as the cavity contracts. This cavitation process was invoked by Iken and Bindschadler (1986) to explain the relationship between basal water pressure and short-term flow variability at Findelengletscher, and by Kamb and Engelhardt (1987) to explain the close association between basal water pressure and flow velocity in mini-surges of Variegated Glacier. These two sets of observations yielded comparable relations between basal water pressure and glacier flow speed, and Jansson (1995) found an equivalent relation at Storglaciären which he expressed as an inverse power relation between sliding velocity and effective pressure at the glacier bed:

$$U_s^E = 30P_E^{-0.4}$$

where U_s^E = sliding velocity predicted from P_E, and P_E = effective pressure (ice overburden pressure minus subglacial water pressure).

Iken (1981) and Schweizer and Iken (1992) took into account that the ice pressure on the bed varies from the upstream to downstream faces of bedrock hummocks, being at a maxi-

maximum on the stoss face and a minimum on the lee face (Kamb, 1970). This minimum pressure, referred to as the separation pressure (P_s), is given for a sinusoidal bed by:

$$P_s = P_o - (\lambda\pi/\tau a)$$

where P_o = overburden pressure; λ = bump wavelength; τ = shear stress; and a = bump amplitude. Schweizer and Iken (1992) argued that the important parameter for cavity formation and sliding was thus not the conventionally adopted effective pressure (the difference between overburden and water pressure) but rather a measure of the difference between water pressure and the minimum basal pressure (P_s) at the lee faces of bumps. Cavitation can begin to occur when basal water pressure reaches this separation pressure. The water pressure at which basal sliding becomes unstable and the base is largely separated from the bed is termed the critical pressure, P_c, and lies halfway between the overburden pressure and the separation pressure:

$$P_c = 0.5(P_o + P_s).$$

Thus even at water pressures significantly below the overburden pressure a cavity system supported by the pressure of water in the cavities can exist, and sliding with cavitation can occur. Stable sliding can occur as cavities grow and unstable sliding is possible when bed separation is extensive. These conditions are independent of the effective pressure ($N = P_o - P_w$), and Schweizer and Iken (1992) argued that the effective pressure should not appear in a realistic sliding law. They suggested instead a sliding law of the form:

$$u_b \approx (\tau_b/P_c - P_s)^n \text{ for } P_w \leq P_s$$

$$u_b \approx (\tau_b/P_c - P_w)^n \text{ for } P_w > P_s.$$

In this formulation the sliding velocity is proportional to the shear stress, as in previous formulations. Here, however, sliding velocity is inversely proportional not to effective pressure, as previously formulated, but to the pressure difference between the actual pressure on the lee of bumps and the critical pressure for unstable sliding. The actual pressure on the lee of bumps is either the separation pressure (when water pressure is equal to or lower than the separation pressure) or the water pressure (when that is greater than the separation pressure).

(e) Friction and sliding of debris-laden ice

None of the models discussed so far have taken detailed account of friction between the glacier and its bed, except in the assumption that it can be counteracted by cavitation or uplifting water pressures at the bed. The impact of debris-rich basal ice on friction has been largely ignored. Iken and Bindschadler (1986) compared measured fluctuations in basal water pressure and surface velocity at Findelengletscher with a theoretical model of sliding over a sinusoidal bed and found that the model predicted too much sliding for any given water pressure. They attributed the discrepancy to the model's failure to address friction introduced by debris. Schweizer and Iken (1992) attempted to develop a sliding law incorporating friction. They considered two types of friction likely to be important at glacier beds: 'sandpaper friction' and 'Hallet friction'.

Sandpaper friction occurs when the basal ice layer in contact with the bed is debris-rich, with debris particles in contact with each other and separated only by interstitial ice. The contact between each particle and the bed is related to the overburden pressure, and the friction between individual particles and the bed can be described by coulomb friction between rigid bodies, but the friction at the bed as a whole is more complicated because the basal layer is not rigid. In this situation, as basal water pressure increases and areas of the base lose contact with the bed, the frictional drag will be controlled by the friction on the areas remaining in contact. The mean pressure on the contact areas will be:

$$P_n = (P_o - sP_w)/(1 - s)$$

where s is the fraction of the bed where separation has occurred. When there is no bed separation, the mean pressure will be the overburden pressure. The frictional stress on the contact area will be:

$$\tau_c = \mu P_n$$

and the frictional drag over the whole area of the glacier bed will be:

$$\tau_f = \tau_c(1 - s) = \mu(P_o - sP_w)$$

The implication of this is that although friction is reduced as water pressure increases, friction is not proportional to the overall effective pressure but to the amount of cavitation and the area of bed contact. As long as separation does not affect the whole bed, frictional drag according to this model will be greater than predicted by a simple model based on effective pressure:

$$\tau_f > \mu N$$

or:

$$\mu(P_o - sP_w) > \mu(P_o - P_w)$$

For a given glacier geometry and hence driving stress, friction along the sliding interface caused by debris in the ice (increasing μ) increases the value of both the separation pressure (P_s, the normal pressure on the lee side of bumps) and the critical pressure (P_c, the pressure at which the sole becomes largely separated from the bed). However, friction has a much greater effect on the separation pressure than on the critical pressure, because much less of the sole is in contact with the bed as the critical pressure is approached. Therefore, when the friction coefficient is high, the separation pressure can approach the critical pressure. In this situation, sliding is steady (albeit slow or zero) for water pressures below the separation pressure, but abruptly becomes unstable as the pressure reaches simultaneously the separation pressure and the critical pressure. High-friction beds can then be characterised by a stick–slip basal sliding regime with low/zero sliding below a water-pressure threshold and unstable sliding above the threshold. Schweizer and Iken (1992) proposed a functional relationship between subglacial water pressure and basal sliding velocity incorporating sandpaper friction as follows:

$$u'_b \sim (\tau_b'/P'_c - P_w)^n \text{ for } P_w > P'_s$$

where τ_b = effective shear stress ($\tau - \tau_f$), and ′ denotes values depending on friction. This illustrates rapid increase in sliding as water pressure approaches the (friction-dependent) critical pressure.

Hallet friction occurs when the basal layer includes isolated debris particles that are not in contact with each other and are surrounded by otherwise clean ice. According to Hallet (1979, 1981), the contact force between the bed and debris particles in the ice in this situation is controlled by the ice velocity towards the bed, and is independent of the overburden pressure, as the particles are supported by a cryostatic pressure. Therefore, friction does not occur on the downstream faces of bed hummocks, as the ice flow there is effectively away from the bed. Since changes in water pressure are mainly important in these lee-face environments where cavitation occurs, water pressure fluctuations are not important controls on basal friction for basal ice with low debris concentrations until the critical pressure is reached. The relationship between debris concentration, water pressure and sliding for Hallet friction can be described as in Figure 7.2.

Shoemaker (1986) expressed Hallet's linear sliding as:

$$\tau_b = \xi \eta u_b + \mu c F$$

where ξ = a bed roughness factor; η = effective viscosity; u_b = ice sliding velocity; μ = coefficient of rock-rock friction; c = projected area of basal rock fragments/unit basal area; and F = average rock-fragment force/projected area of average fragment. According to this relation, for conditions of low bed roughness, the bulk of basal

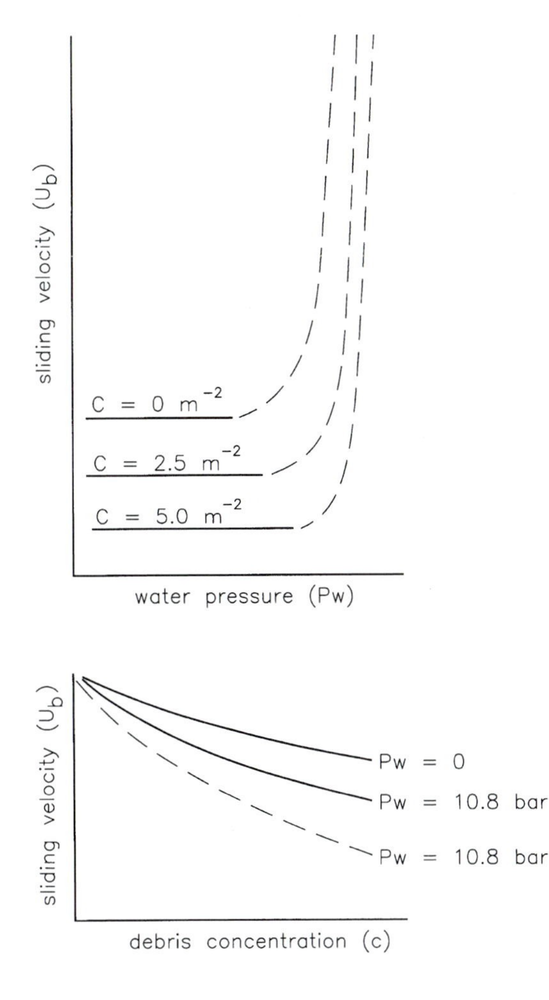

Figure 7.2 Increase in sliding as water pressure approaches the (friction dependent) critical pressure. After Schweizer and Iken (1992).

friction could arise from contact between the bed and debris in the ice. Shoemaker (1986) discussed modification of sliding laws to accommodate the effect of debris on basal drag. He proposed that any such modification would involve a debris-concentration multiplication factor, but allowed that such a factor would, at present, remain generally unknown. The role of debris-bearing basal ice continues to be a major source of uncertainty in modelling glacier dynamics.

(f) Sliding at subfreezing temperatures

No glacier whatever is known to move down any mountain slope without being accompanied by a stream of water . . . gliding over the same part of the river-bed occupied by both of them. In fact, no motion of ice on land can be considered possible without such accompanying water.

(Rink, 1877)

Where there is no basal water and the ice is frozen to the bed, it was traditionally held that sliding could not occur, but in fact bot!. sliding (Shreve, 1984) and other basal processes (Echelmeyer and Wang Zhongxiang, 1987) can operate at subfreezing temperatures. Sliding is made possible because a microscopic liquid layer (1–10 nm in the range –20°C to –0.1°C) can exist at an ice/rock interface even below the freezing point. The thickness of the layer decreases with decreasing temperature, but is increased by the presence of solutes. Subfreezing sliding speeds are thus partially dependent on solute concentration, as shown in Figure 7.3. Fowler (1986) explained subfreeezing sliding by the incorporation of solid friction into the sliding law, and suggested that the sliding law relating basal velocity to basal stress depends also on temperature, such that sliding velocity is dependent on temperature below the melting point. The rates of sliding at subfreezing temperatures are generally very slow, making only a negligible contribution to the overall glacier movement, and this may account for failure of many attempts to detect movement in the field. However, since glaciers can last a long time, speed is not necessarily of the essence. Shreve (1984) showed theoretically that in 10,000 years a glacier might slide 350 m at –5°C and 35 m even at –20°C. Echelmeyer and Wang Zhongxiang (1987) observed sliding rates of ice at –5°C at Urumqui Glacier no. 1, in China, that were two orders of magnitude higher than predicted by Shreve's model. They accounted for this by

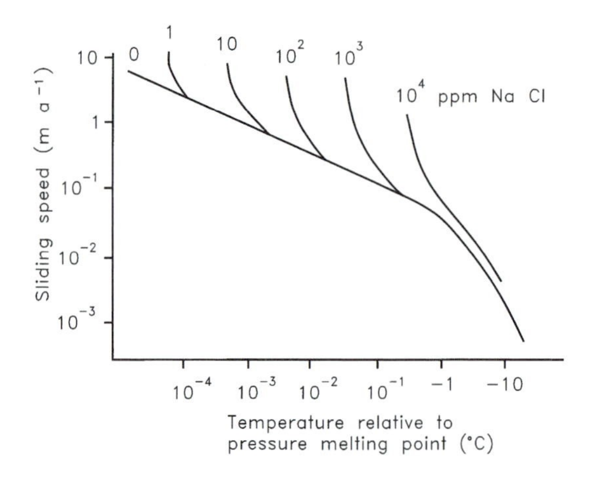

Figure 7.3 Sliding speed at sub-freezing temperatures depends partly on the presence of dissolved impurities in the ice. After Shreve (1984).

lower bed roughness and ice viscosity at their site than in Shreve's calculation. The possibility of subfreezing sliding is very important both for modelling ice-sheet dynamics and for understanding the geomorphological effectiveness of glaciers.

7.2.4 DEFORMING BED

One of the exciting things about glaciology in the last few years has been the way that modellers have taken on board increasingly sophisticated approximations of reality. There has been a growing dialogue between physicists and field observers that has helped to focus field observation and to inform model baselines. One example of this has been the recognition that subglacial till deformation can contribute significantly to glacier dynamics. Boulton and Jones (1979) reported movement of markers in a borehole drilled into subglacial till from an englacial tunnel at Breiðamerkurjökull, an outlet of the Vatnajökull ice cap in Iceland. Over a period of 244 h, 88% of the movement of the glacier was apparently accommodated by movement within the till. This realisation was a milestone in the development of our understanding of glaciers; in the last few years subglacial deformation has been invoked to explain a variety of glacial dynamic phenomena including ice streams and surges. Boulton (1986) referred to deforming bed theory as a new paradigm in glaciology. It has long been clear that large areas of former Pleistocene ice sheets, and significant portions of existing glaciers, were underlain by unconsolidated sediment. Nevertheless, until the 1980s, rigid-bed models of glacier flow were the norm, and the glacier bed was by default referred to as the ice/rock interface. However, when direct observations of subglacial till deformation had been made, and as new geophysical techniques allowed us to find out more about the beds of contemporary glaciers including the Antarctic ice streams (Blankenship *et al.*, 1986), accommodation of deforming bed theory into the glaciological mainstream was relatively, if belatedly, rapid.

Where a glacier rests on a rigid bed the driving stress is accommodated by movement across the bed and within the ice. Where the bed is not rigid but deformable, then (if the shear stress at the bed exceeds the yield strength of the bed material) some or all of the basal shear stress will be accommodated by deformation of the bed. There is effectively a coupling between the ice and the substrate. Most of the theoretical analysis of subglacial deformation thus far carried out has concerned deformation of glacigenic sediment, and deforming subglacial material is generally referred to as till, regardless of its actual provenance. Deformation accommodates movement by several processes: ploughing through the till of clasts attached to the glacier sole; shearing along discrete planes within the till; pervasive deformation of the till; and sliding of till over a rigid substrate (Figure 7.4). In addition, the ice can slide over the upper surface of the till by processes included in classic sliding theory, but this sliding is likely to be limited as till surfaces usually have higher roughness in the

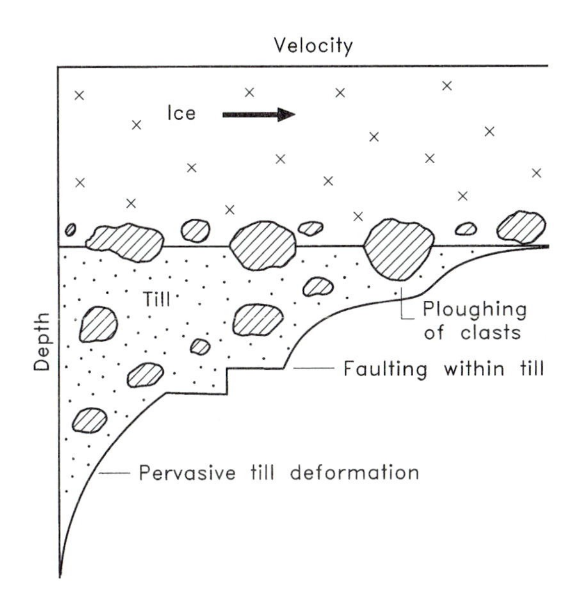

Figure 7.4 Aspects of subglacial deformation involved in glacier motion.

critical range than a glacially eroded bedrock. According to modelling by Alley (1989a,b), in soft-bedded glaciers not dominated by a supply of channelised meltwater, pervasive deformation will account for about 60–100% of the basal velocity.

The yield strength of till depends on pore-water pressure, which in turn depends on water supply and drainage. Significant till deformation occurs only when the till is saturated and pore-water pressure is high. Thus water supply needs to be high, or drainage of water needs to be inhibited either by poor hydraulic conductivity through the till or by the presence of an aquitard layer beneath the deforming layer. Till deformation is thus favoured by high water supply to the bed, an inefficient distributed drainage system, and low-permeability till. When these conditions are met, the till will have low strength and will deform under basal shear stresses much lower than the typical stress of the order of 100 kPa that is required for movement on rigid beds. Yield stresses as low as 2 kPa have been measured in till beneath Ice Stream B (Kamb, 1991).

Flow relations for deforming beds thus take the form (Paterson, 1994):

$$\varepsilon = B(\tau - \tau_0)^a N^{-b} \text{ when } \tau \geq \tau_0$$

$$\varepsilon = 0 \text{ when } \tau < \tau_0$$

where ε = shear strain rate; $(\tau - \tau_0)$ is the excess of shear stress over the yield stress; N = effective pressure (hydrostatic pressure minus pore-water pressure, wp); and B, a and b are empirically derived for local till conditions.

This shows that deformation occurs only when the shear stress equals or exceeds the yield stress, and that deformation is then proportional to the shear stress and to the pore-water pressure but inversely related to the effective pressure. Alley (1989a,b) suggested that the strain rate would be inversely proportional to the square or cube of the effective pressure. The steady-state velocity of the glacier sole due to bed deformation is then given by:

$$u_b = \int_0^T T\, \varepsilon_{xz}[N(z), \tau_b]\,dz$$

where $z = 0$ is the base of the till layer, and $z = T$ is the top of the till layer. This shows that the deformation velocity depends on the strain rate through the thickness of the deforming layer, which depends on the shear stress and the depth-dependent effective pressure. The depth of the deforming layer thus contributes to the overall velocity from deformation. A simple expression of the potential thickness of the deforming layer can be given from the excess of the shear stress over the strength of the sediment. Strength of sediment is determined here by effective pressure and material properties:

$$z_0 - z_t = \frac{\tau_b - (c \tan \Phi N)}{G \tan \Phi}$$

where $z_0 - z_t$ = thickness of the deforming layer; τ_b = basal shear stress; c = cohesion; Φ = angle of internal friction; N = effective pressure ($\rho g h$ – wp); and G = a constant.

This implies that the thickness of the deforming layer is directly related to shear stress and water pressure but inversely related to the viscosity of the sediment. Changes in water pressure, ice thickness, or ice surface gradient could therefore change the thickness of the deforming layer. The geographical distribution of bed deformation beneath an ice sheet is thus likely to be controlled to some extent by longitudinal variation in basal shear stress (Hart *et al.*, 1990) (Figure 7.5). Another limiting control is the supply of debris. For deformation to continue in a steady state, till evacuated by deformation must be replaced from up-glacier. Steady-state deformation would therefore require sediment production by erosion equivalent to the debris flux. Implications of sediment flux in deforming beds is considered further in Chapter 9.

Deforming bed theory is still immature. A major problem is that till constitution and rheology are hugely variable, and a general flow law for till is likely to be even more elusive than one for ice. Most models so far have made the simplifying assumption that till behaves either as a Newtonian viscous material with a linear stress-strain relationship or as a perfectly plastic material with a critical yield stress. In reality, till may well behave in a non-linear fashion. Hooke *et al.* (1997) suggested that till behaved as a frictional Mohr–Coulomb material, such that till strength is correlated with effective pressure, but that the non-linear nature of the mechanical relationships was such that it could not be described by a constant coefficient of friction. One source of non-linearity is that till can dilate and become weaker in response to high strain. A flow law for coupled ice/deforming-bed systems will need to involve more than glaciology alone (MacAyeal *et al.*, 1995). Blake *et al.* (1992) declared that 'bed deformation is a more complex process than previous work has suggested . . . Basal strain rate is observed to change sign, and to fluctuate wildly.' There is also contradictory evidence and counter example to the developing paradigm. For example, Iverson *et al.* (1995) and Hooke *et al.* (1997) showed that, at Storglaciären, the fastest ice flow occurred during periods when the bed was not deforming, and that periods of subglacial deformation corresponded with low overall motion. This was explained by decoupling of glacier and bed at high water pressure leading to fast sliding of the ice over the till, and low water pressure leading to ice/bed coupling and movement by relatively slow deformation. Clearly, local conditions can significantly affect sliding relations for deforming bed conditions, and the detail of how the processes are controlled is as yet poorly understood. Subglacial deformation and rigid-bed sliding could be considered as end-member cases of a continuum of ice flow conditions (Hart, 1995b). Saturated subglacial till can deform rapidly, but frozen till with only interstitial ice does not (Echelmeyer and Wang Zhongxiang, 1987). The liquid/solid state of interstitial water is a critical control in switching from deforming to non-deforming conditions. Fluctuations in the vertical position of the freezing isotherm in a subglacial layer constitute vertical movement of a boundary between debris-rich basal ice and a deforming substrate.

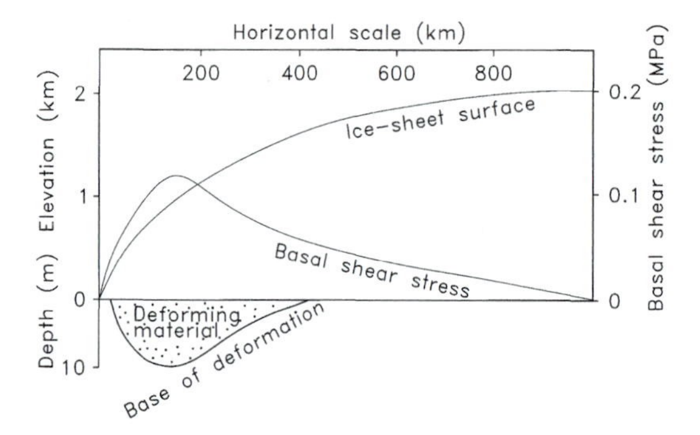

Figure 7.5 Longitudinal variation in basal shear stress is a major control on distribution of subglacial bed deformation.

Deformation of sediment beneath a glacier has several implications for glacier dynamics, sedimentology and morphology. Boulton and Jones (1979) suggested that subglacial deformation could account for many characteristics of former Pleistocene ice sheets, including low surface profiles, rapid disintegration, low isostatic rebound on deglaciation, and glaciotectonic structures in glacial sediments. The basal shear stress will be lower than for a rigid-bed glacier as deformation will begin to accommodate stress at a lower yield stress. Hence the ice-surface gradient is likely to be lower, and glacier motion will be faster for given driving stress. For ice sheets a very low surface profile would imply thin ice extending hundreds of kilometres in from the ice margin, with implications for basal thermal regime, hydrology and geomorphology. Beget (1986) described low yield-strength tills from Illinois, and Alley (1991) considered the characteristics of southern Laurentide till sheets to be compatible with a deforming bed origin. Several others (Matthews, 1974; Boulton and Jones, 1979; Clayton *et al.*, 1985) have suggested that the low profiles of some of the lobes of the Laurentide ice sheet could have been caused by overriding of deformable sediment. Clark *et al.* (1996) modelled the former Laurentide ice sheet, taking account of likely locations of rigid and deforming bed conditions. They found that the ice-sheet morphology emerging from their reconstruction fitted closely with ice-sheet characteristics predicted by the ICE-G4 ice-sheet reconstruction (Peltier, 1994). They concluded that deforming bed dynamics were the most likely control on the Laurentide ice sheet, and that deforming beds provided a mechanism that could explain the size and shape of the ice sheet inferred from sea-level data. Thorpe (1991) attributed reconstructed low gradient profiles with low driving stress (as low as 10k Pa) for outlet glaciers of the Scottish late-glacial Western Grampians Ice Field to deformation of glaciomarine clays with high pore-water pressure. It is probable that ice sheets that extend into infrequently glaciated mid-latitude areas are commonly underlain by unconsolidated sediment, and evidence of subglacial deformation has been found for former ice sheets in Britain, Ireland, northern Europe and North America. Deforming-bed theory also plays a critical role in our current interpretation of ice streams (section 7.5).

7.3 FACTORS CONTROLLING MOVEMENT

The preceding sections on mechanisms of movement indicated that the rate of glacier movement is a function of the dominant mechanisms of flow, the driving force, and the influence of a variety of controlling factors. The dominant mechanism of flow depends largely on environmental constraints, such as thermal regime, topography, and whether the glacier rests on rigid or deformable substrate. Specific mechanisms are controlled by different factors; for example, crystal size and orientation affect the creep rate more than they affect basal sliding. However, some major controlling influences are important in a range of dynamic situations.

Temperature is important, because sliding, regelation and creep are all temperature-dependent. The presence of water is also temperature-dependent, and the basal water pressure is a major control on sliding, on till deformation, and on the transition between surge and quiescent phases of surging glaciers. Roughness of the bed exerts a control on regelation and basal creep, and the sedimentological characteristics of both the debris-bearing basal layer and any unconsolidated substrate exert an important influence on movement by basal or subglacial deformation. Basal debris also increases friction between ice and bed, but stress concentrations resulting from debris-bed friction can cause local strain heating and softening of ice. Ice hardness is a major control on deformation, and is largely a function of temperature, crystal size, dust content and soluble impurities. Pleistocene ice that still exists at the base of many ice sheets is about three times softer than Holocene ice. Table 7.3 summarises how some of these variables affect velocity.

7.4 OBSERVED PATTERNS OF MOVEMENT

7.4.1 OBSERVATIONS OF GLACIER MOTION

Spatial and temporal patterns of glacier motion have been recognised since the earliest

Table 7.3 Some of the variables that affect glacier velocity

Factor	*Key parameters affected*	*To increase velocity*
Gradient	Driving stress	Increase gradient
Ice thickness	Driving stress	Increase thickness
Basal temperature	Presence of water	Increase temperature
Ice temperature	Viscosity/hardness	Increase temperature
Bed roughness	Friction	Decrease roughness
Bed roughness	Creep enhancement	Increase roughness
Bed roughness	Regelation	Decrease roughness
Soluble impurities in ice	Viscosity/hardness	Variable (see Figure 7.3)
Debris in the ice	Viscosity/hardness	Variable
Ice crystal size	Viscosity/hardness	Decrease crystal size
Gravity (constant)	Driving energy source	(Theoretically) increase gravity
Glacier density*	Driving stress	Increase density

*Depends on debris, crevasses, etc.

days of glacier study. Kuhn (1787) referred to measurements of glacier motion in the eighteenth century. Theory and observation of glacier movement have progressed together through time. There now exists a body of observation consistent with theory and with modelled behaviour of glaciers, and there are widely acknowledged patterns of both spatial and temporal variation in movement. Notwithstanding variations between individual glaciers, it is possible to identify some general characteristics, and these are discussed in the next two sections.

7.4.2 SPATIAL PATTERNS OF MOVEMENT

Ice flows from the accumulation area to the ablation area under the influence of gravity. Flow is thus outwards from the centre of an ice sheet and downwards through a mountain or valley glacier (Figure 7.6). The flow centre at the surface of an ice sheet is a point of zero horizontal velocity separating ice flow in different horizontal directions. The flow centre at the bed is not necessarily directly beneath the flow centre at the surface (Figure 7.7) (van der Veen and Whillans, 1992). Surface velocity increases down-glacier towards the equilibrium line, is greatest at the equilibrium line, and decreases down-glacier from the equilibrium line. Flow in the accumulation area of valley glaciers is generally convergent, and in the ablation area generally divergent.

Ice derived from the higher parts of the glacier follows a low-level route through the glacier and emerges closer to the snout, while

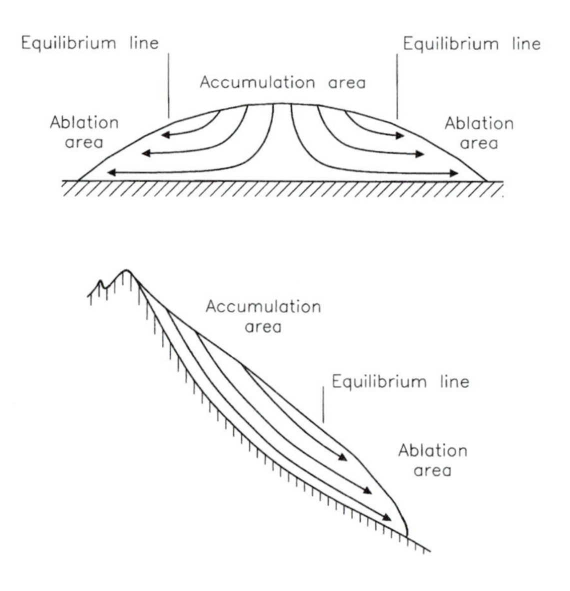

Figure 7.6 Flow is primarily outwards from the centre of an ice sheet and downwards through a mountain or valley glacier.

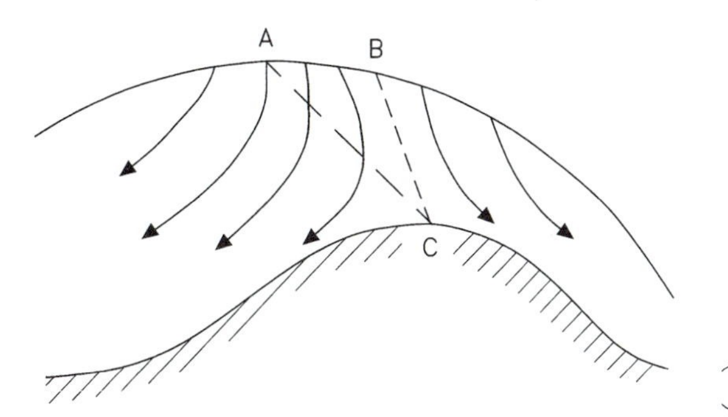

Figure 7.7 The flow centre at the bed is not necessarily directly beneath the flow centre at surface. After van der Veen and Whillans (1992).

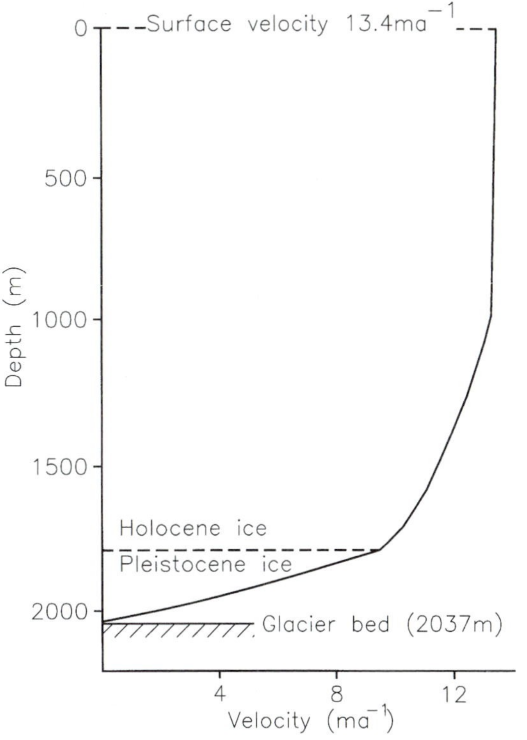

Figure 7.8 Variation of horizontal velocity with depth estimated from bore-hole surveys at Dye 3, Greenland (after Gundestrup and Hansen, 1984.) Velocity decreases with depth, and the transition from Holocene to Pleistocene ice is marked by a distinct rheological change.

ice derived from the lower parts of the accumulation area follows a high-level route and emerges closer to the equilibrium line (Reid, 1896). Flow is dominantly submergent in the accumulation area and emergent in the ablation area relative to the ice surface. Flow close to the margin can sometimes be emergent relative to the bed, depending on compression and basal entrainment.

Ice flows faster close to the surface of a glacier, and more slowly towards the bed. The upper layers of the glacier are carried 'piggyback' style on the lower, so the surface velocity is the sum of the vertical velocity profile (Figures 7.1, 7.8). Most of the movement is accomplished in the lower part of the profile, and the velocity gradient is strongest close to the bed. This is especially true when the lowest ice is of Pleistocene age, which has a softer rheology than Holocene ice. Vertical velocity profiles vary along the length of any flow line through a glacier as temperature, ice thickness and bed conditions change (Dahl-Jensen 1989).

At the surface of a valley glacier the ice flows more quickly close to the centre-line of the glacier and more slowly closer to the lateral margins, due to friction of the ice against the valley wall and to greater ice thickness towards the centre-line. The horizontal velocity gradient is strongest close to the lateral margin. Hanson's (1995) modelling of Storglaciären indicated that the bulk of the resistance to flow came from lateral drag due to the glacier being frozen to the valley walls while probably sliding along the valley floor. In ice streams the lateral boundary to flow is not a valley wall but slower-flowing ice. Echelmeyer *et al.* (1994) measured a transverse velocity profile across Ice Stream B and found the resistive drag to be partitioned nearly equally between the margins and the bed. The flow lines across an ice shelf are determined in part by the relative input volumes and velocities of its contributing ice streams.

Obstructions to flow such as bedrock bumps cause ice to diverge around and over the obstacle. Boulton (1974) described flow streaming around small bedrock obstacles, and Knight *et al.* (1994) described similar processes affecting much larger topographic features. Where the ice is thin, as for example close to the margin, movement is closely influenced by subglacial topography. Knight (1992) showed from fine-resolution strain measurements at the margin of the Greenland ice sheet that the surface strain pattern was closely controlled by ice flexure over subglacial obstacles.

7.4.3 TEMPORAL VARIATIONS IN MOVEMENT

Glaciers rarely move steadily at the same speed for long periods of time. Long-term flow variations are often associated with changes in mass balance and glacier fluctuations (Chapters 3 and 8). Shorter-term fluctuations are associated with surge behaviour and the life cycle of ice streams (sections 7.5 and 7.6). At intra-annual time-scales, variations occur in both horizontal and vertical movement on a seasonal, daily and even hourly basis. Clarke (1991) described diurnal velocity fluctuations of up to 300% at West Fork Glacier, Alaska. Some localised short-term variation can be caused by the dynamics of the sliding process; for example, stick–slip motion results from periodic formation of cold patches during regelation sliding (Robin, 1976). More widespread short-term variation is caused by changes in hydrological conditions at the glacier bed (Hodge, 1974).

Willis (1995) provided a useful review of the relationship between the mechanisms of movement and intra-annual temporal variations in rates movement. For hard-bedded temperate glaciers, variations in basal sliding occur if there are variations in subglacial water pressure, and are often associated with the seasonal evolution of the drainage system. During winter, subglacial water flux is low and is routed via a thin basal film. Basal water pressure is high, and basal sliding can occur. In some glaciers a gradual increase in velocity has been observed, correlated with a gradual build-up of water storage at the bed. During spring and early summer, the water flux increases and is routed predominantly through a linked cavity system; basal water pressure is very high, and rapid sliding and cavitation are likely, possibly with hydraulic jacking as cavities open. Spring commonly witnesses glacier uplift and high velocity. During the summer, the drainage system over much of the bed evolves from a linked cavity system to a conduit system; water pressure drops, and sliding decreases. Only in rare cases where water flux exceeds the capacity of the channel network is enhanced sliding likely during the late summer. During winter, the discharge decreases, the channel network collapses, and flow returns to small values through a basal film. For soft-bedded glaciers, intra-annual variations in sliding occur in response to subglacial pore-water pressure. During winter, subglacial water flux is low and is routed primarily by Darcian flow through the till. Subglacial water pressure is high, and stable bed deformation can occur. During spring and early summer, water pressure and deformation rates increase. As summer progresses, a pipe and channel system might develop at the bed. In that case, deformation will largely cease near to the channels, but might continue if pore-water pressures remain high in areas of the bed away from channels. Temporary water-pressure increases within a channel might cause increased deformation in till close to the channel.

Thus for both hard-bedded and soft-bedded glaciers, water at the bed is a primary control on velocity variations. This has been recognised by field observations in a range of locations. Hooke *et al.* (1989) described a 3-year record of seasonal velocity variations at Storglaciären where gradual decline in horizontal velocity through the winter is followed by abrupt increase in late spring or early summer. In many cases velocity increases have been observed to follow rainfall or melt events after a short time lag reflecting the time taken for water to reach the base of the glacier.

Hanson and Hooke (1994) reported diurnal variations of three times the mean flow in response to water input, with a time lag of about 4 hours between the diurnal melt and velocity cycles. The velocity of Glaciar Solar, Patagonia, is reported to have doubled and halved over periods of less than one day, correlating closely with variations in meltwater output (Naruse *et al.*, 1992). The primary effect of water pressure variations is on sliding and bed deformation, but Hooke *et al.* (1992) also reported temporal variation in internal deformation in response to variation in stress configurations following temporal variation in water pressure at the bed and hence in ice/bed coupling.

There are notable exceptions to the common seasonal pattern of fluctuations in glacier flow. For example, there is no measurable seasonal variation in the motion of Jakobshavn Isbrae in spite of seasonal variation in meltwater input to the glacier drainage system (Echelmeyer and Harrison, 1990). The factors discussed above suggest that this could indicate either drainage entirely within the ice above the bed, or a basal drainage route more than sufficiently capacious to discharge the water supplied without affecting basal water pressure.

7.5 ICE STREAMS

Ice streams are zones of fast-flowing ice within an ice sheet. They can account for the majority of the ice discharge from an ice sheet even though they occupy relatively small areas. About 90% of ice drainage from the Greenland and Antarctic ice sheets is via ice streams, although their outlets represent only about 13% of the Antarctic coastline and an even smaller proportion of the margin of the Greenland ice sheet. Ice streams exhibit a wide variety of characteristics, and examples of different types are discussed in the following paragraphs. A useful review of Antarctic ice streams was provided by Bentley (1987). Ice streams are the focus of a great deal of recent research as they seem to play a major role in controlling the stability of the Antarctic ice sheet.

7.5.1 THE SIPLE COAST ICE STREAMS

The West Antarctic ice sheet drains into the Ross ice shelf through a group of five ice streams (A, B, C, D and E) on the Siple Coast (Figure 7.9). The ice streams carry 90% of the ice flux into the Ross shelf. Ice stream C is nearly stagnant, but all the others have high velocities of 400–1000 m a^{-1}. One of the most closely observed has been Ice Stream B (e.g. Whillans and van der Veen, 1993). Ice Stream B is 500 km long, 50 km wide and about 1 km thick. It has a surface velocity of 1–3 m d^{-1}, and contributes 30 km^3 a^{-1} of ice to the Ross ice shelf. The ice stream is bounded laterally by nearly stagnant ice. Some ice feeds into the stream along most of its sides, but the bulk of ice input is from the up-flow ends of several tributaries. It achieves its high speed in spite of low surface gradient. Beneath Ice Stream B there is a 6–15 m thick layer of till characterised by high porosity and low effective pressure. This deforms in response to shear stress at the base of the ice, accommodating a high

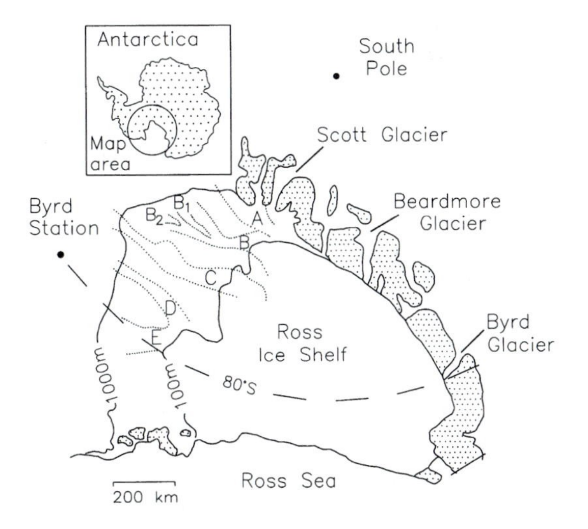

Figure 7.9 Location of the Siple Coast ice streams, Antarctica.

sliding velocity for relatively low driving stress. The drainage system beneath the ice stream is relatively stable, and flow speed is fairly regular. McDonald and Whillans (1992) found no velocity variation in the up-glacier end of the ice stream at scales ranging from 1 day to 2 years.

By contrast, Ice Stream C has virtually stopped. The lower reaches are completely stagnant, the central portions move at about 1–13 m a^{-1}, and the head of the stream moves at about 40 m a^{-1}. Most of the ice stream shows no visual sign of surface crevassing, but radar sounding indicates buried crevasses beneath the surface snow. The depth of accumulation over the buried crevasses reveals the length of time since they were active, and allows reconstruction of the history of the ice stream's stagnation. As recently as 130 years ago, Ice Stream C moved as fast as the others, but it has recently, and quite rapidly, stagnated. Lower sections were affected by stagnation about 130 (± 30) years ago, the upper sections more recently (100 ± 30 years). The very top section of the ice stream has only been affected in the last 30–100 years, with an open crevasse suggesting some remaining activity. Retzlaff and Bentley (1993) refer to a 'wave of stagnation' propagating up the ice stream. The driving stress for Ice Streams B and C is about equal, so the contrasting behaviour is due to controlling mechanics rather than driving stress. Anandakrishnan and Bentley (1993) suggested that the switch from fast to slow flow in Ice Stream C occurred because the till layer increased in strength following a decrease in porosity that prevented it from deforming in a ductile manner. Loss of dilatancy through dewatering in till beneath the ice stream would be expected to lead to stagnation, and the abundance of micro-earthquakes beneath Ice Stream C relative to other ice streams (20 times more abundant beneath C than B) could reflect a new brittle till behaviour. Retzlaff and Bentley (1993) suggested that stagnation occurred because the subglacial water flow became too great to remain stable as sheet flow: R-channels developed that led to a drop in water pressure, which in turn caused dewatering of the subglacial deforming till, thus immobilising it. Upstream migration of R-channels led to upstream propagation of the stagnation front. Anandakrishnan and Alley (1997) presented data to suggest that the stagnation was only affecting the lowest portion of the ice stream, and that it occurred because water piracy, or meltwater capture at the bed, caused loss of lubrication at 'sticky spots' at the bed of the ice stream.

If ice streams are flowing rapidly on a bed of weak, deforming till, the question arises as to what limits the flow velocity and why the flow remains stable for long periods. Kamb (1991) suggested that the apparent stability of ice stream B could be accounted for by the existence of 'sticky spots' at the bed that disproportionately support the basal shear stress. Alley (1993) cited a range of published evidence suggesting spatially variable drag under the Siple Coast ice streams, and Echelmeyer *et al.* (1994) found that while the basal shear stress beneath B was low, it was not as low as would be expected if the whole bed were mobile. At Ice Stream E, MacAyeal *et al.* (1995) identified small, high-friction sticky spots scatted across the bed. These sticky spots comprised about 15% of the bed, and coincided with structural features in the underlying bedrock inferred from satellite imagery. The most likely origin of sticky spots is drag from bedrock bumps penetrating the ice. Discontinuity of lubricating till, another possible cause, is likely to be unstable to the collection of lubricating basal water at high-friction sites (Alley, 1993). In contrast to these results, Whillans *et al.* (1993) found no visible sign of sticky spots at Ice Stream B, suggesting that basal properties as they affect sliding are uniform. Whillans and van der Veen (1997) suggested that nearly all the mechanical control on the flow of Ice Stream B was exerted by the lateral margins, with the effect of basal friction being negligible. There may be substantial differences between the ice streams, or even between parts of each stream, and our knowledge of basal conditions remains sketchy, but it is apparent that wall drag (against non-streaming ice), grounding-line friction (back-pressure from the ice shelf) and sticky spots contribute key resistive forces countering ice-stream movement.

7.5.2 OTHER ANTARCTIC ICE STREAMS

The Siple Coast ice streams, though well studied, are not necessarily typical of Antarctic ice streams. Some, such as the Rutford ice stream that flows into the Ronne ice shelf, share with the Siple Coast ice streams some characteristics such as extensive zones of deforming subglacial sediment (Smith, 1997a,b). A number of other Antarctic ice streams, such as the Byrd Glacier, have been described quite differently. McIntyre (1985) suggested that many of them correspond to fjord-like channels, with the transition from ice-sheet flow to ice-stream flow occurring at the head of the fjord and corresponding with a marked bedrock step. The onset of stream flow is marked by a transition from flow dominated by ice deformation to flow dominated by sliding, but does not seem to involve deformation of subglacial sediments, as the surface slope and driving stress (typically 50–200 kPa) are much higher than could be supported by a deforming till. If these ice streams are topographically initiated, they are unlikely to be susceptible to unstable propagation or surge behaviour, as the bedrock sill will pose a limit to headward propagation of fast flow. The extent to which there are fundamental differences between different Antarctic ice streams remains to be ascertained. As increasing amounts of data about subglacial conditions become available, ideas about the flow mechanisms at each site are still in a state of development. For example, in an early draft of this book I planned to cite the Rutford ice stream as a counter-example to the deforming-bed streams of the Siple Coast (e.g. Frolich *et al.*, 1989), but newly emerging data has indicated that comparison might be more valid than contrast (Smith, 1997a,b).

7.5.3 JAKOBSHAVN ISBRAE

The largest ice stream draining the Greenland ice sheet is Jakobshavn Isbrae. It exhibits substantial differences from its Antarctic cousins. Jakobshavn Isbrae extends up-glacier 85–90 km from its calving terminus. Its length/width ratio of about 15 compares with typical ratios of about 10 for the Siple Coast streams. Jakobshavn Isbrae lies a in deep bedrock trough, with a centre-line ice depth of up to 2520 m, (compared with about 1000 m for the Siple Coast streams, for example) and a bed depth below sea level of up to 1500 m. The great thickness coupled with steep surface slope (0.01–0.03) leads to high driving stress (200–300 kPa), an order of magnitude higher than the Siple Coast streams but only slightly higher than some other Antarctic ice streams. Jakobshavn's lateral transition zone of crevasses and flow change is broader and more gradual than the abrupt lateral transitions of the Siple Coast ice streams (Echelmeyer *et al.*, 1991). Deformational heating at the base and sides of the ice stream affects deformation in accordance with the temperature-dependent flow law, producing meltwater and a thick temperate basal ice layer. The basal temperate layer seems to be critical to the fast flow of Jakobshavn, and it has been hypothesised that the stream is self-perpetuating as its mass is warmed by the heat of internal friction, reducing its viscosity. The temperate layer is thickened by three-dimensional ice deformation as ice is drawn in convergent flow into the ice-stream trough from the head and sides (Iken *et al.*, 1993; Funk *et al.*, 1994). The flow does not exhibit seasonal variability associated with water availability (Echelmeyer and Harrison, 1990), suggesting that basal sliding may not be an important component of motion. Enhanced creep in a thick basal layer of thermally softened ice is currently the dominant hypothesis for the streaming flow of Jakobshavn Isbrae.

7.5.4 ICE STREAMS, TRANSIENT BEHAVIOUR AND ICE-SHEET STABILITY

Our understanding of ice streams remains incomplete. MacAyeal *et al.* (1995, p. 262) suggested that 'the intricate pattern of basal coupling that controls the present flow of Ice Stream E arises from some as yet inscrutable property of the bed that escapes our understanding.' A major issue to be resolved is the question of whether ice streams undergo transient behaviour, like surging glaciers but at a much longer time scale. This is especially important as ice streams are the major route for draining the world's two large ice sheets. Retzlaff and Bentley (1993) discussed a cyclic theory of streaming flow in which an ice stream accelerates by increasing its mass flux, consequently increasing basal melting until subglacial sheet flow of water at its mouth is no longer stable and the stream stagnates. Its drainage basin could then be captured by another stream that would repeat the cycle (Rose, 1979). Patterns of internal folding have been interpreted to indicate headward migration of a zone of changing boundary conditions at the head of streams B and C where ice-sheet flow changes to streaming flow (Jacobel *et al.*, 1993), so streams may both form and then decay by propagation from the nose upwards. Anandakrishnan and Alley (1997), on the other hand, suggested that the water piracy that led to the stagnation of Ice Stream C was due to an accident of basal topography rather than to any internal feedback in ice-stream dynamics. Since ice streams contribute such a major part to the dynamics and potential stability of ice sheets, understanding their behaviour is likely to prove crucial to successful modelling of ice-sheet response to climatic perturbation and the reconstruction of former ice sheets. Ice streams are being incorporated into models of former ice sheets (e.g. Isaksson, 1992), and might help to explain some hitherto puzzling phenomena such as Heinrich events (Chapter 2).

7.6 SURGES

There are real mistakes in the mind of mountaineers, however, with respect to the advance of glaciers. They usually assume that glaciers slide over their substratum, and it is not unusual to hear them tell very amazing stories, about the speed of advancing glaciers, and the jumps they are supposed to make.

(Agassiz, 1840).

7.6.1 GLACIER SURGES

In some glaciers, 'normal' flow is periodically interrupted by episodes of enhanced flow lasting for periods of months or years. These enhanced-flow events are referred to as surges, and the periods between surges are referred to as the glacier's quiescent phase. During a surge, part or all of the glacier experiences accelerated flow velocity; there is a transfer of ice from an up-glacier source or reservoir area to a down-glacier receiving area; the profile of the glacier changes, and there is sometimes a major advance of the snout. In many cases a zone or 'packet' of fast ice-flow moves down the glacier in the form of a kinematic wave. During the quiescent phase, most of the glacier experiences normal flow conditions, except that ice stagnates in areas of the snout that during the surge advanced beyond their pre-surge position, and the ice becomes less thick in the receiving area and thicker in the reservoir area. Many surging glaciers thus undergo a cyclic transition between a long, shallow profile immediately following a surge and a shorter, steeper profile immediately prior to a surge. Surges can last from a few months to several years, and return periods vary from as little as 10–20 years to as much as 500 years.

About 4% of all glaciers exhibit surging characteristics (Sharp, 1988). Post (1969) identified

204 surging glaciers in south-central Alaska and the Yukon Territory alone. Surges have been recognised in all sorts of glaciers: both warm and cold based, large and small, with rigid beds and with deformable substrates, in seismically active and in stable areas. The largest glacier in North America, the 5200 km^2 Bering Glacier, surged in 1993–4 (Fleisher *et al.*, 1995). Surging can be especially important for floating ice tongues, as it can lead to enormous advance of the terminus and extraordinary iceberg production.

The origin of surges has been the centre of considerable debate, and many different theories have been proposed. These have included theories based on external forcing by earthquakes, landslides or volcanoes, and theories based on periodic variations or inherent instability in the glacier's dynamic controls. Sharp (1988) provided a useful review of the field. In the last two decades, detailed observations of glaciers passing through surge cycles have clarified some of the mechanisms that seem to be important in surging. The most extensive observations have been made at Variegated Glacier, Alaska.

Variegated Glacier is a temperate valley glacier 20 km long located on the southern side of the St Elias Mountains in south-eastern Alaska. The glacier has surged five times in the twentieth century, at an average interval of 18–20 years (Tarr and Martin, 1914; Post, 1969; Bindschadler *et al.*, 1977; Kamb *et al.*, 1985). A siege-like research effort monitored the behaviour of the glacier from the mid 1970s onwards in preparation for the surge that was expected in the mid 1980s. The surge occurred in 1982/3, and furnished some clues as to the mechanisms involved. Raymond and Harrison (1988) described the evolution of Variegated Glacier through the last decade of the quiescent period prior to its 1982/3 surge. They reported that the glacier thickened in the upper 60% and thinned in the lower 40% of its length. In 1979, 1980 and 1981, 'mini-surges' occurred in the section of the glacier where the 1982 surge would be initiated, which was also the zone of highest velocity and basal stress increase over the previous decade. These mini-surges were marked by high water pressures, enhanced sliding and turbid water discharge (Humphrey *et al.*, 1986; Raymond and Malone, 1986; Kamb and Engelhardt, 1987). The main surge began in January 1982 and occurred as a zone of accelerated flow that passed down the glacier, with compressive flow below and extending flow above. A surge front developed initially in the upper glacier, expressed at the surface as a steep topographic ramp, 1–2 km long, within which emergent velocities and compressive strain rates were greatly enhanced compared with quiescent flow. This surge front propagated down-glacier into virtually stagnant ice at a mean velocity of about 40 m d^{-1}. Ice velocities in the surging zone were over 20 m d^{-1}. Sharp *et al.* (1988) interpreted the surge in terms of a propagating strain wave, and defined three tectonic zones in the glacier. Furthest up-glacier was a zone of extension experiencing continuous and cumulative elongation and characterised by transverse crevassing. Furthest down-glacier was a zone of compression, characterised by longitudinal crevasses. Between the two was a zone through which the velocity peak passed: a zone of superimposed compression and extension tectonics.

The key to the surge process at Variegated Glacier seems to be related to basal hydrology (Humphrey and Raymond, 1994). The surging region of the glacier was underlain by a basal hydraulic zone of low water velocity and high water storage, inferred to be a distributed flow system. The ice down-glacier of the propagating surge front was underlain by a zone of high water velocity, and low water-storage, inferred to be a conduit system. The volume of water stored upstream of surge front was the major hydraulic control on the surge. Glacier sliding was related to basal hydraulics. During the surge, water discharge was free of turbidity, indicating that no basal water was escaping the glacier. During the termination phase, abrupt decreases in glacier velocity were accompanied by high water discharge with high sediment content, indicating release of large amounts of basal water that had been stored during the surge. Over 1 m of water volume was stored per unit area during the surge, much of it englacially.

On the basis of observations of the Variegated Glacier surge, Kamb *et al.* (1985) and Kamb (1987) proposed the following model of the surge process. The immediate cause of the high flow velocity is high basal water pressure which causes a great increase in basal sliding. The high basal water pressure relates to the breakdown of an efficient channelised drainage system and the initiation and survival of a distributed linked cavity drainage system. Through the quiescent phase, increasing ice thickness and gradient in the reservoir area lead to increasing basal shear stress. This increases the separation pressure (p_s) and hence increases the likelihood of cavities opening in the lee of bedrock hummocks. Basal water pressure (p_w) tends to be highest in the winter, when water flux is low, and so it is in the winter that p_w is most likely to exceed p_s and form water-filled cavities. Once a linked cavity system forms, its characteristics lead to increased sliding, and the high sliding rate mitigates against the unstable opening of orifices between cavities. The high-pressure linked cavity system can therefore remain stable at high sliding speeds. Areas of the bed affected by this condition exert longitudinal stress on adjacent sections to propagate the effect downstream. The surge continues until the water in the cavity system is released. This might be because the water reaches a connection to the proglacial zone, or because high summer water input is sufficient to destabilise the cavity system into a tunnel system (Chapter 6).

This general model of surging with sliding at high basal water pressure has been applied to several glaciers (e.g. Echelmeyer *et al.*, 1987; Harrison *et al.*, 1994). The surge is triggered by disruption of the internal drainage system, which results in large water storage and high basal pressure, with consequent localised high ice velocity. McMeeking and Johnson (1986) emphasised the role of longitudinal stress transfer in spreading the active surge zone down-glacier. This type of mechanism could work both on hard-bed glaciers (by effectively floating the ice) and on deforming-bed glaciers (by causing failure in the bed). Observations have been made at surging glaciers with deforming beds, providing an opportunity to test the applicability of this model. The most detailed observations of soft-bedded surging glacier have been made at Trapridge Glacier in the Yukon Territory, Canada.

Trapridge Glacier is a sub-polar glacier resting on an unconsolidated sediment substrate. It last surged around 1941, and has since been evolving through its quiescent phase towards another surge. The boundary between the reservoir and receiving areas is determined by the position of a thermal boundary; the glacier is cold based in the receiving area, but warm based upstream. This thermal boundary presents a barrier to glacier flow, since basal motion is inhibited in the cold-based region. A large ice-surface step or 'bulge' formed at the position of this thermal barrier, as ice in the reservoir area thickened and slid forwards towards the barrier, and the virtually stagnant ice in the receiving area thinned. In 1980 the step at the face of the bulge was 40 m high. As the reservoir continues to thicken, major changes in basal stress conditions occur beneath the reservoir area.

Clarke *et al.* (1984) proposed the following model for surge behaviour in a sub-polar deforming-bed glacier on the basis of their observations at Trapridge. The premise of the model is that downstream resistance to sliding divides the glacier into a receiving area and a reservoir area. This downstream resistance might be due to a thermal dam, as at Trapridge, or to a downstream increase in the efficiency of the drainage system. A progressive increase in the thickness and surface gradient of ice in the reservoir through the quiescent phase causes an increase in basal stress, and hence increasing deformation of the subglacial sediment. As the rate of deformation increases, it becomes harder for drainage paths to remain open through pipes or channels in the till. Eventually, the deformation rate exceeds the rate at which drainage routes can be maintained, and the drainage becomes distributed and inefficient. Basal water pressure rises, and till deformation is quickly enhanced to the point where the surge occurs by unstable deformation of the substrate. As the surge progresses, surface gradient and thickness decrease until the basal driving stress falls below the threshold at which the rate of regeneration of drainage paths exceeds their destruction; efficient drainage is restored, and the surge ends.

In spite of extensive observation in recent years, details of the surge process remain elusive, even at Variegated Glacier. Controls on the timing of the event, the way in which the surge front propagates down-glacier and the role of different mechanisms of movement are still unclear. The role of tectonic dislocations within the ice during surging have been considered in the context of debris entrainment, but are not well incorporated into models of surge movement. Some dissatisfaction might arise from a determination to find a single unifying surge mechanism. Clarke *et al.* (1984, p. 238) called the idea that several mechanisms can cause surges 'unattractive', but trying to solve the problem by focusing on common characteristics (are they all soft-bed? are they all sub-polar?) might be a mistake. On the other hand, shared hydraulic characteristics seem to offer a useful clue. Surging based on build up of basal water associated with reorganisation of drainage system is at present the ruling hypothesis, and is widely applied to both hard-bedded and soft-bedded glaciers.

7.6.2 ICE SHEET SURGES

It has been suggested that whole ice sheets as well as individual glaciers could surge. Several scenarios for the catastrophic collapse of the Antarctic ice sheet have been postulated (e.g. Wilson 1964, 1969; Hollin 1965, 1972, 1980) and Budd and McInnes (1979) suggested that the Antarctic ice sheet could surge for 250 years at a time on a 23,000-year cycle. If this were so it could have major implications for global sea-level and climate stability. It is difficult to imagine the surge process that has been modelled for smaller glaciers applying to an ice sheet, and evidence that could be cited in support of former ice-sheet surging, such as low surface profiles, can be explained by alternative hypotheses such as the prevalence of deforming beds. Nevertheless, there are a variety of surge-like phenomena that might affect ice sheets. For example, MacAyeal (1993) suggested that ice sheets might progressively thicken until basal thawing is induced, at which point the ice surges, thins, and returns to frozen bed conditions before gradually thickening in advance of the next surge. At a smaller scale, individual components of ice sheets, such as outlet glaciers or ice streams, may show surge-like behaviour. If several ice streams simultaneously switched from non-streaming to streaming behaviour, the effect could in some ways be equivalent to a surge of at least a section of the ice sheet. The periodic discharge of large numbers of icebergs into the Atlantic during Heinrich events could be related to periodic surging of an ice stream through the Hudson Straight. The West Antarctic ice sheet as a whole might experience major dynamic changes if the ice shelves that surround it were to disintegrate in response to warming ocean temperatures. It has been suggested that back-pressure from the ice shelves is a major restraint on flow of ice off the ice sheet, and the loss of this restraining force could allow rapid flow and possible collapse of the ice sheet (Hughes, 1973; Mercer, 1978).

7.7 CONCLUSION

Glacier movement should be placed at the heart of any comprehensive conceptual map of glacial phenomena. Movement is an inevitable consequence of the physical characteristics and environmental context of glaciation, and it is a direct cause of most of the major consequences of glaciation. If glaciers did not move, there would be no glacial geomorphology and a constipated hydrological cycle. The nature of glacier movement is a valuable indicator of englacial and subglacial conditions, but one that we are not yet fully able to interpret. Movement is conditioned by driving stresses, ice rheology and substrate conditions, and especially by water pressure at and beneath the bed. Intra-annual variations in motion reflect seasonally evolving subglacial conditions, and long-term periodcities such as surging reflect long-term changes in subglacial conditions that are as yet incompletely understood. Elucidation of the mechanical background to the differences between 'fast' and 'normal' flow states remains a major goal of glaciology, as does the elaboration of a realistic and comprehensive sliding theory for different bed conditions and the establishment of flow laws for realistic glacier ice and bed conditions. These goals can be approached from a variety of routes, including laboratory experimentation, theoretical modelling, and the interpretation of geomorphic and geologic evidence. This is a huge and exciting area of ignorance in glaciology, the exploration of which may hold the keys to unravelling outstanding problems in all aspects of the subject.

GLACIER FLUCTUATIONS 8

It is remarkable that whereas the Lak Glacier has so greatly shrunk of later years, this Chur Glacier, its immediate neighbour, and which drains another flank of the self-same mountains, should, on the contrary, have greatly swollen. It overflows all its moraines and pours in a broken spreading wave on to the surface of the Hispar.

(Conway, 1894)

8.1 FLUCTUATIONS OF GLACIER MARGINS

For a variety of reasons, glaciers change in size through time. As glaciers expand and contract, their margins advance and retreat. These fluctuations of glacier limits occur at a variety of time scales ranging from days to millions of years. Where margins respond sensitively to local conditions, fluctuations can be identified on very short time scales. For example, calving of large tabular icebergs from floating glacier tongues and ice shelves can very abruptly cause substantial changes in margin position. Calving of 14 large tabular icebergs from the tongue of Jakobshavn Isbrae in the spring of 1982, following the seasonal break-up of sea ice in the fjord and exposure of the tongue to increased wave and tide activity, led to recession of the margin by 2 km in less than 18 days (AGRISPINE, 1983).

Many glacier margins advance and retreat on a seasonal time scale in response to seasonal changes in ablation. During the summer, ablation at the margin outweighs the forward movement of the ice and the margin retreats. Each winter, when ablation is low, the forward movement of the ice exceeds the rate of ablation and the margin advances. Changes in climate over periods of a few years or decades can also be reflected in equivalent periods of glacier advance or retreat. At longer time scales, major fluctuations in glacier extent can be traced over centuries and millennia. Matthes (1939) introduced the term 'Little Ice Age' to describe the period of renewed glaciation following the warmest part of the Holocene, and recognised repeated fluctuations in glacier conditions over the last 4000 years. More recent usage has applied the term 'Little Ice Age' specifically to a period of widespread glacial advance between about AD 1500 and AD 1900 (Grove, 1988). Since about AD 1900, observers world-wide have noticed a trend of glacier retreat from maximum positions achieved during the Little Ice Age. That is why so many glacier margins today lie just behind moraine ridges (Figure 8.1). At the longest time scale, glaciers and ice sheets wax and wane, even to the point of disappearance and reformation, over periods of hundreds of thousands, or even millions, of years. The geological record shows evidence of at least seven major glacial episodes since the Precambrian, and abundant geomorphic evidence testifies to the inception, growth, decay and disappearance of ice sheets such as the Laurentide and Scandinavian during the Pleistocene.

Both advance and retreat can happen either gradually or abruptly, and may be either steady or episodic. Glacier surges can cause extremely fast glacier advance, and also fast retreat. For example, the snout of the Medvezhiy Glacier advanced 1.5 km in less than 2 months during its 1963 surge. However, the areas over which surge-advances encroach typically become ice-free again within a few decades as the

Figure 8.1 At present, many glacier margins stand behind moraine ridges that mark their former Little Ice Age positions. Russell Glacier, Greenland.

ice at the margin stagnates. Floating glacier tongues also demonstrate rapid and substantial margin fluctuations compared with most land-terminating glaciers. For example, Jakobshavn Isbrae retreated ~30 km at average rate of 0.28 km a^{-1} between 1850–1960, and some Alaskan tidewater glaciers have retreated up to 100 km at 0.5–1.7 km a^{-1}. Hughes (1986) reported a retreat of 2 km in just 45 minutes at Jakobshavn in 1985. It is more common for land-terminating glaciers to advance and retreat at a rate of metres, or tens of metres, per year. Different scales of glacier fluctuation can be superimposed upon each other. Long-term fluctuations are frequently overlain by short-term variability. For example, the snout of Solheimajökull, in southern Iceland, has been advancing steadily for about two decades, but interrupts its forward procession each summer when the glacier snout retreats several metres in the face of summer ablation. Long-term glacier retreat is frequently marked by seasonal stand-stills or re-advances of the margin at periods of low ablation, witnessed in the geomorphic record by recessional moraines.

Fluctuations of glacier margins can be reconstructed from geomorphological and sedimentological evidence. Evidence of former glacier fluctuations is provided by a range of indicators including terrestrial and marine sediments deposited by glaciers in different locations at different times, and landforms of glacial erosion and deposition. Fluctuations can also be observed and monitored directly, and can be predicted or modelled theoretically. Schmeits and Oerlemans (1997) simulated the historical variations of Unterer Grindelwaldgletscher by coupling a numerical mass balance model to a dynamic ice-flow model. The coupled model reconstructed the known historical variations of the glacier with reasonable success, and predicted that by AD 2100 only 29% of the 1990 volume of the glacier would remain.

Glacier fluctuations are important for several reasons. Fluctuations result in the formation of specific landform and sediment assemblages, and generate a 'geography' of glacial geomorphology (Chapter 9). Fluctuations reflect, and can be used to identify, changes in climate and in glacier dynamic regimes. In turn, glacier fluctuations influence climate. Surface albedo, and the composition and circulation of the oceans and atmosphere are strongly influenced by glacier extent (Chapter 2). In inhabited areas, glacier fluctuations can have important economic implications (Chapter 10). Nevertheless, it was not until the latter part of the nineteenth century that the extent or significance of glacier fluctuations was widely appreciated (Chapter 12). Detailed historical records of glacier fluctuations exist from many areas for about the last hundred years. World-wide collection of information about glacier variations started in 1894 with the International Glacier Commission (Haeberli *et al.*, 1989). Glacier monitoring at a global scale has been made easier in recent years by satellite remote sensing technology. A satellite atlas of the world's glaciers is being produced with the intention of providing a baseline reference against which to monitor future glacier fluctuations (Williams and Ferrigno, 1988). A new project known as GLIMS (Global Land Ice Monitoring from Space) is scheduled for 1998 onwards, designed to monitor snowline, velocity field, areal extent and terminus position of the world's glaciers.

Our understanding of glacier fluctuations is based on a combination of theoretical modelling of ice dynamics and correlation of past climates with known glacier positions, both of which are poorly constrained (Letréguilly *et al.*, 1991). At the grand scale, our understanding of the causes of climate change that drive glacial and interglacial cycles is also limited. Fluctuations between glacial and interglacial conditions are roughly synchronous at the global scale, but detailed regional correlation is difficult because different glaciers respond to environmental influences in different ways and at different rates. Therefore it is difficult in any direct way either to interpret former ice-margin positions in terms of climate change, or to predict the future response of ice margins to climate change, or to use present and past ice-margin positions to calibrate ice-sheet/climate models. To approach any of these goals it is necessary to identify more closely the links between environmental forces, ice dynamics and glacier fluctuations.

8.2 CAUSES OF GLACIER FLUCTUATIONS

8.2.1 MASS IMBALANCE AT THE MARGIN

Fluctuations of glacier margins are caused by a mass imbalance at the margin. If the amount of ice supplied to the margin by flow from up-glacier is equal to the amount lost at the margin by ablation, then the position of the margin will remain stable. If supply exceeds loss, the margin will advance. If loss exceeds supply, the margin will retreat. The function of ice-margin fluctuation is to maintain mass equilibrium at the margin, by relocating the margin to a position where the local ablation matches the ice supply. This can be achieved either by adjusting the extent of the ablation area so as to control the rate of supply to the margin, or by changing the ablation rate at the margin (for example, by calving). Thus, for example, if ice supply to the margin is increased, the margin will advance until the ablation area upstream of the snout is enlarged sufficiently to negate the increase in input to the margin, or until the margin reaches a position where local ablation is sufficiently high to evacuate the extra input. If ice supply to the margin is decreased, the margin will retreat until ablation is reduced sufficiently to match the reduced input, either by shrinkage of the ablation area or by removal of the margin to a position of lower local ablation. The response of the margin to changes in ice supply or ablation is not instantaneous, and it takes a certain time for the ice margin to retreat or advance to a new equilibrium position in response to any stimulus. If the controlling stimuli are relatively dynamic or frequently changing, then glaciers might be expected to find their margins in a state of disequilibrium, advancing or retreating, for much of the time. Stable, steady-state conditions might be relatively rare. The frequency/magnitude regime of controlling parameters, and the response time or inertia of the glacier, are therefore very important. The areal extent of a glacier is determined by its total mass and its profile, so glacier margins will advance or retreat in response to changes in either of these. Mass imbalance at the margin can thus be driven either by climate-related mass balance changes, or by ice dynamic conditions, each of which is discussed below.

8.2.2 FLUCTUATIONS DIRECTLY RELATED TO MASS BALANCE

Changes in glacier mass balance have a direct impact on glacier fluctuations by controlling the supply of ice to the margin. As explained in Chapter 3, the main controls on mass balance are climatic, and climate-driven variations in accumulation and ablation are a major cause of glacier fluctuations. Oerlemans (1992) modelled the way in which glaciers in Norway would advance or retreat in response to mass balance changes driven by climate variation. He found that equilibrium line altitude (ELA) was sensitive to both temperature and precipitation variations, and that this sensitivity was reflected directly in fluctuations of the snout position. For the glacier Nigardsbreen, the most sensitive of the glaciers that Oerlemans considered, his model predicted that 1 K cooling would lead to a snout advance of 3 km, while 1 K warming would lead to a snout retreat of as much as 6.5 km. The response to a 10% increase in precipitation would be a snout advance of 2 km, while a 10% decrease in precipitation would lead to a 4 km retreat of the snout. Different glaciers are more or less sensitive to climate change. Their sensitivity depends partly on the shape of the glacier. Nigardsbreen has relatively long, narrow tongue fed by a broad accumulation area. A small change in its ELA thus has a large effect on the accumulation area ratio and a major effect on extent of the snout. The impact of changes in ELA depend partly on the altitudinal range of the glacier. For example, a mountain glacier with a large altitudinal range stretching far above the ELA would be less troubled by a rising ELA than would a broad ice dome that did not extend in altitude far above the ELA (Figure 8.2). A 200 m rise in ELA could kill a low-relief ice cap but not inconvenience a tall mountain glacier. Adjacent glaciers that have accumulation areas at different elevations will respond differently to changes in climate. Thompson (1988) described how two adjacent outlets from Vatnajökull ice cap have retreated at different rates during the twentieth century: Skaftafellsjökull has retreated at a rate of more than 17 m a^{-1}, but Svinafellsjökull has retreated at only about 4 m a^{-1} because its accumulation area is at a higher elevation, and is less affected by rising ELA. Bennett and Boulton (1993a) suggested that similar conditions caused asynchronous retreat of adjacent glaciers in Glen Grudie in the Northwest Highlands of Scotland at the end of the Loch Lomond stadial. Climate changes measured during the present century have been correlated directly with glacier fluctuations in many different locations. Some of the longest and most detailed records come from the European Alps, where much of the twentieth century has been characterised by relatively mild conditions and widespread glacier retreat. The retreat has been interrupted in many locations by a re-advance in the third

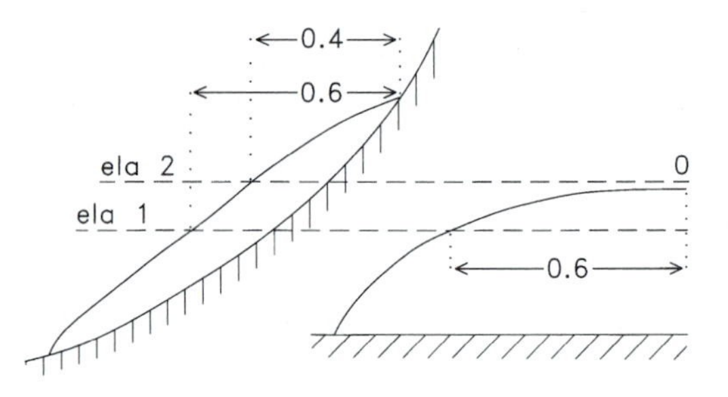

Figure 8.2 The effect of changing ELA on AAR of glaciers of different morphology.

quarter of the century associated, after some time lag, with a period of reduced ablation and increased snowfall that begun in about 1950 (Patzelt, 1985; Pelfini and Smiraglia, 1992). Similar correlations are available from glaciers world-wide (Grove, 1988; Dawson, 1992).

Changes in accumulation generally affect the position of the margin only after a time delay related to the throughput time of ice travelling from the accumulation area to the margin. For large ice sheets, this might amount to a delay of thousands of years between an increase in accumulation at the summit and the corresponding advance of the margin. For smaller glaciers, response times are much quicker, typically of the order of decades. Holmlund (1988) suggested a response time of about 50 years for Storglaciären. A glacier responds to a change in mass balance by changing its profile. If the ice limit is fixed, then mass balance changes are accommodated by changes in thickness. If the edge is mobile, then the ice sheet can respond by changing both its thickness and its horizontal extent. For changes in thickness, Oerlemans and van der Veen (1984) modelled response time (T_r) by:

$$T_r = 1/c(2m + 1)\, L^{m+1}\, H_o^{-2m}$$

where c and m are flow constants, L = length, and H_o = thickness. This implies that the response time will be quicker for thicker ice, for shorter-radius ice sheets, and for faster-flowing (less viscous) glaciers. For a theoretical ice sheet with L = 1000 km, H_o = 2500 m, c = 0.1 $m^{-2}a^{-1}$, and m = 3, the model predicts T_r = 5851 years as a typical value for re-establishing equilibrium following a mass balance perturbation. Because this response time is longer than the wavelength of the forcing fluctuations in mass balance, the ice-sheet response integrates a series of mass balance fluctuations. The total variance is then likely to increase with T_r; so, for a given series of forcing variation, it seems that thin, large-radius and slow-moving ice masses will tend towards greater margin fluctuations than will thick, short and fast-moving ice masses. Jenson *et al.* (1996) suggested that an ice sheet resting on a deformable bed could respond to mass balance changes faster than one on a hard bed, and reach an equilibrium form more quickly. Jóhannesson *et al.* (1989a,b) considered the length of time (T_m) over which a glacier responds to a prior change in climate in terms of the time required for a change in mass balance to supply the difference in ice volume between the steady-state glacier profile before the change and the steady-state glacier profile after the change. They found that:

$$T_m = H/(-b)$$

where H = maximum glacier thickness (m), and $-b$ = negative mass balance (ablation) at the snout (m a^{-1}).

For glaciers 100–500 m thick, with snout ablation rates of 1–10 m a^{-1}, this model suggests response times of the order of decades. The model was tested against data from Blue Glacier by McClung and Armstrong (1993), who found that measured terminus response (advance) lagged a positive mass balance input (area averaged net mass balance) by only about 10 years, in broad agreement with the model. Earlier models of glacier response times, which had been based on theoretical work by Nye (1963), had led to a general belief in a much longer response time – of the order of centuries. Jóhannesson *et al.* (1989b) explained that Nye's theoretical treatment was difficult to apply as it was very sensitive to some potential variables, including the behaviour of kinematic waves in the glacier (discussed below). The new model of Jóhannesson *et al.* (1989a,b) was much easier to apply, was generally robust, and provided a good approximation of response times. A convenient review of the differences between the treatments was provided by Paterson (1994). Table 8.1 shows the response times that have been suggested for different glaciers.

Changes in accumulation high on the glacier can be transmitted rapidly downstream through the glacier as a kinematic wave. A

Table 8.1 Thickness and response time of different sized glaciers

Glacier	*Thickness (m)*	*Response time (years)*	*Reference*
Blue Glacier	1–200	10	McClung and Armstrong (1993)
Storglaciären	250	50	Holmlund (1988)
Greenland ice sheet	3000	3000	Paterson (1994)

kinematic wave (Lighthill and Whitham, 1955) is an area of faster-moving ice within a glacier. It can be initiated when an increase in accumulation leads to a local thickening of ice and steepening of the glacier surface above the equilibrium line. Ice in the thicker, steeper part of the glacier moves more quickly than ice in thinner, gentler areas upstream and downstream, and the zone of fast flow passes down-glacier at a speed several times faster than the flow of the ice itself. The kinematic wave may thus have a physical surface expression as a topographic bulge, although this is not necessarily always so. The topographic manifestation of the wave may dissipate, and the enhanced flow within the wave be maintained by a change in bed conditions rather than by the surface profile.

Kinematic waves have been related to periodic fluctuations of several glacier margins. Lliboutry (1958) and Reynaud (1977) described kinematic waves on the Mer de Glace, and Finsterwalder (1959) attributed a rapid 400 m advance of the Bossons Glacier, Mont Blanc, between 1952 and 1958, to the passage of a kinematic wave to the snout. He suggested that the advance was caused by an immediately preceding increase in accumulation in the firn field, and that the rapid response of the margin was because a pressure wave, or zone of thicker, faster ice, moved down the glacier at about three times the speed of the glacier itself. As the wave reached the snout, the position of the margin advanced in response to the temporarily increased ice flux. Meier and Johnson (1962) reported that a kinematic wave reaching the snout of Nisqually Glacier, Washington, raised the ice surface by more than 30 m. Sturm *et al.* (1986) described thickening of 70 m associated with a kinematic wave that passed down Drift Glacier, Alaska, when the upper and lower portions of the glacier were recoupled after a period of separation caused by volcanic activity. Nye (1960) pointed out that changes in ice thickness brought about by changes in mass balance could propagate in an unstable manner if they occurred in an area where ice flow was longitudinally compressive, which is generally the case in the ablation area. Nye showed that if an ice surface was thickened by a uniform amount, the increase in ice flux caused by the increase in ice thickness would diminish down-glacier because of the decrease down-glacier in total ice thickness. Therefore ice flux at the lower end of a zone of compressive flow would be less than that at the upper end of the zone, and would be insufficient to discharge all of the ice supplied from the upstream end. In that situation, ice would build up in the area between the two ends of the section, theoretically to form a bulge at the surface, and possibly initiating a kinematic wave. The way in which mass balance perturbations propagate through a glacier and influence the position of the margin has been the subject of a great deal of physically based numerical modelling, which has been conveniently reviewed by Paterson (1994).

Changes in ablation at the glacier margin can have an immediate effect on the position of the margin. This is evidenced by the seasonal advance and retreat that is commonly observed in response to seasonal variations in ablation. Small glaciers that terminate at altitudes relatively close to their equilibrium lines are particularly susceptible to occasional seasons of high snowfall or low ablation when the glacier snout remains snow-covered for a whole year. In such a case the margin will advance by an amount equivalent to the annual movement of the glacier. Non-climatic events can also influence ablation and the position of the glacier margin. For example, superposition of debris from extraglacial sources onto the glacier surface in the ablation zone, blanketing the ice and inhibiting ablation, can trigger terminus advances. Several cases have been documented of mass movements or falls of volcanic tephra protecting glacier surfaces from ablation and preventing normal summer retreat of the margin. Sturm *et al.* (1986) described how the deposition onto the piedmont lobe of Drift Glacier of material from the 1966–68 eruptions of Mt Redoubt, Alaska, reduced ablation by a total of 7×10^7 m^3 of ice. Shreve (1966) described the effects of a landslide onto the Sherman Glacier, Alaska in 1964. 3×10^7 m^3 of rock fell 600 m onto the surface of the glacier, and then slid across the ice to form a debris lobe up to 3 km wide and with a thickness of 3–6 m. Within one year the debris cover had prevented the ablation of several metres of ice. If a landslide occurred in mid-glacier, rather than at the snout, the reduced ablation beneath the debris would lead to a locally thick ice zone that could initiate a kinematic wave. If the wave did not dissipate entirely as it propagated downstream from the debris zone, the snout could experience increased ice flux prior to the arrival of the supraglacial debris. Debris cover also serves to modify the response of the margin to climate-driven changes in mass balance. Recession will be least, and re-advance greatest, when the margin is debris covered, and early signs of re-advance will be easiest to identify at debris-covered margins (Rogerson, 1985).

8.2.3 FLUCTUATIONS CONTROLLED BY ICE DYNAMICS

Former glacier limits are frequently used to reconstruct former climates. However, some fluctuations of glacier margins are caused by ice dynamic characteristics that are independent of climate. It is therefore important to be able to recognise margin positions and fluctuations that are controlled by non-climate factors.

(a) Glacier profile

The positions of glacier margins fluctuate as the area covered by a glacier changes. The area of a glacier is determined by its mass and its surface profile. Therefore, changes in surface profile, as well as changes in mass, can lead to changes in the positions of ice margins. Ice-sheet profile is largely controlled by ice-sheet dynamics (see Chapter 5). For a given mass of ice, specific conditions lead to different flow states and hence to different profiles. For example: cold ice flows less readily than warm ice; ice on a frozen bed slides less readily than ice on a warm bed; and a frozen substrate is less likely to deform than a thawed substrate. A glacier in which flow is restricted by any of these conditions is likely to have a steeper profile, a greater thickness, a lower area/volume ratio, and hence less extended margins, than a glacier where the ice is warm and the bed lubricated or mobile. Therefore, changes in the parameters that control ice flow can lead to changes in ice extent and hence to fluctuations of glacier margins.

Many glaciers are characterised by periodic changes in flow state, most clearly so glaciers that surge. Surging glaciers can achieve spectacular advances, and flattening of their profile, in very short times as a result of periodic flow enhancement. The low profile and periodic extension of margins of the Laurentide ice sheet has also been attributed to changes in flow state through time. A periodically deforming subglacial sediment layer or periodic floods of subglacial water could lead to low basal yield stress, rapid flow, profile lowering and marginal extension. Till sheets deposited by the lobes of the Laurentide ice sheet indicate periodic advances of ice lasting between 100 and 1000 years, each contributing to a sequence of till layers. Several models (e.g. Boulton and Jones, 1979; Beget, 1986; Alley, 1991) have associated the Laurentide lobes with deforming bed conditions, and Kamb (1991) explained how deforming bed flow would be expected to be non-linear, or unstable. Ice flow driven by deforming bed conditions is therefore likely to be episodic. This would account for the periodicity indicated by the stratigraphy of the till sheets, and implies periodicity in margin fluctuations. Further details of deforming bed motion are discussed in Chapter 7. An alternative model, put forward by Shoemaker (1992), proposed that periodic flow enhancement and surge-like advance of the Laurentide margin was driven by subglacial floods. He envisaged the ice sheet being floated forwards during each flood and returning to a dormant state between floods. Surges might also affect the Antarctic ice sheet (e.g. Wilson, 1964; Hollin, 1980), causing major changes in ice extent on very long periodic cycles (Chapter 7).

At the scale of large ice sheets, the ways in which ice sheets affect their environments could lead to intrinsic instabilities in ice-sheet development. For example, the influence of a very large ice sheet in reducing regional precipitation patterns could lead to a negative feedback on ice-sheet growth. Several alternative model hypotheses have been proposed for long-term ice-sheet dynamics, but our understanding remains insufficient to allow confident predictions for current ice sheets.

(b) Ice-divide migrations

In ice sheets, ice flows outwards from centres of accumulation in response to a stress gradient imposed primarily by the surface profile of the ice sheet. The ice flows radially outwards from surface domes and laterally away from surface ridges. Ice-sheet summits thus form ice divides that delimit ice-sheet drainage areas in the same way that watersheds delimit drainage basins in terrestrial fluvial systems. The positions of ice divides do not necessarily remain constant through time, and migration of ice divides can involve changes in the surface profile, in accumulation area ratio, and in the catchment area of individual ice-sheet outlets. These changes in turn can lead to the advance or retreat of individual sections of the ice-sheet margin. The surface profile of an ice sheet, and hence the position of the ice divide, will change in response to changes in accumulation patterns. If one side of an ice sheet has a higher accumulation rate than the other, the ice divide will be located closer to the side with the higher accumulation, and so that side of the ice sheet will have a shorter width. Weertman (1973a, 1983) showed that if the margins of the ice sheet are fixed (for example, by a calving coastline) then the change in position of the drainage divide will be an order of magnitude smaller than the change in the geography of accumulation. However, for ice sheets where the position of the margin is flexible, the ice divide can migrate further and can drive changes in the position of the margin.

Geomorphological evidence from several locations has illustrated that significant changes in ice-divide location have been accompanied in the past by reorganisations of ice-flow patterns and margin positions of ice sheets. Dugmore (1989) and Dugmore and Sugden (1991) described the Holocene history of Myrdalsjökull ice cap in southern Iceland. They showed that, as the ice cap grew, local precipitation patterns changed and the accumulation centre shifted towards the southern flank of the ice cap. As the ice divide migrated, the size of the ice-catchment areas of the different flanks of the ice cap changed. Outlet glaciers to the south, including the glacier Solheimajökull, experienced shrinking catchment areas as the ice divide moved south, and outlets to the north experienced an enlargement of their catchment areas. At other times, when the ice cap diminished in size, the reverse occurred, and Solheimajökull experienced an enlargement in its catchment area while northern outlets experienced catchment shrinkage. A consequence of this was that Solheimajökull's fluctuations were anomalous with respect to the ice cap as a whole and to most of its other outlet glaciers. When the ice cap as a whole grew, Solheimajökull's catchment area shrank and its margins retreated or experienced only limited advance compared with other outlet glaciers. When the ice cap as a whole diminished in size, the catchment area of Solheimajökull grew, and its margins advanced, while other outlets experienced retreat. Solheimajökull reached its maximum Holocene extent at a different time from the rest of its ice cap, and Solheimajökull is at present advancing while the ice cap as a whole, and most of its outlets, are in retreat.

Evidence has also been found to suggest that the Laurentide ice sheet experienced ice-divide migration during the course of its existence. Former patterns of ice flow have been interpreted from the distribution of erratics (Prest, 1990) and from the orientation of sediment ridges that formed parallel to the contemporary ice-flow direction (Boulton and Clark, 1990). Boulton and Clark used satellite images to map geomorphological lineations over most of mainland Canada, and reconstructed ice-flow patterns from different times. They showed that the ice sheet was characterised by changes of as much as 2000 km in the positions of centres of mass and flow. These shifts were caused by changes in the distribution of accumulation and ablation that occurred in response to ice-sheet growth.

Both the Myrdalsjökull and Laurentide examples illustrate the importance of coupling between ice sheet and atmosphere. Changes in ice-sheet size affect airflow and precipitation patterns which in turn affect ice-sheet morphology. It is also clear that changes in mass or mass distribution can affect individual sections of a glacier (such as individual outlets from an ice cap) differently. In circumstances such as those where ice-divide migration affects different sections of an ice sheet in different ways, glacier margin fluctuations are clearly not related directly to climate fluctuations alone.

(c) Topographic control, and fluctuations of floating glaciers

A key element in the way that mass balance controls the position of a glacier margin is the rate of increase of ablation towards the snout. For example, for an ice sheet bounded by the edge of a continent, the sudden increase in ablation as ice reaches the sea effectively constrains the maximum extent of the ice to the size of the continent. Any increase in mass that would tend towards expansion beyond the coastline results in an abrupt increase in ablation (by calving) to compensate. For a steep mountain glacier, the rapid increase in ablation with distance down-glacier as a result of change in climate with altitude plays the same role: small changes in the position of the margin accomplish substantial changes in total ablation, and hence offset substantial changes in mass throughput to the snout. By contrast, the margin of a glacier extending onto a flat plain can increase ablation only by increasing its area of ablation. If ablation per unit area is low in the local environment, then a large increase in area is needed to evacuate the excess input caused by a small increase in accumulation, and conversely a slight reduction of input will lead to a substantial retreat of the margin to reduce the ablation area. A small change in mass balance can lead to big response at snout. If no ablation occurs at the altitude of the snout, the theoretical plane-bedded glacier subject to an increase in mass flux will extend indefinitely unless the consequences of changing latitude take effect.

Floating glaciers face a similar difficulty in adjusting their output in response to material provision at margin. They cannot move downhill, and large changes in surface area might be required to make any difference to total ablation. A floating glacier can change its ablation rate most effectively by changing the amount of calving that takes place from its terminus.

Calving is controlled partly by water depth and the length of the calving front (see Chapter 3) and so, for a given ice velocity, the ability to change the calving discharge is controlled by the width and depth of the waterway that the floating tongue or shelf occupies (Mercer, 1961; Meier and Post, 1987). This leads to special conditions of advance and retreat. An increase in mass flux will cause the terminus to advance until it reaches a position where calving increases sufficiently to discharge the extra ice. This will be a position where the fjord increases in width or in depth, so that the calving front expands and the calving rate increases. If a widening or deepening of the fjord does not increase calving enough to accommodate all of the increased flux, the terminus will continue to advance until it reaches a position where the front can expand further, even if this means advancing all the way to the open sea where the ice front can expand unrestricted to a width and depth at which all the ice supplied to the snout can calve. In the reverse situation, a decrease in mass supply will cause the glacier to retreat along the fjord until it reaches a narrower section where the amount of calving is reduced to match the ice supply. In this situation, several important points arise. First, the amount of ice advance or retreat is not related directly to the size of the mass balance perturbation causing the fluctuation. Rather, it is controlled by the geography of the fjord, and will not necessarily be synchronous or of similar scale to contemporary fluctuations of land-based glaciers in the same area. Second, the terminus is likely to become stable only at a limited number of locations, referred to as pinning points, where the width of the fjord changes. It is unlikely to remain stable in the middle of a straight section of fjord.

The fluctuation of floating margins is thus controlled not only by regional mass balance but also by variations in calving dynamics controlled by the topography and bathymetry of the terminal position. Warren (1991) studied 72 land-terminating, tidewater and lake-calving glaciers in west Greenland, and found that the different types of margin responded in very different ways to regional climate forcing. The land-terminating glaciers were controlled dominantly by variations in summer temperature, and 84% of them were stable or in retreat. The tidewater and lake-calving glaciers showed a much more varied pattern, and demonstrated a non-linear response to climate forcing because of the influence of calving dynamics on terminus behaviour. Warren found that topographic pinning points were critical in determining the locations of terminus still-stands, which almost invariably occurred at pinning points irrespective of climatic change or regional trends. Likewise, Motyka and Begét (1996) found that the tidewater Taku Glacier in south-east Alaska sometimes fluctuated synchronously with land-based glaciers in the area and sometimes asynchronously, indicating that individual floating glacier tongues respond to complex interaction of climate-related regional mass balance controls and local calving-dynamics controls. Floating glacier tongues are thus unlikely to be reliable indicators of climatic trends, and moraines or trimlines left by former floating ice tongues should be treated with care in the reconstruction of former climate histories (Mann, 1986).

Although fluctuations of floating glaciers might be climatically initiated, the mechanisms controlling terminus position involve dynamic processes quite independent of climate. Work by Hoppe (1959) and Meier and Post (1987), among others, led to the theory of a tidewater glacier cycle. This cycle involves three stages:

1. A period of slow advance (20–40 m a^{-1}) during which the glacier moves down the fjord through deep water, maintaining a submarine moraine shoal in front of the terminus. This shoal is moved ahead of the terminus by erosion on the proximal side and deposition on the distal side, and greatly reduces calving, hence maintaining the glacier in positive mass balance.

2. A period of relative stability, with accumulation and ablation in approximate balance, with calving suppressed as terminus rests on its moraine shoal.
3. A period of catastrophic retreat (up to 2000 m a^{-1}) begins when the terminus retreats off submarine moraine shoal and finds itself terminating in the deep water behind the moraine. Calving rate suddenly increases and the glacier retreats until the terminus gets back to an area of shallow water, or to the head of the fjord, at which point it becomes grounded and returns to stage 1.

The cycle may be climatically initiated, but at any given moment the position and direction of advance/retreat are controlled by the calving flux, which is determined by fjord morphology. Sturm *et al.* (1991) suggested that this cycle could explain pronounced differences in the observed behaviour of the adjacent Yale and Harvard Glaciers at the head of College Fjord, Alaska. The two glaciers occupy parallel arms of the same fjord, and descend from the same source glacier, but while Yale Glacier has retreated throughout the present century, Harvard Glacier has steadily advanced. Clearly they are not responding to any shared climatic control. It is suggested that Yale Glacier is near the completion of catastrophic retreat phase 3, while Harvard Glacier is in advancing phase 1.

Cycles of advance and retreat might also affect ice-shelf limits, as the calving of sections of ice shelves seems to occur at discrete intervals, rather than as a steady flux each calving event is associated with a retreat of the shelf margin and followed by a period of recovery or advance. Jacobs *et al.* (1986) described more or less uninterrupted advance of sections of the margin of the Ross ice shelf for periods of up to 75 years, implying long intervals between major calving events. Ice shelves respond not only to melting and calving dynamics but also to changes in input from the ice streams that feed them. Changes in ice-shelf thickness caused by changes in ice-stream discharge may cause localised grounding of the shelf, the formation of ice rises, and hence negative feedback to shelf advance. Thickness changes propagate by horizontal advection and do so more slowly than velocity changes, which are rapid (MacAyeal and Barcilon, 1988; MacAyeal and Lange, 1988).

Changes in sea level play an important role in fluctuations of floating glaciers and glaciers that terminate on land close to sea level. Sea-level change is involved in a positive feedback with growing ice sheets, such that growing ice sheets associated with falling sea-levels are provided with extensive coastal areas for expansion before reaching the limiting ocean. Retreating ice sheets associated with rising sea-level are attacked by the encroaching water-line. Change in sea level can lead to a change in the position of the grounding line, and also to changes in the width of a fjord at the water-line, both of which influence calving.

Not only floating glaciers are subject to topographic control of fluctuations. Topographic constraints can also act as pinning points for land-based glaciers. Topographic barriers can pose physical obstacles to ice advance that require ice to thicken before it can make further progress. For example, Burbank and Fort (1985) illustrated the importance of bedrock topography on glacier limits. They found that observed differences of around 400 m between the elevations of the terminal positions of late Pleistocene maximum advances of glaciers in the adjacent Ladakh and Zanskar ranges in the Himalaya were caused not by a climatic difference between the two mountain ranges but by individual valley geometries. Constricted valley morphology prevented glaciers in the Zanskar range from advancing into the lower parts of their drainages. For these areas, any climate reconstruction based on the altitudes of glacier limits would incorrectly identify strong climatic gradients in the region, and give a false impression of mass balance parameters. Topographic control on glacier fluctuations has been recognised from a wide range of different locations (e.g. Sugden, 1991; Warren,

1991; Greene, 1992; Kirkbride 1993). With reference to the interpretation of hummocky moraines from Scotland, Bennett and Boulton (1993a,b) concluded that topography was the principal variable controlling the pattern and organisation of ice decay.

8.3 STYLES OF ADVANCE AND RETREAT

The concept of glacier fluctuations implies a dynamic ice margin – sometimes moving forward, sometimes retreating. Different rates and styles of these migrations have implications for landform creation, and for our interpretation of environmental change. Advance or retreat may be either steady or episodic. In extreme cases it may be unstable or catastrophic, like a surge advance or the retreat of a floating ice tongue along a fjord. Different styles of retreat are especially interesting to glacial geologists and geomorphologists as most of glacial geology is related to deglaciation. It is also relevant to contemporary glaciology as the vast majority of the world's glaciers are presently in a state of retreat from Little Ice Age maxima. Retreat may occur by either backwasting or downwasting. In backwasting, the margin occupies a sequence of increasingly withdrawn positions. In downwasting, a large area of the glacier thins to nothing while the extremity remains in roughly the same place, so that there are no intermediate margin positions to mark a recession. Downwasting happens commonly in the lobes of surging glaciers when dead ice wastes away in stagnation. Similar effects may occur in ice-sheet stagnation. If the ice-sheet profile is relatively gentle, as was the case, for example, for many lobes of the Laurentide ice sheet, then thinning can lead to the rapid disappearance of the ice without protracted recession of the margin. By contrast, a steeply sloping ice margin will, upon thinning, withdraw progressively (Figure 8.3). Clague and Evans (1993) described recent rapid retreat of glaciers in the Saint Elias Mountains in Canada, and suggested that the style of retreat there might serve as an analogue for the decay of the Cordilleran ice sheet in the

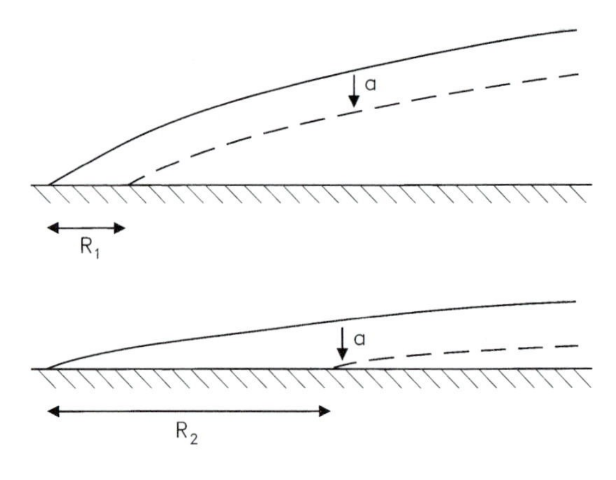

Figure 8.3 The effect of ice thinning on glacier margin position at steep and shallow-gradient glaciers.

Rocky Mountains at the end of the Pleistocene. Over a period of a few hundred years the Grand Pacific Glacier and the Melberne Glacier have lost 250–300 km^3 of ice, more than 50% of their volume. The land-terminating Melberne Glacier thinned by 300–600 m and retreated 15 km. The tidewater Grand Pacific Glacier, which flows through Tarr Inlet into Glacier Bay, Alaska, retreated 24 km between 1879 and 1912. Since 1794, the whole of Glacier Bay has become deglaciated as the floating lobe that filled it has disintegrated and retreated into its tributary fjords (Figure 8.4). In a similar fashion, about 14,000 years ago, tongues of ice rapidly retreated across the British Columbia continental shelf to pinning points in fjords where they temporarily stabilised (Clague, 1985). A traditional view of ice-sheet retreat involved active deglaciation (Charlesworth, 1955) in which ice fields withdrew progressively into valley glaciers and finally into corrie glaciers, all the time maintaining an active flow state, reversing the assumed mode of ice-sheet growth. This model has been challenged repeatedly by the concept of passive stagnation. For example, Bennett and Boulton (1993a) found that the active deglaciation model did not apply to the deglaciation of the Northwest Highlands of Scotland, where corries were unimportant as decay centres, and where ice lingered longest

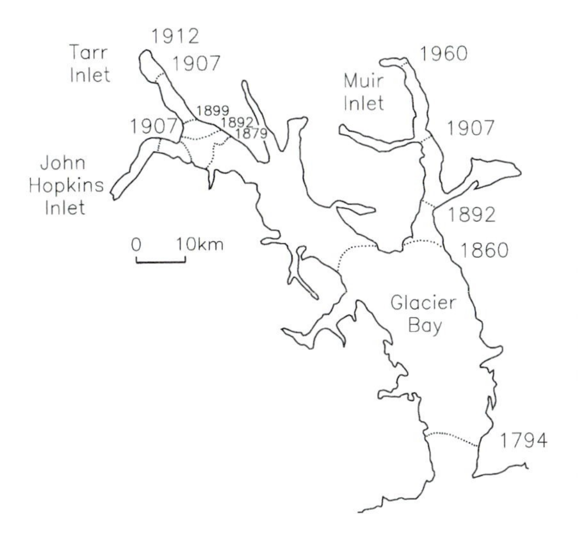

Figure 8.4 Recent retreat history of Glacier Bay, Alaska.

on the northern slopes of east–west ridges. The rate of deglaciation is likely to vary as the process advances. Early on, local microclimate generated by the presence of ice is likely to sustain glaciation in the face of more widespread warming. As the ice mass decreases, the impact of the ice on local microclimate will decrease, and the rate of deglaciation is likely to accelerate progressively.

8.4 CONCLUSION

Glacier fluctuations occur primarily in response to changes in mass balance. The principal driving forces are thus usually climatic, but ice-dynamic and geographical controls strongly influence the observed terminus response to environmental forcing. Responses may be delayed, and may be non-linear with respect to the magnitude of driving forces. In addition, mass balance at the terminus is affected by non-climatic factors such as calving dynamics and ice-divide migration. The positions of glacier margins, and the record of former margin positions, thus reflect a composite effect of several controlling parameters, and the use of former margin positions to reconstruct climate change is fraught with difficulty. Likewise, contemporary advances and retreats of glacier margins cannot always be interpreted as simple indicators of climate change. Especially for large ice sheets and floating glacier tongues, margin fluctuations observed at the present time may have little to do with contemporary climate change. A major consequence of glacier fluctuations is the repeated glaciation and deglaciation of areas of the Earth's surface, and, hence, the origin of glacial geomorphology.

GLACIAL SEDIMENT TRANSFER AND GEOMORPHOLOGY

9

Have you reckon'd that the landscape took substance and form that it might be painted in a picture?

Walt Whitman: *Leaves of Grass.*

9.1 GLACIERS AND GEOMORPHOLOGY

Glacial geomorphology is concerned with the landforms and landforming processes associated with glaciers. It is a broad subject that draws upon aspects of many other disciplines including hydrology, climatology, and geology. This book does not attempt to cover the whole field. However, most of geomorphology can be reduced conceptually to a single issue: the movement of material from one location to another. Glaciers erode bedrock to produce rock debris; they erode and entrain previously existing sediments; they transport material from place to place; and they deposit or release sediment either directly onto the ground, into water, or into the atmosphere. The form of the ground surface is altered both where material is removed and where it is deposited. Glacial landforms thus reflect both erosion and deposition, and glacial sediments reflect the history of the processes by which they were entrained, transported and deposited. Glacial geomorphology is important, therefore, not only in understanding the landforms and landscapes of glaciated environments, but also as a tool for finding out about glacial processes and for reconstructing the characteristics of former glaciers. Shaw (1994) and Kleman *et al.* (1997) provided examples of the symbiotic relationship between geomorphological and glaciological modelling. Shaw showed that models of landscape evolution could be based on conclusions drawn from an understanding of glacial sediments, forms and processes, and Kleman *et al.* showed how geomorphic evidence could be used to reconstruct the evolution of ice-sheet configuration through a glacial cycle. Geomorphology also exerts controls on glacier behaviour. Glacier dynamics and morphology are in many cases largely controlled by their surrounding topography, and rock debris entrained into a glacier affects both the way in which the glacier moves and the geomorphic impact of the glacier on its bed. Furthermore, the geographical distribution of glacial landforms and sediments provides a valuable tool for constraining and testing ice-sheet models. To the glaciologist, geomorphology is much more than the explanation of landforms.

9.2 GLACIAL SEDIMENT PRODUCTION AND TRANSFER

A glacier is a transfer system for ice and water within the hydrological cycle, and also for rock material, or sediment. In the same way that glaciers have an ice mass balance, they have a sediment balance, with inputs, storage, transport and outputs. Glaciers are involved in the production, transport and release of sediment. In geomorphological terms the key processes are erosion, entrainment, transportation and deposition, each of which is discussed in the following sections.

9.2.1 EROSION

Erosion is the removal of material by some moving agent such as wind, water or ice. In glacial geomorphology, erosion occurs at the bed of the glacier, and its principal agents are ice (glacial erosion) and water (glacifluvial erosion). Erosion can involve several different sets of processes: abrasion by rock fragments embedded in, or moved along by, moving ice; crushing or spalling of rock fragments from a rock bed; traction of material from the bed that has become attached to the base of the ice; erosion by mobile till in a deforming subglacial layer; erosion by flowing meltwater; and dissolution into meltwater. Each type creates characteristic landforms that can subsequently serve as evidence of the former subglacial conditions.

(a) Abrasion

Abrasion involves the scratching and polishing of bedrock by rock particles being dragged across it by the glacier. The essentials of glacial abrasion were appreciated by early glaciologists, including Agassiz (1840), who wrote:

> "The lower surface of the ice itself, although smooth and glassy like an artificially polished piece of ice, is generally studded with small sand grains or minute rock fragments which give it a variable degree of roughness. The surface appears similar to a kind of file and may be compared to a sheet of wax which has been strongly pressed against a gravel surface . . . The polish, the striations and the different kinds of grooves which occur over the substratum of all glaciers result from the contact of such a surface with the bedrock, under the action of the movement of the glacier."

The rate of abrasion is controlled primarily by the following factors:

1. The rate at which particles are dragged across the bed. This is function of sliding velocity, which is in turn largely a function of basal thermal regime. Even if sliding does not occur, rock fragments in the basal layer that penetrate upwards a few centimetres into ice that is moving by internal creep can be dragged through the frozen bottom ice to abrade the bed.
2. The concentration of particles at the bed. The more rock particles that are being dragged along, the more scratching and polishing there will be. Debris-rich basal ice is a much more effective abrasive tool than a clean basal layer. However, very high debris concentrations might tend to reduce sliding velocity by increasing friction (Chapter 7), which would have a negative effect on abrasion.
3. The rate of supply of rock fragments to the bed. The contact points of the abrading particles against the bed become worn, so the particles need to be lowered constantly through the ice towards the bed for contact to be maintained and for abrasion to continue. High rates of basal melting, or a vertical strain component derived, for example, from longitudinal extending flow, both lead to downwards trajectories for particles in the basal ice. Rotation of particles during flow because of friction against the bed at their lower edge also helps to bring fresh asperities into contact with the bed. Fresh particles to act as abrading tools can also be supplied by erosion of the bed.
4. The contact force between the particles and the bed. The amount of wear depends partly on how hard the particles are pressed against the bed as they move. Boulton (1974) argued that the contact pressure was determined by the effective normal pressure at the bed (overburden pressure minus basal water pressure). He concluded that abrasion would increase with pressure up to a certain point beyond which the pressure would cause so much friction between particles and bed that the particles would stop moving, and would be lodged at the bottom of the glacier (Figure 9.1). Hallet (1979) took the alternative view that the

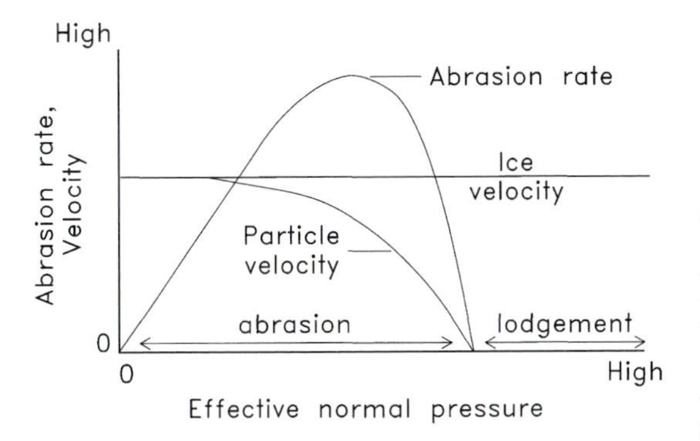

Figure 9.1 Boulton's (1974) theoretical graph of the relationship between abrasion and lodgement. The abrasion rate initially increases with pressure, but then decreases after pressure passes a critical value. The critical pressure depends on the velocity, and the peak abrasion rate depends on the velocity and hardness of the particles.

contact force was independent of the effective normal pressure, because the particles were suspended in the ice. It is a situation analogous to that of a deep-sea diver being flattened against the sea-floor by the weight of water above him (Boulton's model) or being squeezed from all sides and supported by the water pressure (Hallet's model). According to Hallet, the contact force between the particle and the bed is determined by the buoyant weight of the particle combined with the rate of its downward movement. Iverson (1990) conducted laboratory experiments that confirmed Hallet's theory that the velocity component towards the bed strongly influenced the stress between rock fragments and the bed. The Boulton model is probably appropriate where the concentration of rock particles is very high and some particles are in contact with one another so that some particles are pressed down onto the bed by particles above them and are not entirely engulfed in ice. This situation might apply beneath a thin layer of subglacial sediment or beneath a very debris-rich basal ice layer. The Hallet model probably applies where rock particles are more widely dispersed, there is no inter-particle contact, and each particle is surrounded and supported by ice.

Abrasion rates have been measured both in actual subglacial environments (e.g. Boulton, 1974) and in laboratory simulations (e.g. Iverson, 1990), and have also been calculated theoretically (e.g. Lliboutry, 1994). On the basis of field observations, Boulton (1974) reported values ranging from 0.9–36 mm a^{-1}, varying with ice thickness, ice velocity and the lithology of the affected surface. Different particle sizes and concentrations have different abrasive effects. Individual clasts tend to produce grooves (cross-sectional area > 10 mm^2) or striae (cross-sectional area < 10 mm^2) The creation of these features results in the production of fine grained 'rock-flour' and when this flour acts as an abrasive agent it causes polishing of the bed. Boulton (1974) observed that individual striations gradually decreased in depth along their length and attributed this to the build-up of crushed debris between the ice and the rock around the striating clast. Rea (1996) observed polishing of rock surfaces within striations by comminution debris derived from the striating clast. Periodic flushing by water or gradual comminution of the debris could reactivate the grooving process. Practitioners of car bodywork or domestic joinery will be familiar with the problem of an abrasive medium clogging up with abraded dust.

(b) Fracture and traction

Erosion by glacial traction includes a range of processes that have frequently been considered in previous literature under the headings of plucking, quarrying, crushing and joint-block removal. When moving ice becomes attached to material on the glacier bed, if the tractive force exerted by the moving ice is greater than the force holding the material in place at the bed, the material will be entrained and removed (eroded) by the ice. Key issues are: the adhesive and tractive force between the ice and the rock; and the loosening of the rock from the bed. Traditional theory considered both issues together in terms of a bond of frozen water sticking the rock to the ice, and the ice plucking the rock from the bed. However, the tractive force exerted by the ice would usually be insufficient to break solid bedrock; the ice/rock bond would fail before the rock mass failed. A body of theories has thus grown up to explain the weakening of the bedrock that enables the traction to be effective. Rock is predisposed to fracture along existing joints or bedding planes. These may be weakened further by freeze–thaw cryofraction beneath temperate glaciers. Most rock is also afflicted by many incipient weaknesses in its microstructure that might be susceptible to fracture propagation, and it has been suggested that a contribution to loosening bedrock may come from the crushing effect of ice passing over the rock. However, Morland and Morris (1977) showed that the maximum stress generated by ice flow over a bed protuberance would be insufficient to generate fracture in a coherent rock mass. Numerical simulation by Iverson (1991) confirmed this view, but also showed that rock was much more likely to fracture in situations where basal water pressures fluctuated. For example, periodic changes in water pressure in a cavity in the lee of bedrock step would favour the development of vertical fractures and retreat of the cavity headwall in response to stress fields generated in the rock. This effect would be especially pronounced in impermeable rock, where water pressure could remain high in cracks while decreasing rapidly in cavities. Röthlisberger and Iken (1981) suggested that water pressure fluctuations were important also because they could cause hydraulic jacking to force ice away from adhesive contact with the lee faces of bed protuberances. Small-scale fracturing of a rock bed can also be caused by the passage of clasts in the basal ice across the bed to produce friction cracks and gouges. Abrasion itself is no more than a small-scale fracture process involving large stress differences localised beneath the abrading tool.

(c) Deforming layer

Erosion by a deforming layer of subglacial till can take several forms, including cannibalism of till, till erosion of rock substrate, and comminution within deforming till. The very process of till mobilisation beneath a glacier is a form of erosion, as the till is removed from its initial resting place and put into transport towards the glacier margin. For till deformation to continue indefinitely in a steady state, without the eventual exhaustion of the till layer, new till must be supplied at the bed of the glacier to replace that which is being removed. At the down-glacier end of a deforming bed, replacement material is supplied initially from upstream, but without production of new till at the upstream end the head of the zone of mobile till would be expected to migrate down-glacier until the whole glacier rested on a till-free bed. New material could be produced by meltout of debris from overlying basal glacier ice or by the erosive action of the mobile till against the material underneath it. Erosion is probable wherever a mobile subglacial till layer extends all the way down to a sediment/bedrock interface, because of abrasion of the underlying material by the till moving across it. Cuffey and Alley (1996) concluded that accurate theoretical prediction of erosion rates by deforming till were not yet possible, but that sufficient erosion to supply the sediment to replace the flux through a deforming layer was unlikely unless the substrate beneath the deforming layer was very soft. Unconsolidated sediments might provide a rich source of material to re-supply a thinning layer of deforming till. The depth to which a deforming layer penetrates is largely a function of the shear stress at the glacier bed. If the deforming layer penetrates only a short way into a subglacial till body, then till lost to the deformation flux could be replaced by the penetration of the bottom of the deforming layer lower and lower into the subglacial material. The pattern of this lowering is unlikely to be uniform, and Boyce and Eyles (1991) have associated drumlins in Canada with erosion by streams of deforming till which cannibalise the overridden sediments. Hart (1997) identified a continuum of drumlin forms associated with deforming bed processes, and suggested that under conditions of net till erosion, where more till is removed in the deforming layer than can be supplied to it, any obstacle to till flow might be left behind as a drumlin. Erosion also occurs of material within a deforming layer, as a result of particle interaction. Changes in sediment characteristics downstream within a till can therefore give an indication of travel distance and flow characteristics within the till.

(d) Meltwater

Meltwater erosion takes several forms, both mechanical and chemical. In most of geomorphology, chemical processes are usually considered under the heading of weathering, but dissolution of mineral material into water beneath a glacier can also be considered as an erosion process, because material is removed from the site of the reaction by the agent (the water) effecting the reaction. Rates of chemical erosion depend on the subglacial lithology and the characteristics of the glacial water system. Chemical denudation beneath glaciers is more effective than in many other subaerial environments for a number of reasons. First, the solubility of carbon dioxide in water increases at lower temperatures, so glacial meltwater can become more acidic than warmer waters usually do. Second, subglacial environments are usually rich in freshly ground rock fragments that are chemically reactive and also have a high ratio of surface area to volume. Third, many glaciers have rapid flushing rates or throughput rates for water, which means that water in the system does not become chemically saturated, and the hydrological system in the glacier remains chemically reactive. Sharp *et al.* (1995) found that the dissolved load was less than 1.5% of the total load in water draining from Haut Glacier d'Arolla, Switzerland, but that the corresponding rate of chemical denudation was significantly higher than the continental average. Modern glacial environments are characterised by chemical erosion rates and solute yields an order of magnitude greater than the global average. Proglacial environments around the margins of warm-based and melting glaciers present a potentially very active chemical weathering environment, with abundant comminuted rock fragments, copious water and an atmospheric supply of CO_2 (Collins *et al.*, 1996; Gibbs and Kump, 1996; Sharp, 1996).

Chemical weathering by glacier waters, and the relationship between glaciation and weathering, has been an area of growing interest in recent years. There is an important link between glaciation, weathering processes and atmospheric composition. It has been suggested that glacially driven chemical weathering could be a significant factor in carbon cycling and climate change on glacial/interglacial time scales (e.g. Sharp *et al.*, 1995). Tranter (1996) considered the role of glacial meltwater as a sink for atmospheric CO_2 during glacial/interglacial transition, and found that chemical weathering processes associated with glacial runoff at the time of maximum runoff during deglaciation about 10,000 years ago could have made a significant contribution to the observed ~50 ppm decrease in atmospheric CO_2 at that time. Chemical processes in meltwater are considered further in Chapter 6. Mechanical processes of meltwater erosion are broadly the same as fluvial processes in other environments except that the water may flow under pressures higher than atmospheric pressure. The principal processes involved are fluvial abrasion and cavitation.

(e) Erosion by surging glaciers

The foregoing discussion highlights several controlling factors in glacial erosion, such as ice velocity, debris content of basal ice, and basal water pressure, which can vary between locations and between types of glacier. For example, several observers have noted the extreme erosional potential of surging glaciers. Humphrey and Raymond (1994) noted that the erosion rate during the surge of Variegated Glacier, Alaska, was extremely high in comparison with non-surging glaciers, and directly proportional to the high sliding velocity during the surge. The sediment output of the glacier was directly proportional to the basal sliding rate. A dimensionless erosion rate (metres eroded from the bed divided by metres of sliding) for Variegated Glacier is of the order of 1×10^{-4}. Total erosion during the 20-year surge cycle was of the order of 0.3 m bedrock, with two-thirds of this occurring during the 2 years of the surge peak, and the bulk of that during the peak 2 months. High sediment concentrations in meltwater draining the glacier 1 year after the surge implied incomplete flushing of basally produced sediment by the surge de-watering event, and a sediment 'memory' in the system. Some of this delayed release of surge-produced sediment could be due to its entrainment into the basal ice and its gradual release by meltout. Sharp *et al.* (1994) identified different basal ice types formed in surging and quiescent phases of flow at Variegated Glacier. Basal ice that was formed during the surge phase reflected enhanced rock-fracturing processes and efficient flushing of fine comminution products to create a dominantly coarse particle size in the basal debris. Basal ice that was formed during the quiescent phase reflected less extreme fracture processes, less efficient flushing, and hence a more mixed particle size distribution including more fines and less coarse material.

9.2.2 ENTRAINMENT

(a) Erosion and entrainment

Erosion and entrainment are separate processes, but are sometimes confused. Entrainment is the incorporation of material into or onto the glacier. Sometimes erosion and entrainment occur together. For example, if a rock is engulfed by ice, plucked from its resting place, and carried away as part of the basal layer of the glacier, then both erosion and entrainment have occurred. Sometimes erosion occurs without entrainment, such as when the products of abrasion are flushed away by basal water and not incorporated into the basal ice. Sediment is input to a glacier by entrainment from supraglacial and subglacial sources. Supraglacial sources include: material falling or being washed or blown onto the glacier from surrounding land; atmospheric fallout such as volcanic ash; extraterrestrial materials such as micro-meteorites; and particles carried down onto the ice surface in snow or rain. Subglacial sources include eroded material from the glacier bed or valley walls, and, in the case of floating glaciers, material frozen onto the base from the underlying water.

Both supraglacially and subglacially derived debris can be transported into the body of the glacier. Supraglacial material in the accumulation zone is buried by subsequent accumulation and follows an englacial transport path through the glacier. Material entrained near the top of the accumulation zone follows a very low transport path and may be carried right to the bed of the glacier. Material entrained closer to the equilibrium line follows a higher route and emerges onto the surface in the ablation zone. Material entrained at the bed can be raised into the ice to form a basal layer (Chapter 5). The processes by which basal material is entrained and raised into the basal ice are glaciologically very important, partly because the basal ice layer itself is so important, and partly because the entrainment processes reveal a range of mechanisms operating at the glacier bed. The main entrainment mechanisms, involving both attachment of material to the sole and the vertical transport of material from the sole into the basal ice, are discussed below.

(b) Regelation

Localised pressure-melting and refreezing of basal ice around bedrock protrusions (discussed in Chapter 7) is a widely cited mechanism for entrainment of debris (e.g. Kamb and LaChapelle, 1964). Its impact on the basal layer has been reviewed by Hubbard (1991) and Hubbard and Sharp (1993). It is suggested that meltwater is generated on the upstream side of bed obstacles and refreezes on the downstream side. Gas and fine debris are expelled from the ice during the initial stages of refreezing, but trapped in the final stages, generating laminations of clear and debris-rich ice. Because the regelation layer is recycled at each new bed obstacle, then, ignoring deformation effects, the thickness of the layer is limited to approximately the size of the largest obstacles.

A variation on the regelation process has been discussed by Iverson (1993) and Iverson and Semmens (1995). They demonstrated that basal ice can move downwards into a granular substrate by means of vertical regelation. Ice melts above individual grains and refreezes below them at speeds proportional to the pressure gradient downwards across the particles. Ice should intrude downwards to a depth where basal melting matches the rate of regelation, and could generate more than a metre of debris-bearing basal ice.

(c) Water flow through the vein system

Lliboutry (1986) suggested that water formed by pressure-melting at the bed might not flow along the ice/bed interface, but be squeezed away from the bed along a pressure gradient into the ice via the vein network between ice crystals. Lliboutry (1993) predicted that when ice flow involves regelation at the bed, water will be mobile through the vein network in the lowest layers of ice. Knight and Knight (1994) demonstrated that such a flow could be accompanied by transport of sediment from the bed into the basal ice. They suggested that this layer could equate to the dispersed ice facies, with silt-sized debris in the dispersed facies deriving from the vein flow of water and sediment.

Another mechanism for distributing particles through the ice matrix was suggested by Boulton (1967) who proposed that solid particles in ice could be dispersed by shear deformation within the ice. Theoretically, this might result in material from the base being raised into the body of the ice. However, Weertman (1968) modelled the process and found that diffusion would operate only over a distance about one order of magnitude greater than the size of the particles themselves. The mechanism could therefore not account for more than a few centimetres of silty ice. It has yet to be demonstrated that a shear dispersal model might not operate more efficiently if flow of liquid as a 'transmission fluid' through the vein network was incorporated into the theory.

(d) Bulk freezing-on

The freezing to the glacier sole of basal water not derived from localised pressure-melting is sometimes referred to as congelation. It commonly occurs at the boundary between warm and cold basal zones, or where supraglacially derived water penetrates to a cold portion of the glacier bed. Water reaching the bed from extraglacial lakes or streams can import substantial amounts of sediment to the bed, but freezing often results in the expulsion of suspended sediment from the water and the formation of clean ice. The amount of debris likely to be entrained with the congelation of basal water depends on the character of the bed and the sediment load of the water. Where basal water flows through a permeable substrate, debris entrainment is likely to be substantial compared with a case in which water is confined between the ice and an impermeable substrate. Weertman (1961) described entrainment of clean ice occurring when the freezing isotherm was at the bed, and entrainment of debris occurring when the freezing isotherm passed downwards into the substrate. If water can escape from the bed through the substrate, entrainment of ice and debris by temperature fluctuations at the interface will be inhibited (Knight, 1988). Strasser *et al.* (1996) described freezing-on of basal water to the sole of Matanuska Glacier, Alaska, to form a debris-bearing basal ice layer. According to their evidence, water flowing towards the margin through an overdeepened portion of the bed becomes supercooled when the pressure-melting point rises faster than the temperature increases. Ice then nucleates in a variety of forms to create a thick layer of basal ice.

(e) Shearing and folding

Shearing is one of the oldest theories for debris entrainment into basal ice (e.g. Chamberlin, 1895; Salisbury, 1896) and one of the most controversial. Until very recently, the efficacy of shearing or thrusting as a mechanism of debris entrainment was widely disputed. However, recent observations seem to confirm that it is indeed a realistic process. The theory is that sediment is entrained by shearing or thrusting of sediment from the bed along shear zones penetrating upwards into the ice (e.g. Lamplugh, 1911; Goldthwait, 1951, 1971; Bishop, 1957; Swinzow, 1962; Souchez, 1967, 1971). Weertman (1961) and Hooke (1973a) argued that the shear hypothesis was theoretically unsound. Hooke and Hudleston (1978) suggested that, despite the frequency of references to shearing, none of the authors who had previously suggested it seemed to have actually observed it in action. However, Wakahama and Tusima (1981) reported that all the debris bands cropping out as moraine ridges on the ice surface near the margin of McCall Glacier in Alaska were located along thrust faults, and successfully duplicated transport along shear planes in a laboratory model. Sharp *et al.* (1988) were able to make direct measurements of displacement rates of up to 0.1 mh^{-1} along thrust faults in the basal part of Variegated Glacier during its 1982–3 surge. Tison *et al.* (1993) invoked shearing at subfreezing temperatures to explain the characteristics of debris-bearing basal ice from Antarctica. In some situations a freezing-on mechanism might be responsible for initial attachment of sediment to the glacier bed, and shearing for transport to different levels in the glacier. This two-stage process was recognised for example, by Boulton (1967, 1970) in Spitzbergen. Observers in many locations have reported traction of basal transport zone sediments along a new effective bed formed by structural discontinuities passing upwards into the ice. This could be interpreted either as entrainment by shearing or as the thickening of the basal transport zone into a basal ice layer. Shearing might be expected to cause fine-scale changes in the distribution of debris in the ice. For example, Johnson (1995) suggested the attenuation by shearing of a band of massive and laminated debris into a diffuse layer of suspended particles in basal ice in Greenland, and proposed this as a possible origin for part of the dispersed or clotted basal ice facies (Chapter 5). Hart (1995a) proposed that the entire sequence of stratified and dispersed basal ice can be attributed to progressive differential deformation of material entrained at the bed, although it is not clear how this mechanism could explain the chemical and isotopic differences that occur through the sequence.

Folding of both glacier ice and substrate at the base of the glacier can incorporate subglacial sediment into the basal ice. This is especially likely close to the margin, where compressive flow in thin ice overriding marginal debris accumulations can lead to the formation of cavities in the lee of subglacial moraine ridges. The higher pressure and velocity on the up-glacier side of the cavity can cause squeezing of basal sediment into the cavity, and the rolling of the cavity into a debris-filled recumbent fold. In warm ice, such folds are a likely focus for thrusting, and can lead to the formation of debris layers within the basal ice. Incorporation of subglacial material into folds and thrusts is especially likely where the substrate is deformable. Boulton (1974) described how folds could develop in basal ice as a result of flow deformation around basal bumps, and how such folding could raise material from the bed into higher positions in the ice during flow.

Combined folding and thrusting has been described from several locations, and seems especially pronounced in surging glaciers and in sub-polar glaciers with thermally induced compression at the margin. Clarke and Blake (1991) identified a recumbent fold/thrust debris structure penetrating upwards from the bed at the boundary between warm and cold basal zones beneath Trapridge Glacier, Alaska, and illustrated the development of basal folds and thrusts as a mechanism for generating debris layers within the basal ice. Sharp *et al.* (1994) described a variety of fold and thrust structures from the basal layer of Variegated Glacier.

(f) Entrainment from cavities

Tison and Lorrain (1987) considered in detail the processes of attachment and transport of ice formed as floor-coatings in subglacial cavities. At Glacier de Tsanfleuron, Switzerland, they observed that the rock floors of natural subglacial cavities were coated with ice layers 1–10 cm thick. These floor coatings incorporated layers of debris that had fallen or been washed onto the cavity floor. The floor-ice was in places contorted and detached from the floor, and these twisted sheets of ice were picked up by the glacier sole where it rejoined the rock bed at the down-glacier limit of the cavity. The sheets were stretched and dragged along by glacier movement to progressively form a basal ice layer. This basal layer was characterised by an alternation of ice and debris layers, which could be either inherited from the banded structure of the original floor coating, or produced during folding and attenuation during flow. Knight (1987) suggested this as a major source of stratified ice, and Hubbard and Sharp (1995) as the source of their 'interfacial' and 'dispersed' facies. Figures 5.20 and 5.21 show aspects of these processes. Where cavities occur in areas of deformable subglacial sediment, the sediment can be squeezed into the cavities and become enfolded into the ice flow. This is frequently observed where glacier margins advance over moraine ridges, and can lead to the intercalation of debris bands into non-basal ice as shown in Figure 9.2.

(g) Apron overriding

Ice and debris already existing beneath or in front of the glacier can be attached to the glacier sole to form part of the basal layer. Proglacial ice sources include ground ice, lake ice and glacier-margin accumulations of snow and ice. Souchez *et al.* (1994) have attributed the origin of the silty ice forming the basal layer of the GRIP ice core to non-glacial ice formed in front of the developing young Greenland ice sheet. They infer from the isotope characteristics of the silty ice that it probably formed as wind-drift or ground-surface ice outside the limits of the ice sheet and was incorporated into the basal ice as the ice sheet advanced over it. Shaw (1977a,b), Hooke (1973a,b), Evans (1989) and others have recognised the incorporation of ice and debris into the basal layer by the glacial overriding of proglacial accumulations of snow, superimposed ice and ice-cored debris. Evans suggested that the incorporation of alluvium-dominated proglacial debris

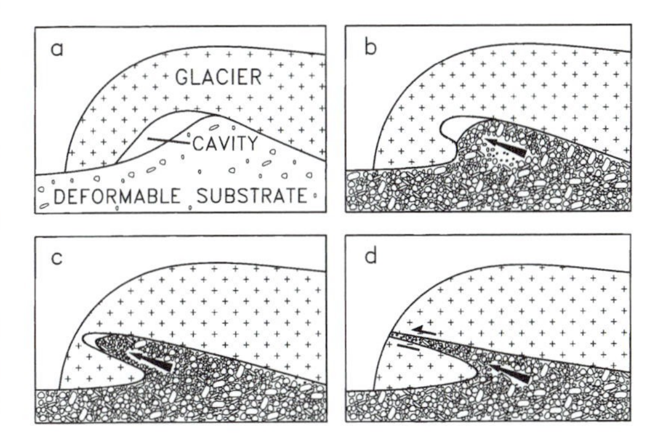

Figure 9.2 Subglacial sediment can be squeezed into cavities and enfolded into the ice flow.

fans or 'aprons' is a major source of basal debris where glaciers re-advance over extensive areas of dead ice. Sharp *et al.* (1994) recognised apron entrainment as an important process at Variegated Glacier. Where glaciers advance over frozen ground including ground ice or permafrost, the subsequent subglacial entrainment of material from the substrate would also involve attachment of pre-existing ice, which would have distinctive chemical, isotopic and structural characteristics.

9.2.3 TRANSPORT

Material can be transported horizontally and vertically through glaciers by a variety of mechanisms. The main agents of transport are: the movement of the ice itself carrying the material with it; gravity-driven flow or mass movement of supraglacial debris; water transporting sediment rapidly through the glacier drainage system; and glaciotectonic deformation of subglacial and proglacial sediments. The transport route depends largely on the point of entry of the material and the flow characteristics of the glacier. A distinction is sometimes drawn between low-level (active) transport, where the material is brought into contact with the bed, and high-level (passive) transport, where the material remains in the supraglacial or englacial zones. Material entrained at the surface in the accumulation zone is usually buried by snowfall and travels downwards into the glacier to follow either an englacial or a basal transport route. The higher in the accumulation zone the material is entrained, the greater the chance of it following a basal transport route. Material entrained in the lower part of the accumulation zone follows an englacial route before re-emerging onto the surface in the ablation zone and following a supraglacial route. Material entrained onto the surface in the ablation zone follows a supraglacial route to the margin. Supraglacial debris can fall or be washed down to the englacial or basal zones via crevasses or moulins. Material entrained at the bed may be transported along the bed in the basal transport zone, or travel upwards into the basal ice layer. Where folding or thrusting occurs in the basal ice, or where successive layers of material are attached to the bed by freezing-on of water and sediment, basal material might penetrate several tens of metres upwards into the ice. Close to the margin where the ice is thin, basally derived material might emerge at the glacier surface. Subglacial material, such as in a deforming subglacial layer, may follow an entirely subglacial transport route. Fine material flushed through the substrate by groundwater driven by glacial pressure also follows a subglacial transport route, although it is not generally considered as glacial material. Non-rigid subglacial and proglacial material can also be transported as a result of stress imparted by a glacier. In recent years increasing attention has been given to sediment deformation as a means of sediment transfer (e.g. Croot, 1988; Boulton, 1996). In its broadest definition, glaciotectonic deformation can include deformation of a subglacial layer, pushing or bulldozing of proglacial sediments by an advancing ice margin, ploughing of an underwater surface by icebergs, and even debris flows on the ice surface (van der Wateren, 1995). These processes can lead both to sediment transfer and to the creation of distinctive sediment structures and landforms. Tectonic deformation of subglacial and proglacial material is favoured by poorly consolidated substrate materials, high water pressures and high horizontal stress gradients.

Although debris in transport can be found at any location in a glacier, it is not evenly distributed. In ice sheets nearly all of the debris is transported in the basal layer. In glaciers surrounded by ice-free mountains, much more englacial and supraglacial transport occurs. Hunter *et al.* (1996) measured debris flux through three Alaskan tidewater glaciers and found that basal ice accounted for 43–74% of the debris flux. They also noted a fourfold difference in englacial debris concentrations between glaciers in their study, owing to different amounts of sediment input to high levels in the trunk glacier from tributary glaciers and to the different significance of lateral sediment sources in broader and narrower valleys. Englacial transport occurs mainly in specific debris-rich structures such as medial moraines. These can occur either when the lateral moraines and basal transport zones of two valley glaciers coalesce at a glacier confluence, or when debris is supplied continuously or periodically from a point source such as a nunatak and carried down-glacier by ice flow. If the nunatak is in the accumulation zone the moraine may not emerge at the surface until a point below the equilibrium line. Hunter *et al.* (1996) found that 39% of the debris carried by the Grand Pacific Glacier was in medial moraines. Vere and Benn (1989) described transport of debris in medial moraines and recognised three categories of debris routing: discrete concentrated longitudinal septa containing subglacially comminuted debris derived from basal entrainment; diffuse longitudinal septa containing passively transported debris entrained supraglacially above the firn line; and supraglacially transported debris spreads entrained as rockfall debris below the firn line.

The physical characteristics of debris in transport are determined partly by the debris source and the characteristics of the debris at the time of entrainment, and partly by processes that operate during transport. In supraglacial transport and in englacial transport where particles are not in mutual contact, little alteration of the debris might occur. By contrast, where material is transported in contact with the bed, processes of erosion and comminution at the bed tend to create characteristic particle shapes and particle size distributions that depend on particle lithology, on basal conditions, and on distance of travel. In glacial sediments, characteristics of particle size, shape and orientation can be used to reconstruct former transport pathways and glacier characteristics (Boulton, 1978; Haldorsen, 1981; Dowdeswell *et al.* 1985). The distribution of clasts of different lithology in tills can also be used to reconstruct ice flow and sediment deformation paths down-glacier of bedrock marker lithologies (e.g. Boulton, 1996).

9.2.4 DEPOSITION

The ice sliding down from the highland is loaded with gravel and stones, partly spread over its surface, partly imbedded in its layers. On being wasted by melting, it leaves all these transported matters on the ground, whereupon during the period of progress they are pushed forward by the edge of the ice, and piled up in front and on the sides of it.

(Rink, 1877).

Deposition occurs when material is released from the ice or from glacial transport at the margin or the base of a glacier. Release of material onto the glacier surface is sometimes referred to as supraglacial deposition, but if the glacier is still moving this is often only a temporary stage in the sediment transport process. Deposition may occur directly onto the ground or through water. Many glaciers release sediment into water, and glacimarine sediments form an important part of the glacial sediment record. A narrow definition of glacial sedimentation would include only primary sedimentation directly from the ice to the position of rest, but broader definitions include secondary processes such as deposition through water, and resedimentation of glacigenic materials by flowage. The characteristics of glacial sediments can reflect both the processes of their deposition and also the processes of glacial erosion and entrainment by which the material was produced (e.g. Haldorsen, 1981). A substantial literature exists on the classification of glacial sediments and the processes by which they form (e.g. Schlüchter, 1979; van der Meer, 1987; Goldthwait and Matsch, 1988) and several convenient summaries of depositional processes have been produced (e.g. Whiteman, 1995). Following a broad definition, the main mechanisms of deposition include:

- release of debris by melting or sublimation of the surrounding ice;
- lodgement of debris by friction against a substrate;
- deposition of material from meltwater (glacifluvial deposition);
- chemical precipitation;
- flow and resedimentation of deposited material;
- glaciotectonic processes.

Different processes of sedimentation are dominant in different parts of a glacier. In a supraglacial location, material can be released by ablation of the ice surface. This occurs most commonly by melting, and is referred to as meltout, but sublimation can make a significant contribution in some environments (Shaw, 1977b, 1988). Englacial debris, and debris in basal ice penetrating to the surface, is then exposed on the ice surface and can contribute to a supraglacial sediment layer. This supraglacial sediment is liable to redistribution by flow, wash and mass movements on the ice surface as ablation continues. Resedimented material derived from flow of supraglacial debris, sometimes referred to as 'flow-till', can make up a substantial proportion of glacial deposits in environments where supraglacial sedimentation occurs (Boulton, 1968; Lawson, 1979). The extent of supraglacial sedimentation depends on the debris content of the ice, the ablation rate, and the rate of removal of sediment from the surface by processes such as wash, deflation and mass movement.

In an ice-margin location, material can be released by ablation as in a supraglacial environment and dumped directly into the proglacial environment. Reactivation and re-sedimentation of this material, both by movement of the ice margin and by fluvial and mass-movement processes, is common. The characteristics of ice-marginal sedimentation might depend more on the local climate than on the thermal regime of the glacier. On the basis of depositional processes and sediments at the margin of the Antarctic ice sheet in the Vestfold Hills, Fitzsimons (1990) showed that the influence of the polar maritime climate on sedimentation outweighed the influence of the cold basal thermal regime. Meltwater produced at the glacier surface during a short summer period played a major role in the formation of deposits, reworking ablation tills by flowage and mass movement. Glacier margins can also experience deposition of sediment emerging by extrusion from beneath the ice when a subglacial sediment layer is mobilised by glacial pressure (Boulton *et al.*, 1995).

In a subglacial environment, material can be released by ablation either in cavities or in contact with the bed, or can be lodged against the bed by moving ice. Lodgement of sediment against a rigid bed occurs when the friction of the clast against the bed outweighs the tractive power of the ice. In lodgement, material is released from the ice usually by either pressure-melting or plastic deformation of ice around clasts. Release of basal sediment by meltout or sublimation beneath moving ice can produce conditions conducive to lodgement, to the development of thick basal till sequences, and to the possibility of subsequent subglacial deformation of that released material. Lodgement can also occur against the upper surface of a deforming bed if the overlying ice is moving faster than the deforming layer. Deposition associated with deforming layers is commonly considered to involve the cessation of deformation either throughout the layer or for a certain thickness at its lower margin, the material that comes out of transport being effectively deposited. As Hart (1997) expressed it, deposition within a deforming subglacial layer occurs when more material is supplied to the layer than can be removed by deformation within the layer.

Subglacial deposits can also include chemical precipitates, although these are not always considered in the context of glacial sediments. Chemical precipitates such as calcite can be deposited when solute-rich waters freeze. In carbonate environments, such as where limestone bedrock is present, comminuted carbonate rocks contribute to a highly reactive rock flour. Meltwater produced at the bed can take carbonate into solution, and if the water subsequently refreezes the carbonate can be released as a precipitate onto bedrock or basal debris (Souchez and Lemmens, 1985; Sharp *et al.*, 1989; Fairchild *et al.*, 1993).

Release of sediment into water can produce different effects from terrestrial sedimentation. Sediment can be released directly from the ice, by discharge of sediment-rich meltwater, and by decay of icebergs. The characteristics of the sediment and subaqueous landforms reflect aqueous as well as glacial processes and conditions (Drewry and Cooper, 1981; Powell, 1981; Powell and Molnia, 1989).

The mechanisms of deposition impart specific characteristics to the deposited material. Till fabric and consolidation have both been used to infer depositional processes and glacier characteristics from glacial sediments. According to van der Wateren (1995), the deformed substratum is the fingerprint of the various strain regimes of an ice sheet, and the distribution of structural styles in deformed sediments can be used to reconstruct former ice sheets. Boulton and Dobbie (1993) suggested that consolidation characteristics of formerly subglacial sediments can be used to infer basal melting rates, subglacial groundwater flow patterns, ice overburden, basal shear stress, ice-surface profiles and the amount of sediment removed by erosion.

9.2.5 SEDIMENT BUDGETS

It is clear from the foregoing sections that different glaciers would be expected to produce different amounts of sediment depending on the processes that operate beneath them and the nature of the material over which they flow. Denudation rates beneath glaciers can vary from 0.01 mm a^{-1} to more than 100 mm a^{-1} (Hallet *et al.*, 1996). Quickly moving warm-based glaciers generally produce more sediment than slow cold-based glaciers. Glaciers on deforming beds commonly discharge more sediment than glaciers on rigid beds. Larger glaciers tend not to be very variable in their sediment production from year to year, whereas small glaciers can be characterised by extreme fluctuations because periodic flushing of high proportions of the total bed area can lead to sediment exhaustion. Surging glaciers are characterised by fluctuations in sediment production between their surge and quiescent phases. Also, the locations of sediment input, storage and output vary from glacier to glacier. Basic sediment budget components for glaciers are illustrated in Figures 9.3 and 9.4, but huge differences in the relative importance of different components are likely to occur between glaciers. A number of studies have attempted proglacial (fluvial) sediment budgets, into which glacial sediment constitutes an input (e.g. Maizels, 1979; Warburton, 1990) and there have been some studies of discrete components of the glacial sediment budget, such as the moraine sediment budgets described by Small (1983, 1987) and Small *et al.* (1984). There have also been some attempts to characterise, or at least to delineate, the glacier sediment system as a whole, but quantitative sediment budgets for glaciers are difficult to achieve. Even in the smallest glaciers, large parts of the sediment transfer system are inaccessible; and in larger glaciers, measuring sediment flux through all parts of the sediment system over the whole area of the glacier is virtually impossible.

Typical values for glacial sediment production, averaged over the area of the glacier, are of the order of a few thousand tonnes per square kilometre per year (kt km^{-2} a^{-1}). For typical values of rock density, 3 kt km^{-2} a^{-1} would be equivalent to about 1 mm a^{-1} of denudation. Hallet *et al.* (1996) reviewed data on sediment flux and effective erosion from literature on more than 60 glacier basins and found values for erosion rates ranging from 0.01 to 60.07 mm a^{-1}. They also found that sediment

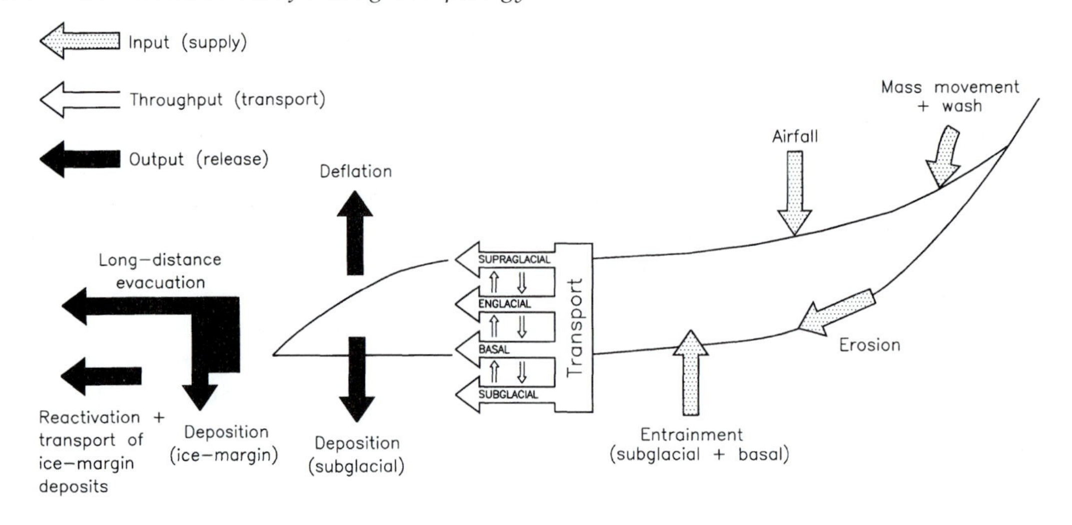

Figure 9.3 Schematic representation of the main elements of a glacier sediment budget.

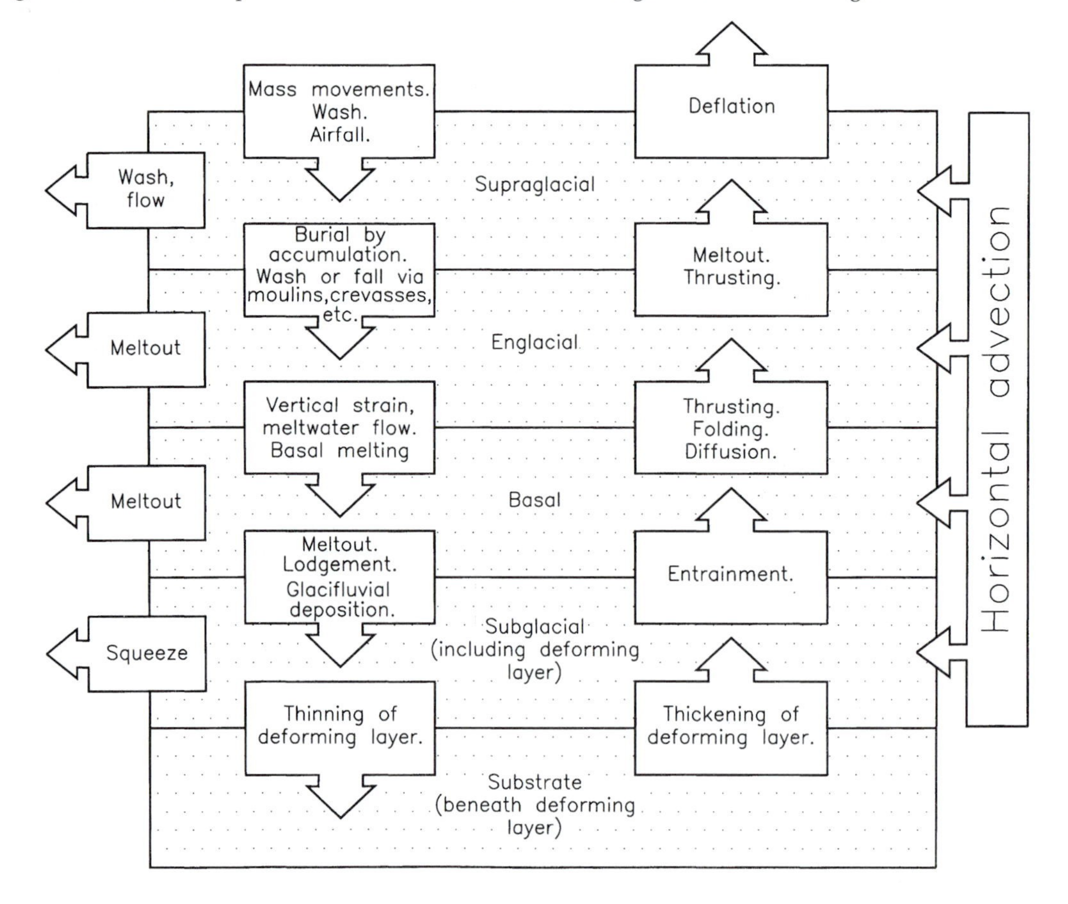

Figure 9.4 Diagrammatic representation of the main interrelationships in a glacier sediment budget.

yields increased with the extent of glacier cover within a basin, basins with more than about 30% glacier cover demonstrating erosion rates an order of magnitude higher than glacier-free basins. According to Hallet *et al.* (1996, p. 232) 'erosion rates from extensively glaciated basins are essentially unsurpassed'. Some values that have been obtained for different glaciers are shown in Table 9.1.

The sediment discharge from the glacier, which constitutes the output element of a glacial sediment budget, is not necessarily reflected directly in downstream proglacial sediment fluxes. The proglacial fluvial sediment budget, and the glacigenic input to ocean sediments, vary both with glacial sediment production and with periodic storage and release of sediment by sediment traps at the glacier margin. Key controls on sediment throughput past the marginal zone are topography and meltwater routing. Marginal sediment storage varies with glacier advance and retreat due both to the interaction of glaciers with fringing moraines and to variations in meltwater discharge and competence.

Where glaciers terminate in areas without topographic obstruction, sediment released from the ice margin flows directly into the distal proglacial zone. Where glaciers lie behind fringing moraine ridges there is a high potential for sediment storage to occur in traps such as lakes and other closed basins in the proximal proglacial zone, and large amounts of the material released from the ice are held in storage between the ice margin and the moraine. Moraine ridges also focus meltwater discharge from the proximal proglacial zone, localising fluvial processes. Merrand and Hallet (1996) described the situation at Bering Glacier, Alaska, where the bulk of the glacier's sediment is routed through the ice-marginal Vitus Lake, from which there is only one main stream outlet. The denudation rate beneath the glacier is estimated to be in the range 10–100 mm a^{-1}, but the sediment discharged at the distal end of the lake is equivalent to no more than about 2 mm a^{-1}, indicating that huge amounts of glacier-derived sediment are being placed into storage in the lake basin. When glaciers advance across their fringing moraines, not only is the material in the moraine itself activated and released into the forefield, but much of the material that was held in storage behind the moraine is either moved forward, re-entrained into the glacier, or carried beyond the position of the former moraine by meltwater flow. When the glacier margin lies on the distal side of a fringing moraine, material released by the

Table 9.1 Erosion rates beneath selected glaciers

Glacier	*Erosion rate*[a] *(mm a^{-1})*	*Source*
Nansen, Greenland	0.01	Andrews *et al.* (1994)
Nigardsbreen, Norway	0.15	Hallet *et al.* (1996)
Erikbreen, Svalbard	0.19	Sollid *et al.* (1994)
Engabreen, Norway	0.41	Bogen (1989)
Gornergletcher, Switzerland	2.22	Bezinge (1987)
Fedchenko, Central Asia	4.36	Chernova (1981)
Johns Hopkins, S.E. Alaska	47.24	Cai (1994)
Margerie, S.E. Alaska	60.07	Hunter (1994)
Breiðamerkurjökull, Iceland (AD 1730–1890)	700.00	Bjornsson (1996)

[a] Values have been standardised from original references following the suggestion of Hallet *et al.* (1996) that a sediment yield of 1 kt km^{-2} a^{-1} is equivalent to an erosion rate of 0.37 mm a^{-1}.

glacier can more easily pass directly into the proglacial zone, with less potential for storage close to the margin. Both wash and mass movement processes contribute to the evacuation of material away from the margin and also to the re-activation of sediments in the moraine. Meltwater discharge, sediment production and sediment storage control the flux of material into the proglacial zone. The readvance of a glacier over a set of fringing moraines is therefore likely to be associated with a characteristic sequence of change in downstream sediment flux. Maizels (1979) developed a model where the occurrence of aggradation and degradation on the proglacial valley train during glacier advance or retreat depended on a balance between meltwater competence and sediment loads which in turn depended on combined glaciological, geomorphological and hydrological changes associated with glacier fluctuations.

There is a complex and locally variable balance between water power, sediment availability and the openness of sediment routeways between the glacial and proglacial environments. If this component of the glacial sediment budget were more fully understood there would be the potential for reconstructing former glacier and ice-sheet behaviour on the basis of periodic variations in sediment characteristics in the geological and geomorphological record.

9.3 PROCESS ENVIRONMENTS AND GLACIATED LANDSCAPES

9.3.1 LANDSCAPES AND GEOMORPHIC PROCESS ENVIRONMENTS

A landscape is an assemblage of landforms. Landforms, and hence landscapes, are generated by the operation of geomorphic processes. Geomorphic processes are in many cases environment-specific. The nature of geomorphic processes varies from place to place depending on a range of environmental conditions, and varies also through time at any one location. The term 'process environment' is used to denote a set of environmental conditions that can be associated with particular geomorphic processes and hence particular styles of landscape development. Landscapes that cover substantial areas, and develop over considerable spans of time, reflect the operation of a range of different processes, in a range of different environments, past and present. This is especially evident in the development of glacial landscapes, which evolve in the most transparently time-transgressive of process environments. Consider, for example, the formation, growth and decay of an ice sheet. A single location may, over a period of time, be first proglacial, then ice-marginal, then subglacial, then marginal again and then proglacial and eventually paraglacial before becoming effectively postglacial. During its subglacial period, the location will be subject to a whole sequence of different process environments through time associated with varying ice thickness and the changing thermal and dynamic regime of the glacier. Towards the end of the sequence, the location could retain a record of several episodes of erosion and deposition by glacial and fluvial processes under different subglacial conditions. The landscape is thus like a palimpsest: a parchment which has been overwritten many times but which retains part of the image of each writing.

One aspect of glacial geomorphology that is of particular interest to the glaciologist is the geography of geomorphic processes. The distribution of landforming processes within the glacier system is controlled by the distribution of environmental conditions. If we can understand the way in which the environment controls the processes, then we can use landform evidence to reconstruct former glacier environments. For example, it is only possible to interpret features such as striations because we know the basal stress and thermal conditions under which striations form. From our point of view it is therefore valuable to consider glacial geomorphology from the point of view of glacial process environments.

9.3.2 GLACIAL PROCESS ENVIRONMENTS

Glacial landforms are commonly discussed in terms of the processes by which they are created: landforms of erosion and landforms of deposition, for example. It is also useful to consider landforms in terms of the locations within a glacier where they are formed. This is especially useful if landforms and sediments are to be used as indicators of former glacier characteristics. The distribution of landforms reflects the geography of the glacier that created them. This is true because different parts of a glacier are commonly associated with specific sets of geomorphic processes creating specific assemblages of landforms and sediments. We can therefore speak of glacial geomorphic process environments, where a set of related processes occurs to create a distinctive set of landforms and sediments. The sediment–landform assemblage created in a specific process environment is sometimes referred to as a landsystem, which Eyles (1983) defined as a recurrent pattern of genetically linked land facets. The major glacial landsystems and process environments where landforming processes are important are supraglacial, subglacial and ice-marginal. In a broader geomorphic context, proglacial and paraglacial environments are also relevant to glaciology. Supraglacial features only survive as landforms when the ice on which they rest wastes away, lowering the supraglacial material onto the ground. Supraglacial material on an active glacier will be carried forwards to the glacier margin and deposited as part of the ice-marginal landsystem. Subglacial and ice-marginal landforms may or may not survive the deglaciation process. Features associated with the major glacial process environments are discussed in the following paragraphs. Convenient reviews of glacial geomorphology include those by Sugden and John (1976), Drewry (1986), Bennett and Glasser (1996) and Benn and Evans (1998).

9.3.3 SUBGLACIAL GEOMORPHOLOGY

Processes of erosion, entrainment, transport and deposition that operate in the subglacial environment give rise to recognisable landform and sediment assemblages. Features of subglacial geomorphology include eroded bedrock, eroded and deformed sediment, deposited till, and fluvioglacial phenomena. Erosional phenomena include 'negative' (recessed) features such striations, grooves and troughs, and 'positive' (protuberant) features such as roches moutonées and whalebacks. Landscape assemblages incorporating such features include regions of areal scouring, where the ground is affected by widespread erosion, and regions of selective linear erosion, where local variations in subglacial conditions lead to abrupt local differences between areas of erosion and areas of no erosion (e.g. Sugden, 1968, 1974). Depositional and deformational phenomena can be identified at a range of scales including vast till sheets, lineated features such as flutes, drumlins and Rögen moraines, and even individual sediment structures and clast characteristics. In addition, many subglacial landforms are controlled by solute and sediment transfer in meltwater at the bed. Although glacial geomorphology is a field that far exceeds the scope of this book, the important point from a glaciologist's perspective is that subglacial geomorphology reflects glaciological conditions. In particular, the basal thermal regime, the debris characteristics of the basal ice layer and the pressure distribution at the bed have a major controlling impact on the geomorphic effectiveness of glaciation.

The characteristics of specific subglacial environments control the nature of the processes that operate, and hence the landforms and landscapes that are produced, in the subglacial zone. The basal thermal regime has long been recognised as playing a crucial role in controlling erosion and deposition. For example, Boulton (1972, 1974) associated specific processes of erosion and sedimentation with different thermal regimes, and Sugden (1977,

1978) identified geographical zones of different geomorphic characteristics that would be associated with different thermal zones beneath an ice sheet. Cold-based areas are likely to experience little sliding and minimal erosion, although more erosion can occur if there is a debris-bearing basal layer. Warm-based areas are more likely to be associated with sliding and significant abrasion. Substantial melting at the bed where debris-bearing basal ice is present is likely to be associated with till deposition. Transition zones between warm- and cold-based ice are likely to be characterised by erosion and sediment entrainment. Postglacial landscape types such as landscapes of areal scouring or selective linear erosion can thus be used to reconstruct areas of different subglacial thermal regimes. Kleman and Borgström (1994) identified a landform assemblage characteristic of glaciers where parts of the bed are frozen and parts are thawed. Frozen patches of the bed are characterised by the preservation of preglacial surfaces with little glacial modification, while the boundaries between warm- and cold-based areas are marked by specific thermal boundary landforms that comprise a 'frozen-patch landform assemblage'.

Other important parameters controlling subglacial geomorphology include the basal pressure regime, the morphology of the glacier sole and the competence of the substrate. These are in part related to water pressure which is, in turn, dependent partly on temperature. Patterson and Hooke (1995) reviewed published descriptions of drumlins to arrive at a characterisation of the physical conditions under which drumlins form. They found that specific glaciological characteristics were associated with drumlin formation, and that parameters such as regional topography, substrate material and substrate thickness were not relevant. They found that drumlins form in a restricted zone up-glacier from, and separated from, the ice margin, in which strain is longitudinally compressive and transversely extending, pore-water pressures are high, shear stress may be low, and the ice is relatively thin but still thicker than some critical threshold below which drumlin formation is inhibited. Ice thickness seems to be very important through its control on basal pressure distributions. Wintges (1985), for example, found a direct correlation between ice thickness and the length of crescentic fractures on subglacial bedrock. In discussing the evolution of roches moutonées, Sugden *et al.* (1992) suggested that plucking beneath ice sheets is favoured by two contrasting sets of conditions. One is where melting occurs and ice velocity is high beneath major ice streams. The other is close to the ice margin where the ice is relatively thin, and overburden pressure is therefore low, permitting relatively easy removal of loose blocks from the bedrock. Mooers (1989a,b, 1990) found that distinct subglacial process zones existed in the near-margin areas (the outer 100 km) of the Laurentide ice sheet. The spatial organisation and process characteristics of these zones were determined largely by thermal regime, and involved water-pressure variations driven by basal thermal conditions. Mooers studied landforms produced beneath two adjacent lobes of the ice sheet. In the Rainy Lobe, a frozen toe 1–2 km wide existed at the margin and subglacial water drained through sediments at a high pore-water pressure. This facilitated thrusting of subglacial sediment in the freeze-on zone at the boundary between thawed and frozen bed conditions. The sediment was then transported to the ice surface by compressive flow in the marginal zone, leading to substantial supraglacial sedimentation. This in turn led to formation of large proglacial outwash fans and, during ice retreat, extensive hummocky stagnation topography. By contrast, the margin of the Superior Lobe was characterised by temperate conditions. Water flowed at the bed in a channel system, creating tunnel valleys, and drumlins formed in a time-transgressive zone within 20–30 km of the retreating margin. During retreat, closely spaced recessional moraines were formed, with little evidence of subglacial thrusting or the development of extensive supraglacial deposits or stagnation topography.

Even a cursory review of subglacial geomorphology thus indicates that it should be possible to reconstruct the nature of former subglacial conditions on the basis of landform and sediment characteristics. However, our understanding of the controls on glacial geomorphic processes is incomplete, and many landforms remain ambiguous. This includes even the most common of glacial landforms, such as drumlins and roches moutonées, that might be familiar to the most novice glacial geomorphologists.

9.3.4 GEOMORPHOLOGY AT THE GLACIER MARGIN

(a) The glacier margin

The glacier margin is perhaps the most exciting of glacial environments, because the whole gamut of glacial processes can here be observed directly without the hazard or inconvenience of subglacial adventure. The ice margin is like a coastline around the edge of the ice; a border between the glacier and its surroundings. The frontal margin of a glacier is the limit between the glacial and proglacial zones. It marks the position where the gradual ablation of the glacier finally outweighs the accumulation of mass; all of the ice has been lost, and all of its sediment load released. It is the end of a chapter in the hydrological cycle. At the margin, the observer can peer a very short way into the interior of the ice to observe subglacial and englacial processes, can view a portion of the glacier surface to witness the supraglacial realm, and can survey the proglacial zone with relative ease. The glacier margin itself also offers its own particular menu of processes and phenomena. Processes specific to the glacier margin relate primarily to the release of debris and water, to the physical manipulation of surface materials by the movement of the ice front, and to the transition between the thermal, pressure and material regimes of subglacial and subaerial environments. The principal geomorphic features can be glacial, fluvioglacial, or glacitectonic.

(b) Release and deformation of sediment

Sediment can be released directly from the ice at the margin, or brought to the margin by water flowing from the interior or surface of the glacier. Sediment can also be transferred to the margin by deformation of the subglacial material. The amount of sediment supplied to the margin is a function of the size and speed of the glacier, the glacier's erosive capability, the erodibility of the substrate, and the input of sediment from extraglacial sources such as rockfalls or tephra. Sources of sediment, and mechanisms of transport and deposition, were discussed earlier in this chapter. The geomorphic features produced at the margin depend on whether the margin occurs on land or in water, and the processes and environments of their formation are reflected in their morphology and sedimentological structure.

On land, principal features of the ice margin include moraine ridges and outwash fans. Ridges form by sediment deformation in front of an advancing or oscillating ice margin (glaciotectonic or push moraines), by dumping of sediment from the ice front over a protracted period at a stationary margin (dump moraines), or by squeezing of sediment from the subglacial deforming layer (extrusion moraines). Sediment dumped at the margin can be englacial or basal sediment released by meltout at the margin or can be supraglacial debris carried to the margin by the forward movement of the ice. When thick supraglacial debris covers the margins of retreating glaciers, ablation can be inhibited and an ice-cored moraine, sometimes referred to as an ablation moraine, can form in front of the retreating glacier margin (section 9.3.6). When moraine ridges are created at a terminal position from which the glacier subsequently retreats without advancing further, the features can remain as a long-term record of the position of the margin. Glaciers that retreat episodically with periods of still-stand or minor re-advance can leave a series or recessional moraines marking positions of the margin during retreat. Glaciers in continuous retreat, without extended still-stands or re-advances, tend to leave unconsolidated sheets of debris behind with no ridges. Advancing glaciers generally demolish previously existing ridges that they override, and produce ridges at their margins only if pushing of proglacial sediment is possible. Small push moraines can be formed by bulldozing of proglacial sediment on a seasonal basis where an ice margin oscillates with seasonally varying ablation. Larger push moraines can form by the superposition of several seasonal moraines or by a more substantial advance of the margin into deformable materials. Dump moraines grow in size for as long as an ice margin remains *in situ* to supply sediment, their rate of growth depending on the rate of sediment supply and ablation. The structure and sedimentology of moraine ridges reflect the processes of their formation. Material carried out of the glacier by water, or washed out of the fringing moraines by water, tends to form a low-gradient debris fan in front of the glacier. Ice-contact fans can thicken over time in front of a stationary ice margin so that when the ice subsequently retreats a steep ice-contact slope is left to mark the former position of the margin, and the surface of the fan stands higher than the exposed ground up-glacier of the former margin.

Where glaciers terminate in water, depositional processes are somewhat different. Where a glacier is grounded on its bed beneath the water, glacial deposition can occur beneath the margin by lodgement and meltout. However, material that is dumped from the front of the glacier into water, or from the base of the glacier where the glacier is floating, forms a deposit that is not strictly a glacial deposit, as its character upon settling will be controlled largely by aqueous sedimentation processes. Principal features of ice margins in water can include subaqueous moraines caused both by pushing of proglacial sediments and by release of sediments from the glacier, subaqueous grounding-line fans formed from material emerging from beneath the glacier into the water at the grounding line, ice-contact fan deltas that form when grounding-line fans grow and emerge at the water surface, and a distal proglacial zone in which sediment settles out from suspension in the water and rains out from icebergs drifting away from the ice margin (Figure 9.5). The relative importance of sediment evacuation routes varies from glacier to glacier, depending on hydrological and sedimentological conditions. At Coronation

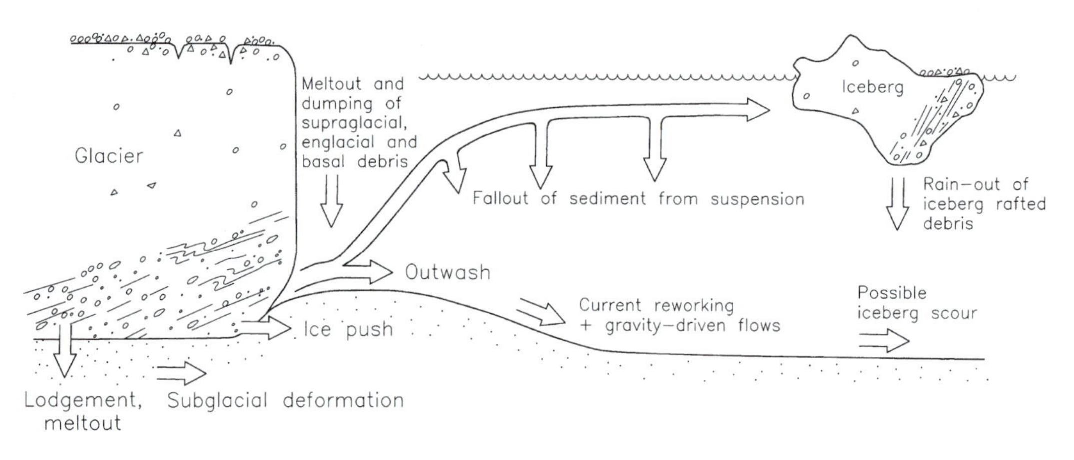

Figure 9.5 Key characteristics of glacial sedimentation processes at marine and lacustrine ice margins.

Glacier on Baffin Island, Syvitski (1989) identified the principal elements of the sediment routing out of the glacier as glacifluvial discharge (86%), supraglacial dumping (9%), subglacial deposition (3.7%) and ice-rafted deposition (0.8%). Similarly, Hunter *et al.* (1996) suggested on the basis of observations of tidewater glaciers in Alaska that up to 98.6% of the total sediment yield could be delivered through the glacifluvial discharge. They also suggested that the sediment flux due to subglacial deformation could be an order of magnitude higher than the flux of sediment out of the glacier ice. Useful reviews of sedimentation at marine and lacustrine glacier margins include those provided by Dowdeswell and Scourse (1990), Powell and Molnia (1989), Hambrey (1994) and Bennett and Glasser (1996).

One of the most important aspects of the glacier margin is the way in which it controls the release of water and sediment to the proglacial zone. Material that issues from the glacier at the margin can pass directly into the proglacial zone, or can be stored in the ice-marginal zone as discussed in section 9.2.5. This storage comprises marginal kame and moraine formations, including both material deposited directly from the ice and material carried out of the ice by water but trapped within the moraine system. Sediment traps include closed intra-moraine basins, channels and ponds adjacent to the glacier, and debris fans on the distal side of the moraines. When glaciers re-advance over previously formed moraines, large amounts of sediment from the moraine system can be released from storage. Regional sediment flux measured downstream within a glaciated basin is thus very sensitive to the position of glaciers relative to their fringing moraines. When glaciers are in positions behind fringing moraines, the bulk of sediment produced at the margin can go into storage in the moraine belt and not reach the proglacial region. When glaciers have no such fringing moraines, sediment is more likely to transit into the proglacial system. Huge pulses of sediment might be input to the proglacial system as advancing glaciers break through fringing moraines. Major events such as jökulhlaups can have a similar effect, flushing stored sediment out of the ice-margin area into the proglacial zone.

(c) Release of water

Water is released by the melting of ice at the margin, and, more significantly, by the outflow of water derived from melting in the supraglacial, englacial and subglacial zones throughout the glacier. Apart from what escapes into the groundwater zone, all of this water (combined with water input to the glacier from rainfall, from inflowing surface streams or from the inflow of draining ice-marginal lakes) eventually exits the glacier at the margin (Figures 9.6 and 9.7). Both the water and the sediment that it carries have a major impact on the ice margin. The morphology of the ice front is in many areas determined by the nature and volume of the water output. Sugden *et al.* (1985) showed how discharge variations in a river skirting the snout of Russell Glacier in Greenland caused variations in the calving rate from the ice-margin cliff and variations in flow velocity up-glacier. The geomorphology of the marginal zone is strongly affected by meltwater erosion and deposition. In many areas the margin is characterised by large amounts of water stored in topographic hollows formed by the position of the ice margin in the landscape. During the 1996 jökulhlaup from Skeiðarárjökull, Iceland, 60–100 $\times$ 10^6 m^3 of water was temporarily stored in a lake that formed between the ice margin and a set of fringing moraines (Andrew Russell, personal communication). The morphology of the ice front and the post-flood topography of the proglacial area were both substantially modified by the existence of the lake. Ice marginal lakes are a major potential store for glacially derived sediment, and also offer the potential for periodic release of temporarily stored water into the proglacial

Figure 9.6 Water emerging from a small subglacial channel at the margin of the Greenland ice sheet. Russell Glacier, Greenland.

Figure 9.7 Large portal formed by melting and calving where a subglacial channel emerges from the snout of Solheimajökull, Iceland.

drainage system. The huge magnitude and significance of lake drainage during the retreat of the Laurentide ice sheet was discussed in Chapter 2. The periodic release of stored water from Glacial Lake Missoula was responsible for creating the distinctive eroded landscape of the channelled scablands in the north-western USA and dumping substantial volumes of sediment onto the continental shelf and slope (Bretz, 1969).

In glaciated areas where jökulhlaups are common, the periodic high-magnitude discharges associated with them can be the dominant control on proglacial landform evolution. For example, Russell (1991) and Maizels and Russell (1992) showed that the sedimentology and morphology of the channel of the Watson River in the proglacial area of the Russell Glacier, West Greenland, remained virtually unchanged over periods of several years between major jökulhlaup events, and that the size and shape of the channel was adjusted to the jökulhlaup discharges rather than to the more common, lower magnitude, melt-dominated discharges. This is a function partly of the recovery time required for a channel to readjust to lower flows after a high flow, and partly of limited sediment supply following scouring of the channel by the flood. Jökulhlaups often carry with them large amounts of sediment and often blocks of ice derived from the glacier. When deposition occurs in the proglacial area, the combination of sediment and ice can lead to characteristic features such as kettle holes, scour marks and boulder rings that can be used to reconstruct the hydraulic and sedimentological characteristics of the event (e.g. Maizels, 1992; Russell, 1993b).

9.3.5 PROGLACIAL GEOMORPHOLOGY

The proglacial environment is the area that is beyond the edge of the ice (and therefore largely beyond the scope of this book), but still subject to the direct effects of the glacier's presence. The geomorphic and geological processes that operate here are conditioned by the activity of the glacier: these include the passage of water and sediment from the glacier through the proglacial zone, and deposition of that sediment, or erosion by that water. In particular the unusually high sediment yield of glacierised areas, and the unusually variable discharge regimes of rivers fed by glacial meltwater lend distinctive characteristics to proglacial areas. At the broadest scale, the proglacial area could be considered to extend almost globally, as fine-grained products of glacial erosion are distributed by air and water to areas far removed from glaciers. The impact of glaciers on eustasy and isostasy also lend a global aspect to their sphere of geomorphic influence. At the scale of individual drainage basins, the effects of glaciation in part of a basin can reach to the extreme distal limit of the basin. Coastlines of glaciated areas have different characteristics from those of non-glaciated areas in terms of geomorphic materials, diversity of wave-energy environments, and range of different process environments per unit length of shoreline (Fitzgerald and Rosen, 1987). At a more restricted scale, most (if not all), of the world's glaciers at present occupy positions of relative retreat compared with their former maximum extent. In other words, the proximal proglacial areas of most glaciers have in the past been subglacial. The proglacial zone thus offers the opportunity to study a whole range of time-transgressive glacier environments in one location.

9.3.6 DEGLACIATION GEOMORPHOLOGY AND THE SUPRAGLACIAL LANDSYSTEM

Deglaciation involves both the exposing of former subglacial landscapes and the creation of specific new landforms of deglaciation. Different styles of deglaciation lead to the creation of different types of landforms. In an active glacier, material on the ice surface is carried forwards by the movement of the ice and dumped at the glacier margin. The previous section described ice marginal landforms created in such situations. However, where glaciers are stagnant or collapsing, surface material may be lowered vertically downwards onto the ground as the ice melts away beneath it. In situations where the supraglacial debris cover is thick, the melting of the underlying ice might take a long time, and the deglaciation process can be 'retarded' substantially. This is especially likely when the mean annual temperature is low, and the glacier ice can be preserved by permafrost. Thick beds of buried Pleistocene glacier ice that are still preserved by permafrost beneath ablation sediments have been described from several locations including Siberia (Astakhov and Isayeva, 1988) and the Canadian Arctic (Lorrain and Demeur, 1985). Downwasting of the ice is the only way in which supraglacial features can survive in tact upon deglaciation, but in most situations the character of landforms and sediments left behind by the lowering of supraglacial materials onto the ground are strongly conditioned by meltwater action and mass movement during the disintegration of the ice. The irregular melting of the ice core, and the sorting of surface materials by water and mass movement, characteristically generates a hummocky topography. The postglacial surface is often an inversion of the supraglacial topography, as low-points on the ice are characterised by thicker accumulations of sediment. McKenzie and Goodwin (1987) described ice-collapse topography in an area of Alaska currently experiencing rapid deglaciation. A hummocky and trenched surface is formed by mass-wasting processes caused by differential melting of buried ice, and the landscape changes progressively as the deglaciation proceeds. In areas of retarded deglaciation landscapes can develop over thousands of years as ice cores gradually decay beneath thick sedimentary coverings.

9.3.7 PARAGLACIATION

The term paraglacial was first defined by Church and Ryder (1972) to describe non-glacial processes that are directly conditioned by glaciation. They used the term to refer both to proglacial processes, and to processes that occur around and within the limits of former glaciers, which are the direct result of the former presence of the ice. A more restricted definition would be to confine the term to areas that have undergone a change from glacial to proglacial, or from proglacial to non-glacial, and where the surface forms and processes are in a state of transition or disequilibrium. A paraglacial environment is one in which geomorphic processes are directly conditioned by the effects of a former glacial environment although the area in question is no longer directly affected by that environment. The key issue is that the geomorphic process environment has changed over time, and that the landscape of the area is in a period of readjustment during which the surface sediments are out of equilibrium with the processes affecting the surface. For example, a subglacial meltout till, deposited to form a relatively stable sediment beneath the debris-rich basal layer of a warm-based glacier, will upon deglaciation find itself in a subaerial fluvial environment where erosive processes are very different from those beneath the glacier. The till may not remain a stable deposit in the proglacial area, and will be subject to rapid erosion. At the transition from glacial to proglacial conditions, sediment supply to the drainage system will be enhanced because of the availability of material that is unstable to erosion in a proglacial environment. The deglaciation process will therefore involve a sediment pulse that lasts until a new equilibrium is established between the surface materials and the process environment. The period of disequilibrium is the period of paraglaciation. While proglacial areas are those still affected by a glacier although they are beyond the glacier margin in terms of space, paraglacial areas are those still affected by a glacier although they are beyond the limits of glaciation in terms of time.

9.4 CONCLUSION

Glacial geomorphology arises from, and can therefore be used as an indicator of, glacier processes. Many students and researchers approach glaciology largely as an aid to the understanding of glacial geomorphology. It is perhaps a shame that more glaciologists do not entertain the study of glacial geomorphology as an aid to the understanding of glaciology. Key glaciological issues such as glacier sliding, the development of basal ice and deforming bed motion are intimately connected to geomorphic processes, and landforms are the calling cards of glacier-bed processes. Palaeoglaciological reconstructions such as that of Kleman *et al.* (1997) point the way to collaborations between glaciology and geomorphology that could be of enormous value. The sediment transfer that constitutes the heart of glacial geomorphology is also significant at a much broader scale than that of individual landforms or even landform regions. The sedimentological concomitants of glaciation and deglaciation induce continental-scale tectonics and associated sea-level change, as well as changes in the chemical composition of oceans and atmosphere as a result of changes in global chemical cycles that are associated with the glacial and deglacial sediment weathering processes. Sugden and John (1976) called for a more glaciological type of geomorphology and a more geomorphological type of glaciology. The former, a glacial geomorphology with a sound glaciological basis, is now becoming well established, but the latter, a glaciology that exploits the potential of geomorphological evidence, is only showing the first signs of emerging.

GLACIER HAZARDS AND RESOURCES 10

The snow from heaven . . . waters the Earth and makes it bring forth.

Isaiah: 55(10)

10.1 GLACIERS AND HUMAN ACTIVITY

Glaciers affect humans directly and indirectly in a variety of ways, both good and bad. The direct impact of glaciers on society can be spectacular and devastating, and the indirect impact can be wide-ranging and pervasive. Impacts can be identified in terms of both present glaciers and former glaciers and their effects. Glaciers have a direct and immediate effect on only a small proportion of the world's population; only a tiny fraction of the world's glaciated area is close to human settlement. However, a much larger number live close to rivers that are fed by glaciers, or live in houses built on sediments that were deposited by glaciers. All of us live with an atmosphere, climate and oceans, the condition of which is directly related to glaciers (Chapter 2).

As with all natural impacts on human activity, the nature of glacial hazards and resources is conditioned by the relationship between the natural environment, human activity and the social perception of natural phenomena. Whether a natural process or feature is a hazard or a resource depends very much on the point of view of the people regarding it. High meltwater discharges are a problem for the bridge-builder but a boon for the hydro-power engineer. Icebergs are a hazard to shipping but a potential source of fresh water to arid lands. The extent to which hazards can be mitigated or resources exploited depends partly on the level of technological development of the society concerned. Glacial resources such as ice for refrigeration that were valuable to low-technology societies are largely irrelevant in a high-technology environment. The unpredictability and sediment yield of glacial meltwater streams only become problematic when the opportunity arises to exploit the hydro-electric potential of such streams. Economic penetration into increasingly remote areas has increased human exposure to glacier hazards in spite of widespread glacier recession. Tourism, especially skiing, brings an increasing number of people into contact with glaciated environments, providing evidence of some of the potential value of glaciated areas, but also providing potential victims for the natural hazards of those areas.

In all cases, the key to the exploitation of resources and the mitigation of potential hazards is prediction of the location, timing, magnitude and quality of phenomena. Prediction is based on understanding of: glacial processes; the needs of resource users; and the vulnerabilities of hazard victims. The following sections consider some of the aspects of glaciation that have been considered as hazards or resources in different circumstances. It is indicative of the nature of hazards and resources that some phenomena are listed both as hazards and as resources.

10.2 GLACIER-RELATED HAZARDS

Glaciers and glacial environments offer a daunting array of hazards to inhabitants and travellers. Tufnell (1984) listed a range of afflictions that have beset the Chamonix valley through history, including advance of glaciers into the village, floods devastating properties, rocks and icy debris from glaciers showering onto habitations below, the inundation of land by encroaching ice and the loss and destruction of property and livestock. Some hazards are exclusive to the area of the glacier itself, such as crevasses which, especially when bridged by a covering of snow, can prove fatal to unwary travellers. Others affect areas well beyond glacier limits. For example, glacier outwash streams are notoriously variable both in discharge and planform, and proglacial areas are subject to devastation by course changes in such rivers, which can result in the washing away of areas of land, or covering of the ground in gravel and boulders.

10.2.1 ICE AVALANCHES

Where glaciers terminate on steep slopes, as is the case in many mountain areas, ice calving from the glacier may plunge downslope under the influence of gravity before coming to rest at a lower elevation. Data from such events in glaciated areas around the world have been reviewed by Alean (1985) in an effort to characterise the events and predict their occurrence. In some cases, ice avalanches occur in conjunction with landslides or snow avalanches, but the largest known 'pure' ice avalanche, from Mt Iliamna, Alaska, in 1980 involved $\sim 2 \times 10^7$ m^3 of ice. More typically, avalanches are of the order of 10^4–10^6 m^3. Although less common in most locations than snow avalanches, ice avalanches have nevertheless accrued a substantial death toll. At least 124 people were killed in ice avalanches between 1901 and 1983 in the Swiss Alps alone, and individual events have wiped out whole settlements, causing large numbers of deaths. In 1965 an ice mass of $\sim 10^6$ m^3 fell from the tongue of the Allalingletscher in Switzerland and killed 88 people who were working on the construction of a hydroelectric power plant in the valley below. In 1597 an entire village near the Simplon Pass was buried beneath an icefall. The village of Randa in Switzerland is so subject to periodic ice falls from the Weisshorn that evacuation of the village has been a recommended precaution at times when the rate of flow of the glacier is observed to increase (Röthlisberger, 1978). Ice avalanches from Nevado Huascaran, the highest Peak in Peru's Cordillera Blanca, have twice this century caused massive devastation. In 1962 about 4000 people were killed by an ice avalanche in which about 2×10^6 m^3 of ice fell about 3000 m from an unnamed glacier into the populated valley below. The cascading ice mass reached speeds exceeding 105 km h^{-1} and doubled its volume entraining debris of all sorts as it plunged downhill. The avalanche travelled 16 km in eight minutes and buried nine villages in an icy rubble up to 20 m thick. Bodies were found as much as 160 km downstream at the Pacific mouth of the Santa River. In 1970, a combined rock and ice avalanche from the same mountain buried the town of Yungay, killing about 18,000 people.

Like snow avalanches or landslides, ice avalanches are characterised by a starting area, a transit zone or avalanche track, and a deposition area. Important parameters for prediction are: conditions required for the onset of avalanching; the timing of the event; the total volume of the avalanche; and the runout distance or reach. These parameters seem to vary with the nature of the starting area. One type of starting area occurs where the glacier bed steepens abruptly, for example over a cliff, and ice calves in sheets from an ice wall as it advances to the lip of the break of slope (Figure 10.1a). The timing of avalanches in this situation can be predicted from the rate of flow of the glacier, and the volume of the avalanche can be predicted from the cross-sectional area of the glacier and the amount of overhang that is likely to occur before the ice breaks off. A second type of starting zone occurs where glaciers rest on steep but regularly sloping surfaces (Figure 10.1b). The critical gradient at which the glacier in this case becomes susceptible to avalanching depends largely on the basal temperature. Glaciers that are frozen to their beds remain stable at greater slope angles

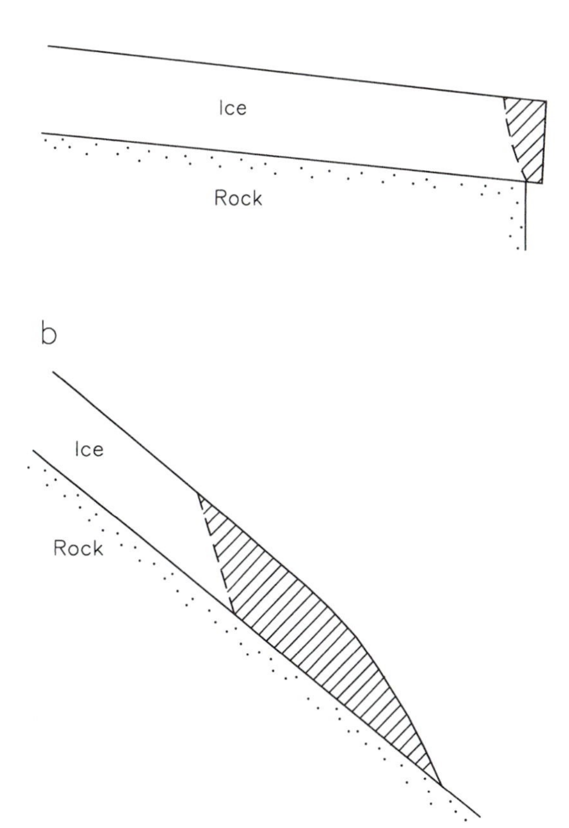

Figure 10.1 Two different types of starting situation for ice avalanches, (a) with and (b) without a break of slope. After Alean (1985).

than those with temperate beds. Starting zones of this second type can produce larger avalanches than the break-of-slope type, since ice can build up over a larger area before the fall occurs. For glaciers frozen to their beds and for break-of-slope type events, avalanching is controlled by mechanical failure within the ice. For glaciers with unfrozen beds and regular slopes, seasonal flow variation seems to play a more important role.

The hazard from ice avalanches stems not only directly from the ice, but also from associated events. For example, in designing reservoirs that are likely to be in the path of ice avalanches, care must be taken to ensure that the retaining dam is sufficient to accommodate waves caused by ice falling into the water. Catastrophic breaching of moraine-dammed lakes often results from waves caused by ice falls into the lakes. Floods from ice-dammed and moraine-dammed lakes have been described as the single most important glacier hazard in China, and the majority of lakebursts that have been studied there have result from icefalls into lakes (Ding Yongjian and Liu Jingshi, 1992).

10.2.2 FLUCTUATIONS AND SURGES

Glacier fluctuations pose a direct threat of overriding areas of human occupancy, and indirect threats of changes to water supply, flood routing, and disruption of transport routes. The significance of the problem varies from time to time as rates of change in the area of glaciation vary. During the Little Ice Age (a period of lower temperatures affecting most of the globe between about the thirteenth and nineteenth centuries) substantial glacier advance was a dominating feature of life in many glaciated areas. Grove (1988) reports historical records from Europe, Asia and the Americas of glacier advances in historical times that have overrun villages and farmland. For the area around Breiðamerkurjökull in southern Iceland, for example, which is now a bleak and largely uninhabited place with ice reaching close to the sea, records from the earliest settlement of Iceland (AD 870–930) through to the eighteenth century describe a landscape of woodland and prosperous settlements progressively disappearing beneath the advancing ice. The settlement of Fjall, prosperous in 1660, was abandoned by 1695 and was beneath the advancing ice by 1709. Similarly in the Alps, the village of Le Chatellard was destroyed by the advancing Mer de Glace between 1600 and 1610. Few human installations nowadays are situated very close to the margins of glaciers because we are in a period of history where most glaciers have recently retreated from more advanced positions and are currently re-advancing over recently exposed land. The same was not true for early settlers in places such as Iceland, who populated glaciated areas at a time when the phenomenon of substantial glacier advance was less apparent from the landscape, and utilisable farmland extended right up to the margins of glaciers that were encroaching on land that had not been glaciated for a long time.

Glacier surges represent a special case of glacier fluctuations that are largely unrelated to widespread climate change, and are often associated with other hazards such as floods. While climate-related advance is usually predictable and slow to take effect, surge advances can be very abrupt and can surprise inhabitants at some distance from the glacier. The Solda Glacier in Switzerland is reported to have advanced 1200 m in about a year. The mechanisms of surging were discussed in Chapter 7.

10.2.3 FLOODS AND LAHARS

Outburst floods, and the mass movements that they sometimes initiate, are among the most devastating of glacier hazards, and also among the most frequent in occurrence. For example, in the Swiss Alps, damage from glacier floods occurs on average every 2 years. Some of the glaciological controls on the initiation of outburst floods or jökulhlaups were discussed in Chapter 6. Jökulhlaups occur when water stored in, or dammed up by, the glacier is suddenly released. Haeberli (1983) reported that 60–70% of glacier floods in the Swiss Alps were caused by the breaching of ice dams of the outburst of ice-marginal lakes, and 30–40% by the rupture of water pockets within glaciers. Björnsson (1992) described floods of various origin in Iceland. In Iceland there are at present about 15 periodically draining ice-marginal lakes with peak discharges of 1000–3000 $m^3 s^{-1}$, and in addition floods are produced regularly from six subglacial geothermal areas (e.g. Gudmundsson *et al.*, 1997). The subglacial lake Grimsvötn has drained at regular 4–6-year intervals for over half a century, producing peak discharges of 600–10,000 $m^3 s^{-1}$. In addition, at least 80 subglacial volcanic eruptions have been reported since the settlement of Iceland. Some of the largest Icelandic floods, 'Katlahlaups' with peak discharges of up to 300,000 $m^3 s^{-1}$, have been generated by eruptions of the volcano Katla, beneath the Vatnajökull ice cap. A jökulhlaup from Myrdalsjökull in 1918 had a discharge equivalent to three times that of the Amazon. Some jökulhlaups are easy to predict because they occur on a regular basis in known locations. In some circumstances, lake formation and drainage depends on the position of glaciers in the landscape, and periods of frequent jökulhlaups are separated by long periods when no events occur. For example, disastrous floods afflicting the Cordillera Blanca in Peru restarted in the 1930s after a period of some 200 years without floods, as the flooding depended on the glaciers retreating behind the position of Little Ice Age moraines to form unstable moraine-dammed lakes. The likely damage or hazard can be predicted not on the basis of water volume alone, but on the mechanism of drainage, the peak discharge, and the local topography and lithology. In many cases, water released from the glacier combines with loose sediment from moraines or other proglacial materials to produce mudflows. These are prevalent especially from draining of moraine-dammed lakes where moraine is incorporated into the flow, and from volcanically triggered jökulhlaups where the water combines with ash and unconsolidated volcanic materials to produce mudflows known as lahars.

In most cases, the magnitude of major jökulhlaups is such that attempts at engineering defences are useless. Nevertheless, engineering measures can be taken to reduce the magnitude or likelihood of flooding. For example, the Gietroz Glacier in Switzerland advanced from a tributary valley to block the Val de Bagnes to create an ice-dammed lake in 1549, 1595 and 1818. In 1595 the outburst flood caused when the lake overflowed killed 500 people. When the lake formed in 1818, engineers cut an artificial channel across the dam, and succeeded in draining about one third of the lake water before the lake burst. Nevertheless, 50 people perished in the resulting flood (Grove, 1987). In 1892, a subglacial water pocket of more than 100,000 m^3 broke out from the Tête Rousse glacier killing 175 people. Engineers proceeded to drill a tunnel through rock to the bed of the glacier to drain other possible water pockets (Vivian, 1977). Recession of glaciers in the Cordillera Blanca of Peru since 1927 has resulted in the formation of substantial lakes between retreating glacier tongues and the fringing Little Ice Age moraines (Lliboutry *et al.*, 1977). As the lakes filled, the dams became unstable, and liable to collapse either because of the pressure of water or because the water level rose to a point where waves caused by ice falls into the lake could overtop the dam. Some of these lakes drained catastrophically, incorporating material from the dam to course downhill as mudflows. In 1941 the town of Huaraz was devastated by a flood that killed 6000 people. Immediately thereafter a commission was established to mitigate the hazard of the Cordillera Blanca lakes. The floods have been successfully prevented by artificially restricting lake depths through the installation of artificial drains and canals through the moraines.

Jökulhlaups commonly occur in unpopulated areas and pose little threat to human activity, but some have affected human population centres and installations. In southern Iceland on 5 November 1996, meltwater that had been produced as a result of a subglacial eruption beneath the ice cap Vatnajökull emerged as a jökulhlaup from the snout of the outlet glacier Skeiðarárjökull. Discharges of around 45,000 $m^3 s^{-1}$ swept away the roadway and bridges from a section of Iceland's main national ring-road, causing Icelandic Prime Minister Oddsson to say that 4 hours of flooding had knocked back the road-building programme by 20–30 years. The incursion of economic development into remote areas has rendered jökulhlaups that used to be relatively insignificant much more important. For example, one of the largest periodically draining ice-dammed lakes in North America, the $\sim 8 \times 10^8$ m^3 Strandline Lake in Alaska, used to drain through an area of uninhabited wilderness. However, that area now contains oil and gas installations, roads, bridges and power lines leading to Anchorage, Alaska's largest city (Sturm and Benson, 1985)

Even more notorious than jökulhlaups are the glacio-volcanic mudflows, or lahars, that are commonly associated with volcanic activity on glaciated mountains. Many volcanoes are capped with glaciers, and eruptions can lead to the melting and dislodgement of huge volumes of ice, to generate floods, ice falls and lahars. The eruption of Mt Redoubt, Alaska, between 1966 and 1968 blasted, melted, scoured and/or washed away about 6×10^7 m^3 of glacier ice from the upper part of Drift Glacier (Sturm *et al.*, 1986). Surface and subsurface melting of the ice generated a series of jökulhlaups, one of which flooded the site of the oil-tanker terminal on Cooke Inlet at the mouth of the Drift River. The same volcano erupted again between 1989 and 1990, and $11–12 \times 10^7$ m^3 of ice was dislodged by flows that required evacuation of the oil terminal (Trabant and Meyer, 1992). One of the most deadly and widely known glacier-related disasters ever to have occurred was the lahar caused by melting of snow and ice during the eruption of the volcano Nevado del Ruiz in Colombia in 1985 (Naranjo *et al.*, 1986; Williams, 1989–90). About 10% of the volcano's ice cap melted and combined with volcanic ejecta to create lahars that flowed down several separate channels off the volcano. Peak flow in one channel reached 48,000 $m^3 s^{-1}$, with a velocity of 38 kmh^{-1} and a wave front about 40 m high. The lahar flooded through the town of Armero at about 30 $km\ h^{-1}$ killing more than 20,000 people. The lahar had been predicted, and the threat to Armero was well understood. The huge death toll has been attributed to the unwillingness of government authorities to risk the economic and political cost of premature evacuation (Alexander, 1993).

10.2.4 ICEBERGS

Icebergs pose a threat to shipping in both hemispheres. In the northern hemisphere the most problematic icebergs have been those that drift as far south as the North Atlantic shipping lanes. The most prolific source of icebergs affecting this area are glaciers draining to the west coast of Greenland. Icebergs calving from Jakobshavn Isbrae and about 20 other glaciers in West Greenland drift with the West Greenland Current first northwards and then south with the Labrador Current towards the area of the Grand Banks, a shallow area of sea extending into the Atlantic east of Newfoundland (Figure 10.2).

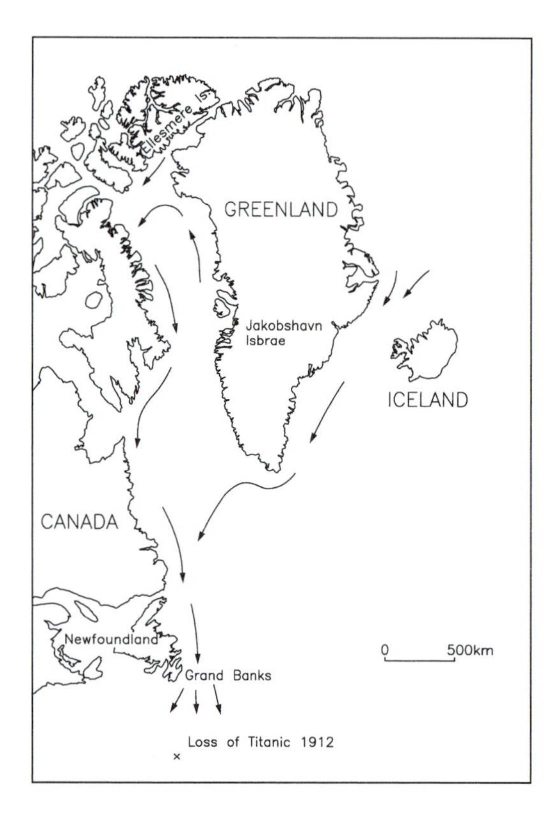

Figure 10.2 Main drift paths of icebergs in the North Atlantic Ocean.

The area is sometimes referred to as 'iceberg alley'. In addition, icebergs calving from Ellesmere Island, Novaya Zemlya, and Spitzbergen, when released from the polar gyre, commonly drift through the Denmark Strait between Iceland and Greenland and then around the southern tip of Greenland to join the West Greenland route. West Greenland icebergs typically survive for up to 2 years, but larger bergs, and ice islands, can last for as much as 10 years. On rare occasions icebergs survive into very low latitudes. Icebergs were sighted close to Bermuda in 1907 and 1926, and close to the Azores in 1921. In the Grand Banks area alone, between 1882 and 1890, 54 ocean liners were reported sunk or damaged in collisions with icebergs. The most famous victim of a North Atlantic iceberg was the SS *Titanic* which in 1912 struck an iceberg of ~200,000 tons and sank with the loss of 1503 lives. On a more individual scale, the explorer Gino Watkins is thought to have drowned when an iceberg 'rolled' (shifted its vertical orientation) in the water close to where he was kayaking. Submarine installations such as cables and pipes, as well as shipping, are subject to iceberg damage. Large icebergs can ground on the sea-bed even at substantial distances from land, carving furrows in the sea-bed and ripping up cables and pipelines. Barrie (1980) found that in the northern Labrador Sea, where up to 2500 icebergs per year drift southwards in the Labrador Current, iceberg scour marks were gouged on the sea-bed in water-depths of more than 180 m. Scours had an average width of 30 m and an average depth of 5 m, but much larger scours, up to 17 m deep and up to 200 m wide, occurred in shallower water.

In 1914, following the loss of the *Titanic*, an organisation now known as the International Ice Patrol was established under the administration of the US Coastguard to find and announce positions of hazardous icebergs. In recent times monitoring has also been carried out by satellite remote sensing (e.g. Vinje, 1980; AGRISPINE, 1983), and various theoretical models have been proposed to predict the paths of icebergs (Marko *et al.*, 1988). Iceberg production varies through time at individual glaciers; for example, rapid retreat of Columbia Glacier near the Alaskan oil shipping port of Valdez has resulted in increased iceberg production threatening oil-tanker routes. Attempts to destroy bergs by bombing and other means have largely been unsuccessful, but attempts to tow bergs for short distances to divert them from collision courses with fixed installations such as oil-rigs have met with some success. The first iceberg-proof fixed oil-rig, the Hibernia, was established in the Grand Banks area in 1997 at a cost of about $5 billion. The rig is designed to withstand the impact of bergs up to 6 million tonnes, but any bergs approaching within 6 miles of the rig will be subject to towing to minimise the risk of collision.

10.3 GLACIER-RELATED RESOURCES

People have made use of glaciers and glacier-related phenomena since prehistoric times, and glaciers continue to be a resource in a variety of ways today. Both the perception of what resources glaciers offer, and the ability to exploit those resources are technology-dependent, and the history of human use of glaciers is almost certainly not yet complete.

10.3.1 REFRIGERATION

One of the earliest uses of glacier ice, and one that continues today in some areas, was as a refrigerant. Caves within glaciers have been used as stores for perishable provisions, and ice cut from glaciers, or collected from calved fragments, has been used commercially and domestically for preserving food and cooling drinks in areas far removed from naturally occurring ice (Ellis, 1982). Before refrigerators were invented, Norway exported ice to other European countries, and a substantial trade in North American lake-ice flourished in Europe in the second half of the nineteenth century (Proctor, 1981). Ice houses built to supply the kitchens of the European aristocracy were stocked with ice from distant glaciers. In the 1850s, ice was transported from Alaska to California, and small ice-bergs from southern Chile were transported as far north as Peru for use as refrigerant. In most parts of the world the trade in glacier ice perished in the early years of the twentieth century with the advent of mechanical methods of refrigeration, but the traditional market in glacier ice continues to the present day in areas such as the Cordillera Blanca of Peru, and in the Hunza region of the Karakoram Mountains (Smith, 1989). Ice is cut from glaciers in the mountains and transported to market by mule and truck, packed in grass, sacking or sawdust for insulation. The drinks and flavoured ices that are made from glacier ice are probably safer for consumption by tourists than are refreshments derived from some other local water sources.

10.3.2 WATER

Glaciers are an abundant source of the most basic of human resources: water. Meltwater from glaciers is the source of many of the world's major rivers, and the root of both secular and spiritual resources: the Gangotri glacier at the source of the Ganges has become such an attraction to tourists and pilgrims that human activity has been blamed for accelerating the retreat of the glacier (Jayaraman, 1996). Meltwater can be used for irrigation, for industry, to produce power (discussed in more detail below) and, when clean, for drinking and domestic purposes. Many arid areas, including the Thar desert and the Atacama desert, receive water for irrigation from meltwater streams draining glaciers in adjacent mountains. The glaciers of China produce an annual runoff of 5.64×10^{10} m^3 of water, which is equivalent to 2% of the country's whole surface-water resources. Glaciers are especially valuable as a water source since they produce most water in hot dry weather, at times when other sources, such as precipitation, are at a minimum. Some major cities, such as the Bolivian capital La Paz, derive their water supply almost entirely from glacier-related sources. In the USA the Arapaho Glacier produces about 260 million gallons of drinking water per year for the city of Boulder, Colorado.

Several attempts have been made to change water supply from snow and ice melting by artificial means (e.g. Higuchi, 1969). Traditional farming methods in parts of Asia have involved spreading dirt on spring snow to speed its melting, and a similar procedure can be used to increase summer melting of glaciers. A thin layer of a dark, heat-absorbing material such as powdered coal can enhance surface melting by as much as 55%, while a thick coating of material, or use of a thermally insulating material, can reduce melting.

Glaciers can supply water not only directly but also indirectly, via icebergs, the use of which is discussed below.

10.3.3 ICEBERGS, ICE ISLANDS AND ICE SHIPS

Icebergs, being composed of fresh water, offer a potential water resource if they can be transported to areas where water is needed. The cost of processing water from a captive iceberg would be less than the cost of recovering fresh water from ocean water by desalinisation. This is a good example of how the definition of a resource depends on human perception: if the need is great enough and a market exists, the cost and technical difficulty of exploitation can be faced. In the late 1970s Prince Mohammed al Faisal of Saudi Arabia established Iceberg Transport International Ltd., and galvanised the international scientific community to consider issues related to iceberg utilisation. Faisal's ultimate goal was to transport a 100 million ton iceberg 9000 miles from Antarctica (the continent that holds the bulk of the world's freshwater reserves) to Saudi Arabia (a country without a single perennial stream). Weeks and Campbell (1973) and Weeks (1980) reviewed the possibilities of iceberg utilisation and showed it was broadly feasible, but that it was likely to be a realistic possibility only for short transport routes in the southern hemisphere. One major problem would be to transport the iceberg. Not only would the physical manipulation of an iceberg be an awesome task, but the iceberg would be subject to melting and fracturing *en route*. For relatively short routes such as from the Antarctic to Australia, it should be possible to transport a iceberg with about 50% loss, but for the journey to northern hemisphere sites, the iceberg would be in danger of disappearing completely during the voyage. Icebergs can thin by as much as $3\ m\ d^{-1}$ in adverse conditions. The problem could be overcome to some extent by insulating the iceberg, but it would be technically difficult to prevent melting from the sides and bottom. De Marle (1980) produced specifications for a project in which icebergs would be towed from the Weddell Sea to South Africa via the cold Benguela current to minimise melting. Problems could also arise from thermal pollution of waters *en route* and at the destination by the cold fresh water produced by the melting iceberg. Local microclimate changes at the destination site have also been postulated. Problems would exist at the destination for docking and processing the iceberg. Large bergs would have a draft of as much as 200 m, and could not be towed close to shore. Around most of the south coast of Australia, icebergs could be brought no closer than about 35 km. The iceberg would have to be moored and then processed at sea. Ice could be mined from the iceberg and either floated, or piped as a slurry, to the shore. Both the water and the energy released by the melting of the ice could be utilised. Weeks (1980) estimated that a $1\ km^3$ iceberg could, in melting, produce as much as $120 million worth of electricity.

Large tabular icebergs that calve into the Arctic Ocean are sometimes referred to as ice islands. They are usually smaller versions of the large tabular icebergs that are common in the Antarctic. The main source of ice islands seems to be calving of the Ellesmere Island ice shelf, the snouts of the large Pettermann and Jungersen Glaciers in North Greenland, and a few other locations. In the Arctic Ocean, ice islands can remain in circulation, frozen in the sea ice, for years. Tracking these islands has provided information on circulation of the circumpolar gyre ocean current, and several have been put to use as mobile laboratories. One such island, labelled ARLIS II (Arctic Research Laboratory Ice Station No. 2) measured about 15 km^2 and was used as a scientific base from 1961 to 1965 before drifting out of the circumpolar gyre into the southward East Greenland current and eventually breaking up in the North Atlantic.

Ice islands can make convenient floating landing strips for aircraft, and artificial ice islands have been proposed as strategic devices. During the Second World War, the British Government considered plans by inventor Geoffrey Pyke to build an aircraft carrier from a material to be known as Pykrete, composed of ice strengthened with wood-pulp. The plan was abandoned as the cost and difficulty would be even more than that of constructing a conventional carrier. However, the ice ship would have had advantages: for example, it would probably have proved very resistant to attack, as indicated by the limited success of attempts to destroy hazardous icebergs by bombing.

10.3.4 POWER

One of the major uses of water derived from glaciers is in power generation. In some areas, such as Greenland, the total human power requirement could theoretically be met by hydroelectric power fed by glacier meltwater. Norway derives all its electricity from hydroelectric power schemes, many of which are supplied by glacier water. Glaciers are especially valuable as a water source since they produce most water in hot dry weather, at times when other sources, such as precipitation, are at a minimum. The Massa hydroelectric power station near Brig in Switzerland runs largely off summer meltwater from the Grosser Aletschgletscher. Meltwater produced in summer is stored in reservoirs and released to the hydroelectric power station in winter. In some cases, water is fed into the power schemes directly from the glacier. For example, Wold and Østrem (1979) described how engineers overcame the problem that water discharging from the snout of the Bondhusbreen, an outlet glacier of the Fogelfonni ice cap in south-west Norway, emerged at a lower elevation than the reservoir that fed Mauranger power station. When the power station was designed it was decided to build a subglacial water intake at the bed of the glacier, well above the level of the snout. A tunnel was cut through the bedrock beneath the glacier, and water intakes were drilled upwards to the glacier bed beneath ice 160 m thick. Volumes of about 60×10^6 m^3 a^{-1} were captured and fed to the reservoir via a tunnel system. However, there are problems in utilisation of meltwater streams: the discharge is variable, difficult to predict, and highly sediment charged. Sediment in the water can block or erode water intake structures, can fill reservoirs quickly with sediment, and can cause rapid wear of pumps and turbines. For these reasons sediment traps must be incorporated upstream of the water intakes, designed to trap all of the bedload and much of the suspended load (Bezinge *et al.*, 1989).

10.3.5 WASTE DISPOSAL

Recently, the idea of using glaciers as natural dustbins has been the object of some debate. The issue revolves especially around the problem of disposing of radioactive industrial wastes. These wastes need to be kept out of contact with the biosphere for periods of up to 250,000 years. In the early 1970s, the International Atomic Energy Agency considered the possibility of burying waste beneath the Antarctic ice sheet (Zeller *et al.*, 1973; Weertman *et al.*, 1974), but the idea was rejected on the grounds that the necessary isolation from the biosphere could not be guaranteed. The potential consequences of accidental leakage from glacier storage are unsettling because of the direct link between glaciers and the rest of the global hydrological system. Further research into the issue has continued, however, as the problem of long-term storage of waste remains unsolved. Philberth (1977) proposed that all of the world's high-level radioactive waste products for a 30-year period could be disposed of in an area of 15 km radius, at a depth of between 20 and 100 m beneath the surface of either the Greenland or Antarctic ice sheets. He proposed that the waste should be made insoluble by being fused into glass, and then loaded into 30 million lead-shielded spherical containers of 0.4 m diameter. He argued that the deposit would not affect the stability of the ice sheet and that even the most 'upsetting' natural ice-sheet instabilities or climate changes could not cause radioactive contamination. The stored material would need to be sufficiently difficult to retrieve that no potential malefactor would be able to gain access, but not so difficult that the material could not be retrieved if necessary.

Zotikov (1986) has considered the problem as an application of the thermophysics of glaciers, and described three styles of storage. The first method is to allow the waste to melt its way through the ice and then reside at the bed of the ice sheet. A cylindrical waste container 30 cm in diameter and 3 m long would melt its way through 1.5 km of ice at −50°C in 3 or 4 years. The bed of a glacier is probably not a stable environment for storage, however, and leakage would communicate quickly with the outside environment via basal meltwater. An alternative solution would be to allow the waste to descend very slowly through the ice so that it remained above the bed, surrounded by ice well below the melting point, for a sufficiently long period. This could be achieved either by diluting the waste with an inert filler so that each waste receptacle would emit less heat, or by fixing the receptacle via a cable to an 'anchor' frozen into the ice above. In the latter case, the container would descend at the rate of vertical movement of the ice itself, which, if the container were wisely located, would allow as much as 250,000 years storage time. The weak link in this strategy would be the mechanical link between the container and the anchor, which would have to remain intact throughout the period. A third approach would be to pack the waste in containers with a density less than the density of water, so that they would float in a water-filled englacial chamber generated by the heat of the waste. The container would descend from the surface through the firn layer and come to rest in a water bubble in the uppermost layer of impermeable ice at a depth (about 50 m) where meltwater is unable to escape through the porous firn. As the ice moves vertically downwards through time, the water bubble will remain in place at the top of the ice column, the waste container effectively trying to melt its way up into the firn. Zotikov (1986) considered the implications of these possibilities for the thermal state of the glacier containing the waste, and concluded that none of the methods could be recommended in view of potential hazards that were impossible to evaluate with our current level of understanding of the behaviour of ice sheets.

10.3.6 MATERIALS, MINERAL TRACERS AND LANDSCAPES

Glacial activity produces as major by-products materials such as sand and gravel. Many formerly glaciated areas, such as Britain, are rich in glacier-related sand and gravel deposits that are a major industrial resource. In Britain alone the building industry uses nearly 300 million tonnes of sand, gravel and aggregate each year (Merritt, 1992). The distribution of these materials reflects the geography of former glaciers and the geography of glacial geomorphic processes. For example, fluvioglacial landforms such as eskers, kames and outwash fans are easily identifiable sand and gravel sources (Crimes *et al.*, 1992).

The distribution of glacial deposits can also be used in the location of mineral resources (e.g. Shilts, 1976; Evenson *et al.*, 1979; Eyles and Menzies, 1983). In areas where bedrock is obscured by a surface covering of glacial sediment, mineral prospecting by direct investigation of bedrock is difficult. However, the location of specific lithologies or mineral deposits in the bedrock can be inferred from the distribution of material in the overlying deposits. The concentration of a particular material in till usually peaks within a few kilometres of the source, and then declines in a negative exponential curve. Locating the head of this dispersal fan by sampling the surface material serves to focus subsurface prospecting efforts.

In the last few hundred years, especially since the growth of the romantic movement, glaciers have been a source of aesthetic interest. Both active glaciers and formerly glaciated landscapes form the basis of tourist industries in many parts of the world. Both the USA and Canada have 'Glacier' National Parks. The role of ice and glaciers in the imagination, and the cultural perception of polar phenomena were considered by Spufford (1997) and are discussed further in Chapter 12.

10.4 ENGINEERING PROPERTIES OF GLACIAL SEDIMENTS

One of the major consequences of glaciation from a human perspective is the creation of new land surfaces by the deposition of sediment by ice and meltwater. A substantial proportion of the earth's total ice-free land area is surfaced with glacigenic deposits, and the characteristics of these deposits are therefore important in an engineering and construction context. One of the main characteristics of glacial deposits is their immense variability, even over very short distances. This is related to the variability of glacial processes over short distances, and the time-transgressive superposition of processes and deposits. The deposits themselves are physically highly variable because of the range of materials involved and the range of different processes by which they are deposited. The glaciological processes and conditions that give rise to specific sediment types also produce specific landforms, so landform evidence can be used as an indication of sediment type. Geomorphology therefore has a role in the prediction of the distribution of surface materials and ground properties.

Fookes *et al.* (1975), Boulton (1975b) and Boulton and Paul (1976) discussed relationships between the geotechnical properties of tills and the processes of till formation. Grain size distribution, consolidation, jointing and clast fabric influence properties such as bearing capacity, plasticity, settlement, slope stability, ease of excavation and value as fill or construction material. Deposits related to specific glacial processes can thus be predicted to have specific properties. For example, lodgement tills are likely to be highly consolidated, clayey and unsorted, and are thus likely to be difficult to excavate but to have good bearing and stability characteristics. They often have a strong clast and jointing fabric that will affect their shear strength, and high clay content that will affect their hydrological characteristics. By contrast, ablation tills and fluvioglacial materials are likely to be less consolidated and hence easier to excavate but less stable. Waste disposal in formerly glaciated areas also requires a knowledge of the properties of glacial sediments. Discussing the siting of waste tips, Gray (1993) described how the heterogeneous nature of glacial till, incorporating both impermeable clay-rich sediments and permeable lenses of coarser material in very small areas, makes it unreliable as an aquitard, and thus unsuitable as a substrate for a waste tip without an artificial lining.

10.5 CONCLUSION

Glaciers have a significant impact on human activity, for both good and bad, both locally and globally. Glacier hazards are tragically familiar in glaciated areas, and glacier resources of various kinds have been exploited in the context of different technological perceptions and capabilities. Although technology has played a role in the mitigation of glacier hazards, hazard mitigation usually relies heavily on hazard prediction, which the theoretical status of glaciology is not yet able to provide at the level of sophistication that it one day almost certainly will. The exploitation of resources depends on a perceived need as well as a technological capability, and, hitherto, the perceived potential value of glaciers has been relatively limited. It is exciting to speculate what glaciers might offer if any effort was expended on exploring their full potential. Glaci-electric power stations driven by the slow but hugely forceful motion of ice streams, or by the phase change of billions of tonnes of ice at the melting point, are not yet, as far as I know, even science fiction. However, the future of applied glaciology, like the future of theoretical glaciology, will be filled with things that we have not yet even imagined.

ICE CORES 11

Ever since scientists first took an active interest in alpine glaciers, they have been overcome by an irresistible need to drill into them.

(Rado *et al.*, 1987)

11.1 ICE CORES

Drilling through glaciers and extracting ice cores is a major source of information for glaciologists and for other environmental scientists. From a scientific perspective, ice cores are a major resource offered by glaciers and ice coring is a major practical application of glaciology. Coring provides access to the interior and base of the glacier, affording a view of the ice, of the hydraulic system, and in some cases of the substrate. Ice sheets are depositional environments that preserve layered sequences of material precipitated from the atmosphere, so the extracted cores of ice provide one of the main sources of data for palaeoenvironmental reconstruction. Problems exist in acquiring cores and in the interpretation of the material the core contains: the cores themselves and the information within them are difficult to extract! Nevertheless, cores drilled through polar ice sheets can provide a record of snowfall and climate stretching back more than 250,000 years. Shoji and Langway (1989) have suggested that the thickest parts of polar ice sheets might contain a record stretching back as much as 3 million years. Ice coring has been at the heart of several major environmental reconstruction initiatives, and the logistical, financial and scientific requirements of deep coring programme continue to encourage major multinational collaborations. A by-product of ice-coring programmes has been the substantial amount of general glaciological research carried out in order to plan and accomplish the coring programmes, as well as the research based on the data that such programmes have acquired. Study of the problems facing different types of coring operations, and of the difficulties surrounding the use of core data, can highlight a range of glaciological issues.

11.2 ICE-CORING HISTORY AND PROCEDURES

Even the early pioneers of glaciology engaged in glacier coring. Louis Agassiz drilled holes up to 60 m deep into Unteraargletscher between 1840 and 1842, and drilling operations by Adolf Blümcke and Hans Hess at Hintereisferner between 1895 and 1909 achieved the first successful drilling to the bed of a glacier (Clarke, 1987). Since then a large number of holes have been drilled into small glaciers and the upper sections of thick ice sheets, but very few cores have been retrieved that penetrate right through from the surface to the bed of thick ice sheets. These few cores, details of which are summarised in Table 11.1, have achieved a critical status in the glaciological literature. The first deep core to penetrate the total thickness of an ice sheet was retrieved from Camp Century in north-west Greenland in 1966, and the second from Byrd Station, Antarctica, in 1968. The bed of the Greenland ice sheet was reached again at Dye 3 in 1981. A number of shorter cores reached the bed through thinner ice, and some of these penetrated ice of similar age to the ice at the bottom of the deeper cores. For example, a core at Taylor Dome, Antarctica, reached the bed through about 540 m of ice, with an estimated

Table 11.1 Details of major cores that have reached through more than 1000 m thickness of ice to the base of the Greenland and Antarctic ice sheets

	Camp Century, Greenland	*Byrd Station, Antarctica*	*GISP2, Greenland*	*Dye 3, Greenland*
Date of core	1966	1968	1993	1981
Location	77° 11′N 61° 08′ W	80° 01′ S 119° 31′ W	72° 58′ N 38° 48′ W	56° 11′ N 43° 82′ W
Surface elevation (m)	1885	1515	3208	2490
Ice thickness (m)	1388	2164.4	3053.4	2037
Age of bottom ice (thousand years)	100–130	65–90	250	100–130
BIL thickness (m)	15.7	4.83	13.11	24.5
Bed	Till	Unconsolidated	Rock	Rock
Basal temperature (°C)	–13°	–1.6(est)	–9.22	–13.2
Surface temperature (°C)	–24.4	–28	–31	–20
Accumulation rate (m a^{-1})	0.38	0.13	0.24	0.56
Surface ice velocity (m a^{-1})	5.5	12.8	1.63	12.5

age of about 160,000–170,000 years at the bottom. At Law Dome, an independently flowing ice cap at the edge of the east Antarctic ice sheet, basal ice close to the bed was achieved at the bottom of an 1196 m core (Morgan *et al.*, 1997). Other deep cores, such as that at Vostok in Antarctica, have penetrated to great depths without reaching the bed. Details of some of these cores are given in Table 11.2. In 1989 two complementary drilling programmes were initiated at Summit, the highest point on the Greenland ice sheet, with the aim of penetrating the thickest, least disturbed part of the ice. The European Science Foundation's Greenland Ice Core Project (GRIP) and the American Science Foundation's Greenland Ice Sheet Project Two (GISP2) established separate drilling sites 28 km apart to facilitate cross-checking of results and analysis of local flow patterns. Both cores penetrated more than 3000 m of ice, and reached ice approximately 250,000 years old. The GISP2 core reached bedrock at 3053.4 m in 1993, becoming the longest core so far recovered. Differences between the bottom sections

Table 11.2 Details of some ice cores that have not reached the bed, and some that have reached the bed through ice less than 1000 m thick

	GRIP, Greenland	*Vostok, Antarctica*	*Devon Is., Canada*	*Renland, Greenland*	*Dome C, Antarctica*
Dates of drilling	1992	1974 onwards	1972–1973	1987	1978
Location	72° 58′ N 37° 64′ W	78° 28′ S 106° 48′ E	75° 20′ N 82° 30′ W	71° 18′ N 26° 44′ W	74° 30′ S 123° 10′ E
Surface elevation (m)	3238	3500	1800	2350	3240
Ice thickness (m)	3030	3800	299.4	325	3400
Length of core (m)	3028.8	3500 (1997)	to bed	to bed	905
Age at core bottom (thousand years)	250	450	100–130	100–130	40
Surface temperature (°C)	–32	–57	–23	–18	–53
Accumulation rate (m a^{-1})	0.23	0.027	0.20	0.50	0.023

of the GRIP and GISP2 cores indicated some disturbance of the lowest layers of ice by flow, and a new drilling programme was subsequently initiated in an attempt to retrieve an equivalent core through undisturbed ice. Drilling for this North GRIP (or NGRIP) core started in 1996, 316 km north of the original GRIP site (Dahl-Jensen *et al.*, 1997). To provide a southern hemisphere equivalent to the Summit cores, a new Antarctic coring programme has also been established. This EPICA (European Project for Ice Coring in Antarctica) core, drilled through a greater thickness of ice, could provide an even longer environmental record than the Greenland cores.

Blümcke and Hess at the turn of the century, used a hand-powered rotary drill and achieved a drilling rate of about 4–5 m per hour. The technology of drilling improved relatively slowly. Subsequent drilling efforts developed motorised rotary drills (e.g. Miller, 1951) and electro thermal drills that bored through the ice by melting (e.g. Stacey, 1960), but drilling remained a problematic exercise. The logistical problems of transporting and powering motorised equipment, the difficulty of penetrating debris-laden ice, and the slow rate of penetration all hindered operations. From the 1960s, improvements in drilling and transportation technology opened the way for drilling operations to attempt penetration of the ice sheets. A major step forward was the development of hot-water drills (e.g. Gillet, 1975). With such equipment, drilling rates of 100 m h^{-1} became possible, and debris within the ice was less of an obstacle to the drill. Rado *et al.* (1987) described a self-flushing hot-water drill that penetrates the ice by spraying warm water through the tip of a probe. The water spray both melts the ice and disturbs debris accumulating in the bottom of the hole, to prevent it from obstructing the drilling process. Splettstoesser (1976) and Langway (1967) provided details of various ice-coring procedures, and Zotikov (1986) discussed the physical background to thermal drilling in some detail.

Varying conditions during drilling programmes have often required that several different coring methods be used for different stages of drilling a single hole. For example, in the drilling of the Byrd station core, the first 227 m was cored with a thermal corer and the remainder with a rotary mechanical drill. For the Camp Century core, the first 285 m of drilling was accomplished with a thermal drill in a dry hole, the next 265 m was accomplished with a liquid-filled hole, and the lower sections were penetrated with a mechanical drill. Variations in ice characteristics, especially temperature, have a major effect on both the ease of drilling and the quality of the recovered core. Warmer ice in the lowest 300 m of the Camp Century hole permitted a much higher quality of core recovery than had been achieved in the middle sections of the hole.

Some of the technicalities of deep drilling are illustrated by the procedures employed by the Greenland Ice Core Project (GRIP). GRIP used an 11 m long electromechanical drill to accomplish its 3028.8 m core. The drill was named ISTUK, from the Danish word for ice (*is*) and the Greenlandic word for drill (*tuk*). The drill was constructed in 1978, and originally used in the GISP1 project to drill the Dye 3 core in 1981. The drill was set up in a shallow excavation on the ice-sheet surface, protected by a temporary building and connected by tunnels to underground laboratories and core-packing facilities. During drilling the drill was suspended on a thin steel cable, and cored the ice in sections 2.4 m long. After coring each section the drill was raised to the surface, tilted to a horizontal position, and the section removed. For the lower sections of the hole it took up to 50 minutes to raise the drill and 50 minutes to lower it. During the period of raising and lowering, an electric current was passed to recharge batteries in the drill, and these batteries powered the actual drilling. The use of batteries and a small recharging current, rather than powering the drill directly from the surface, prevented the need for thick power cables to connect the drill to the surface. The cable used was only 7.2 mm in diameter, and carried both the recharging current and the electrical signals sent from the surface to control the operation of the drill itself. Drilling was accomplished by the rotation of cutting knives at the tip of the probe, and the drill barrel retained both the core and the ice chips generated by the drilling. Coring was finally stopped when debris in the ice close to the bed destroyed the cutting knives and prevented further progress. The drill hole was prevented from closing by being filled with an oil of the same density as the ice. Drilling proceeded 24 hours a day, with an average progress of 150 m per week. A summary of the programme was provided by European Science Foundation (1995).

An important aspect of any drilling programme is site selection. The choice of site for drilling an ice core depends on the purpose to which the core is to be put. Specific purposes may require that the core be taken at a specific locality – for example, to measure ice thickness at a particular point. If the purpose of the core is to reconstruct former climates from the ice record, then several factors influence the choice of location so as to maximise the potential record contained in the core and to minimise the possible disturbances to the record. Drilling at the flow centre is ideal as the flow centre is not affected by horizontal flow, so the ice retrieved from the core should have originated at the surface directly above. The GRIP and GISP2 cores were deliberately located at the very summit of the Greenland Ice Sheet so as to avoid many of the problems of recovery and interpretation that beset earlier cores such as the core at Camp Century. However, the flow centre at the base of an ice cap may not be directly beneath the topographic summit, and determining the location of the ice column with the least disturbed vertical profile can be problematic (van der Veen and Whillans, 1992). When drilling takes place other than at the flow centre, ice lower in the core is derived from snowfall at higher elevations up-glacier along the ice flow-line. This was one of the problems encountered in dating and interpreting the Camp Century core. Changes in altitude affect the isotopic signature of ice accumulating at the surface, so the ideal site should have remained close to a constant elevation throughout its history, or should have a known pattern of historical elevation change. Drilling should take place at a location that has not experienced surface melting. Melting, percolation and refreezing of water disturb the climate record. Runoff of meltwater generates gaps or lacunae in the record. Melting at the bed likewise can destroy sections of the record, so an ideal site should have been cold-based throughout the history of the ice sheet. The basal layers of ice can be disturbed by folding and shearing, and this can be minimised by drilling at a site where the bed is relatively even. The observed differences between the adjacent GRIP and GISP2 cores demonstrate the impact of basal folding on the preserved record.

11.3 USES OF ICE CORES

Drilling through glaciers facilitates access to the interior and base of the glacier, and extraction of ice cores. There are many applications of drilling both for glaciological studies and for environmental reconstructions. Access to the interior and base of ice makes possible the direct inspection of englacial and basal conditions, and the emplacement of monitoring equipment. For example, recent investigations have involved borehole video (Pohjola, 1994) and monitoring of phenomena such as basal water pressure, thermal conditions, hydraulic conditions and strain. Extraction of ice cores facilitates reconstruction of past environments and environmental change on the basis of changing physical and chemical characteristics of ice with depth through the core. Drilling through glaciers has also been put forward as a possible way of disposing of, or putting into long-term storage, dangerous industrial waste. Zotikov (1986) discussed possible ways of safely entombing waste in ice sheets (Chapter 10). The most important use of ice cores has been in environmental reconstruction, which is discussed below.

Ice cores provide a wealth of different types of data that can be used to reconstruct former environmental conditions. An ice core provides a profile or cross-section through the material that has accumulated on the glacier, disturbed by flow of the material. The precipitation that accumulates on the glacier surface preserves a record of atmospheric characteristics at the time of its transformation into ice. The core thus provides a record of climate, atmospheric conditions (including gas, dust and anthropogenic influences), rates of accumulation and the nature of glacier flow. Glaciologists have therefore measured a wide range of characteristics of ice in cores, including stable isotopes, gas content and composition, dust and dissolved impurities, and structural and mechanical properties of ice. It has even been shown that the temperature profile through the ice at depth contains information about past climatic conditions, as surface temperatures are carried downwards in the accumulating ice and modified by flow and thermal equilibration in a predictable way (Dahl-Jensen and Johnsen, 1986).

Analysis of the stable isotope composition of ice from deep cores has been one of the most productive approaches to environmental reconstruction. The isotopic composition of ice in glaciers was discussed in detail in Chapter 4. It is controlled largely by the composition of the snow that is precipitated to form the glacier. This in turn is controlled by the position of glaciers in the hydrological cycle. When water is evaporated from the ocean, the initial vapour is isotopically lighter than the source water, and subsequent condensation and precipitation from the vapour tends to make it increasingly light as heavy isotopes are preferentially released during these processes. Consequently, precipitation falling as snow on land is isotopically very light relative to the ocean water from which it originally derived. The precise isotopic composition of precipitation varies with a range of environmental controls, including altitude, latitude, season and temperature (Dansgaard *et al.*, 1973). It is not always easy to isolate which effects are dominant in an ice-core record, but Cuffey *et al.* (1994) discussed the calibration of the so-called isotopic palaeothermometer, and found that $\delta^{18}O$ is a useful proxy for surface temperature. The isotopic composition of each layer of snowfall accumulating on a glacier surface is preserved as the snow turns to ice, and the glacier can thus offer a record of environmental conditions at the time of the original precipitation. Stable isotope records in $\delta^{18}O$ have been a major tool for reconstructing histories of climate change on the basis of ice-core data.

Gas content and composition in ice can also be valuable tools in environmental reconstruction. The air that is trapped in the ice during the firnification process can be retrieved from cores and analysed. The composition of the air samples is fixed at the time the bubbles are cut off from the atmosphere at the firn/ice transition. The air is thus a little younger than the ice around it (Schwander and Stauffer, 1984). At Vostok, owing to the slow transformation of snow to ice in cold dry conditions, the difference between the date of precipitation and the date at which the air is finally trapped in the ice is about 2500 years. The bubbles in ice cores preserve a sample of the atmosphere from the time the bubbles formed, including gases such as CO_2, CH_4 and N_2O. However, the gas composition can be affected by a range of processes that mask the original atmospheric composition. For example, the CO_2 content can be affected by reaction of CO_2 with water in the snowpack and in the ice, and by the reaction of carbonate dust in the ice. Nevertheless, ice-core measurements compare well with available historical records of the gas composition of the atmosphere (Raynaud *et al.*, 1993). The total volume of air trapped in the ice is a function of the volume of pore space and the pressure of the gas in the pores. The gas pressure is partly a function of atmospheric pressure at close-off, which is in turn a function of altitude. Variations in total air content have therefore been used as an indicator of variations in surface altitude, or ice thickness. If successful, this provides some control on our interpretation of the isotope record, as it gives an indication of how much of the observed δ variation is due to climate change and how much to change in ice-sheet elevation. Martinerie *et al.* (1992) found a good correlation suggesting that total air content in ice decreases with elevation by about 0.17 g mm^{-3} 100 m^{-1}, but the elevation control on air content is not unequivocal, as air content is also controlled by other factors, including temperature (which varies with elevation) and wind strength. A change of 1 K can change air content by 0.2 g mm^{-3}. Seasonal variations in temperature alone, therefore, can generate differences in air content as great as several hundred metres of elevation change.

Dust and dissolved impurities can also be retrieved from ice cores to give an indication of former environmental conditions. Particles of various types are deposited onto ice sheets from the atmosphere. These may be of terrestrial or extraterrestrial origin, and include both natural and anthropogenic material. Ninety-five per cent of the microparticles in ice cores are derived from dust carried from ice-free land. Concentrations of this dust vary significantly in ice cores. Ice formed during the Pleistocene contains dust concentrations up to 70 times greater than those in Holocene ice, with the highest dust concentrations correlating with the coldest periods. The material can be deposited in precipitation as 'wet' fallout, or as isolated particles in 'dry' fallout. Different types of material are susceptible to different kinds of deposition, and so tend to be deposited in different geographical distributions. Some areas of ice-sheet surface, such as the East Antarctic, experience predominantly dry deposition, while others, such as Greenland, are dominated by wet deposition. Some materials remain in the atmosphere, without being deposited, for much longer than others. For these reasons, a global picture of the distribution of material in the atmosphere cannot be produced reliably from ice-core data alone (Davidson, 1989). Impurities in ice are the main control on the electrical conductivity of the ice. Electrical conductivity measurement (ECM) thus provides a way of identifying down-core variations in impurities, and can reveal patterns of variation that can be correlated across cores. Measurements of impurities in ice cores have been a major source of information about pollution, including evidence of industrial activity, the growth of fossil fuel use, and fallout from nuclear activity. Even nuclear pollution can have its value: Chernobyl fall-out preserved in ice sheets has been used as a reference horizon for dating ice cores (Pourchet *et al.*, 1988).

Structural features of the ice can be used to reconstruct former surface environments. For example, the thickness of annual layers can be used to reconstruct accumulation rates. Features such as seasonal melt layers can be used to reconstruct the extent of surface melting, percolation and refreezing, and hence the characteristics of climate and surface snow. Buried wind crusts can indicate former wind frequency and magnitude. Some of these features were discussed in Chapter 5.

11.4 DATING OF CORES

Because of the importance of dating ice cores, and the difficulty of doing so, a wide range of dating methods have been attempted or proposed (e.g. Hammer *et al.*, 1978). Some, such as thermoluminescence dating and neutron activation of tephra layers, have not been widely adopted. The four most commonly applied approaches are discussed in the following sections.

11.4.1 DATING BY FLOW MODELS

Ice-sheet flow models can be employed to predict the age of ice at any horizontal and vertical position in an ice sheet on the basis of the direction of ice flow, the speed of flow, and hence the time taken to travel from the accumulation site to the dating site. The date predicted is as accurate or otherwise as the reconstruction of the flow. In the simplest case, at an ice-sheet summit where horizontal movement can be ignored, the age of ice at any depth is a direct function of the accumulation rate or the rate of burial. This in turn depends on the rate of precipitation, the rate of compression with burial, and vertical thinning associated with horizontal advection of ice. In more difficult cases where the ice to be dated is not at the flow centre, the route and travel time for horizontal flow needs to be taken into account. Major problems can arise here, because the flow pattern was almost certainly different at different times during the period of flow being considered. The thickness of the ice sheet, the directions and speeds of flow, the accumulation rate, ice temperature, the depth of the Holocene/Pleistocene transition, the rate of vertical thinning and many other parameters vary through time and from place to place through the ice sheet. Different flow models vary primarily in the extent to which they neglect or simplify this variability. Some models are 'time-dependent' in that they accommodate changes in ice sheet configuration over time. Others are 'time-independent', in that they neglect the probability of change through time, and model ice flow under constant conditions. Some models, known as 'sandwich models', assume that ice layers do not vary horizontally, while others, known as 'flow-line models', take account of variations in conditions along the route from the accumulation site to the sampling site (e.g. Stolle and Killeavy, 1986; Reeh and Paterson, 1988). Flow-line modelling can be based either on mathematical approximations derived from fluid dynamics, or on finite element methods. Finite element flow-line models have successfully predicted observed flow-line conditions in several cases and are widely used (Hooke *et al.*, 1979). The assumptions of constant flow conditions made by time-independent models become increasingly untenable when modelling older ice, and time-dependent models need to be adopted. Reeh (1989) suggested that steady state models could achieve reasonable dating precision only back to ice about 10,000 years old in large ice sheets, and only for ice up to about 1000–1500 years old for smaller ice masses. Flow modelling is the only dating tool available prior to collecting a core, and thus is the primary tool for site selection. However, the problems of flow modelling dictate that for precise dating of a core after collection, flow modelling should be combined with other dating methods.

11.4.2 DATING BY SEASONAL VARIATIONS

Ice cores, like many other phenomena that grow incrementally by natural processes, can be dated by counting the occurrence of annual layers or patterns of seasonal variation. There are many types of seasonal variation preserved in ice cores, several of which have been used for dating. For relatively shallow cores, annual accumulation layers can be counted to provide an age for different depths in the core. This method is limited, as the layers become too thin to measure below a certain depth. For shallow cores, and in the surface layers of deeper cores, characteristics of seasonal variations in the snowpack can be interpreted. These include variations in density, air content, grain size and chemistry, and a range of surface-crust and melt phenomena. Table 11.3 shows some of these seasonal indicators.

Some of the same features can be traced in deeper ice cores, along with other seasonally variable phenomena. Seasonal density variations can be identified up to 100 m below the surface. Less dense layers reflect loosely packed, coarse-grained summer snowfall. Dense layers reflect fine-grained, hard-packed winter snowfall. Seasonal melt layers can be traced for hundreds of metres into the ice, often throughout the entire thickness of a glacier. These are layers of ice formed by refreezing of meltwater within the firn and are recognisable as bubble-poor layers in the ice. Seasonal melt surfaces are often also visible, where lower summer snowfall and summer melting leads to a relative abundance of dust at the surface.

Table 11.3 Some of the seasonal indicators that can be identified in ice cores

Phenomenon	*Summer layer*	*Winter layer*
Density	Lower	Higher
Hardness	Lower	Higher
Grain size	Coarser	Finer
Air content	High	Low
Light transmission	Lower	Higher
Radiation crusts	Present	Absent

These surfaces remain visible as dust layers in the ice. Ice cores also record seasonal variations in isotope values, acidity and concentrations of microparticles. Seasonal variations are generally best preserved where accumulation is higher, and temperature lower. For example, molecular diffusion through firn and ice, which tends to cause a blurring of seasonal variations in isotope ratios, is minimised where individual seasonal layers are thicker. Melting and percolation of meltwater rapidly degrades the record of seasonal variations.

11.4.3 DATING BY MARKER HORIZONS

Individual layers with recognisable characteristics can serve as marker horizons to correlate dates between different cores, and even to correlate between the ice-core record and the non-glacial record. Marker horizons from dated terrestrial and marine records can sometimes be recognised in ice cores and used to date the core. Exceptional layers of various types have been used as marker horizons to date ice cores. The most direct use of marker horizons for dating is where historical events of known date have left an imprint in the environmental record. For example, since 1952, radioactive impurities in precipitation have been derived from nuclear weapons tests and the nuclear industry. Radioactive products of nuclear weapons tests are a well used source of dating marker horizons. Comparing specific ice profiles with records of precipitation samples allows a tree-ring type matching. For example, total beta-activity increased mainly due to weapons tests in 1954 and 1961/2. The effects of the tests carried out in the northern hemisphere occur in the ice record after about 1 year in northern hemisphere ice and after about 2 years in southern hemisphere ice. Thus, for example, the main ^{3}H emissions associated with thermonuclear tests in 1961/2 produce a peak in northern hemisphere ice core records in 1963. Fallout from the Chernobyl nuclear accident in 1986 has also been identified as a reference horizon in ice accumulated since that date (Pourchet *et al.*, 1988). Historically identifiable volcanic eruptions that have left an acidity peak in the ice record have been used to date ice layers (e.g. Hammer, 1977; Clausen and Hammer, 1988). For example, the eruption of the Icelandic volcano, Laki, in 1783 has left a clear record in ice cores in Greenland and Canada. Where volcanic acidity peaks are geographically widespread and occur in many cores, if they can be dated by layer counting in one core they can then be used as reference horizons in other, undated cores. Other natural events such as influx of extraterrestrial material could also provide such horizons. Ice of Pleistocene and Holocene ages have different rheological characteristics such that the boundary between the two forms an important marker horizon in ice cores that penetrate to that depth. Several phenomena change markedly at this horizon, including crystal size, microparticle content and ice hardness (Chapter 4).

11.4.4 RADIOMETRIC DATING

Radioactive isotopes in glaciers occur in the ice itself, in aerosols deposited on the glacier

Table 11.4 Characteristics of unstable isotopes that have been used frequently in glaciology

Isotope	*Half-life (years)*	*Location*	*Origin*
^{137}Cs	30.17	Aerosol	Nuclear weapons
^{85}Kr	10.76	Bubbles	Nuclear industry
^{3}H	12.26	Ice	Natural or weapons
^{210}Pb	22.3	Aerosol	Terrestrial decay
^{90}Sr	28.8	Aerosol	Nuclear weapons
^{32}Si	172	Aerosol	Cosmic radiation
^{39}Ar	269	Bubbles	Cosmic radiation
^{14}C	5,730	Bubbles	Cosmic radiation
^{10}Be	1,500,000	Aerosol	Cosmic radiation
^{36}Cl	301,000	Aerosol	Cosmic radiation
^{81}Kr	213,000	Bubbles	Cosmic radiation

surface, and in air trapped in the ice (Stauffer, 1989). Radioactive isotopes are derived from several sources, including terrestrial sources, cosmic radiation that causes radioactive isotopes to be produced in the atmosphere, and anthropogenic nuclear activity. Table 11.4 lists some of the main radioactive isotopes used in dating glaciers. Dating from radioactive isotopes can be achieved either by using the radioactive signatures of specific events as markers in the ice, as mentioned earlier, or by decay counting, although this may require large samples. It is also necessary to know the original concentration of the radioactive isotope in the ice, as well as the present concentration, to achieve dating. Concentration of most radioactive isotopes varies temporally and spatially by substantial amounts; for example, natural concentration of ^{10} Be varies naturally by a factor of 8. Ice can be dated by ^{14}C dating of carbon in CO_2 gas in the ice (Andrée *et al.*, 1986). The procedure only works for ice less than about 30,000 years old, and requires large amounts (several tonnes) of ice to be sampled to provide sufficient carbon for dating. Other radiometric dating methods have also been employed on glacier ice; for example, Nijampurkar *et al.* (1985) reported collection of more than 4 tonnes of ice and water for ^{32}SI and ^{210}Pb analysis. These volumes of material would be impractical for core samples. Radioactive isotopes derived from human activity can also provide a basis for decay-counting dating methods. For example, ^{85}Kr has steadily increased in atmospheric concentration due to emissions from the nuclear industry.

11.5 CONCLUSION

Paterson (1981) famously argued that a handful of mathematical physicists who may never have set foot on a glacier had contributed more to glaciology than had all the measurers of ablation stakes and terminus positions. It is tempting to see remote sensers and ice-core drillers as the inheritors of the ablation-stake tradition. Ice cores provide data that can be used to calibrate models, to test hypotheses, and to inspire new ones. Following Paterson's argument, it is especially important in the case of ice-coring programmes that the practical programme of data collection is carefully attuned to the theoretical research context. Ice cores give us one of the few windows that we have into what is actually down there beneath the ice surface. Unfortunately it is a window somewhat narrower than the sole of my boot, and it needs to be carefully placed.

RESEARCH DIRECTIONS IN GLACIOLOGY 12

While an abstract insight wakes / Among the glaciers and the rocks / The hermit's sensual ecstasy.
W.H. Auden: *Lay your sleeping head*

12.1 GLACIOLOGY

Glaciology in its narrowest sense is the study of glaciers. More broadly the term can be used to describe the science of all aspects of snow and ice. This is a broad field that draws extensively on other sciences such as physics, chemistry, meteorology, oceanography, geography and geology. *The Journal of Glaciology*, published since 1947, has included papers on topics ranging from ski design to ice on other planets, written by authors with institutional affiliations as diverse as Marine Dynamics, Medical Physics and Jet Propulsion. A dominant theme in glaciological research, however, has been the formation, distribution, properties, behaviour and effects of glaciers. If all the books and papers that have been written about glaciers were piled into a huge mound it would be of such proportions that the items at the bottom would be compressed and deform under the weight of the great mass above. An item selected at random from the heart of the pile would probably be dense, convoluted, and difficult to understand. Like owners who grow to resemble their dogs, a science mimics its subject. The purpose of this chapter is to identify something of the history of glaciology so as to provide a context for the position of the science that has been set out in the previous chapters, and to describe the current condition of glaciology so as to provide a context for the developments that might be anticipated in the future.

12.2 THE HISTORICAL DEVELOPMENT OF GLACIOLOGY

12.2.1 THE HISTORY OF GLACIOLOGY

Like most sciences, the history of glaciology has been highly non-linear. Rather than recording steady progress, the discipline has seen sudden major advances followed by periods of settlement and consolidation. Even the two concepts that now form the fundamental core of glaciology – namely that glaciers are dynamic agents and that glaciation is a time- and space-transgressive phenomenon – rose to scientific respectability in the mid-nineteenth century only through a revolutionary overwhelming of formerly accepted beliefs. In more recent times, major steps forward in glaciological thinking have accompanied the advent of new ideas or new technologies that have opened up new lines of research. Cunningham (1990) bemoaned the fact that there is no standard history of glaciology, but several convenient reviews of aspects of the history of glacier studies are available, including those by Chorley *et al.* (1964), Clarke (1987), Walker and Waddington (1988) and Cunningham (1990).

12.2.2 EARLY GLACIOLOGY

A cursory review of the forerunners of modern glaciology might suggest that the scientific understanding of glaciers began in the 1840s, with the work of the Swiss scientist Louis Agassiz. Certainly in the first half of the nineteenth century the scientific understanding of glaciers by the majority of the scientific establishment was, from a modern perspective, virtually zero. It is shocking to the modern reader how little was understood about glaciers until so recently. In the eighteenth century only a tiny number of scientists had any familiarity at all with glaciers. The age of adventurous tourism and romantic appreciation of wilderness had not yet arrived, and, in Europe at least, few people other than shepherds and hunters ventured high enough into the mountains to encounter glaciers. Until the first half of the nineteenth century, the conventional wisdom of Earth Sciences was that glaciers were static features, neither changing their position through time nor causing geomorphological effects in the landscape. It was not even generally accepted that the ice in a glacier moved. However, by the third quarter of the nineteenth century many of the fundamentals of glaciology that are still in place today were understood; the entire basis of modern glaciology established itself into the scientific world between about 1830 and 1870.

Agassiz has acquired a huge reputation as the father of glaciology and as the initiator of the 'Glacier Theory': that is, the idea that glaciers have been more extensive in the past and that the landscapes of presently unglaciated areas owe their form to their glacial legacy. However, Agassiz was only one of several scientists who, at more or less the same time, were developing a set of glaciological ideas that had been emerging through the observations of other scientists, primarily in the European Alps, for more than a century before them. Agassiz was a character of great energy, ability and ambition, and it was he who was at the crest of a wave of scientific conversion, but the image of Agassiz 'introducing' the glacier theory to an ignorant world is a myth. By the time Agassiz visited Scotland in 1840, on the famous trip which is commonly credited with introducing the glacier theory to Britain, 20 years had passed since Robert Jameson had accepted the idea of a Scottish glaciation, and the scientists of Edinburgh were able to take Agassiz on a glaciological tour of their area, showing him the glacial features of which they were already aware (Cunningham,1990).

The earliest recorded glaciologists were not scientists but mountain dwellers and travellers who observed glaciers on a regular basis. Early scientific observers commented on the fact that people living near glaciers were familiar with their behaviour, and even had an understanding of their properties. Kuhn (1787) reported that a Grindelwald shepherd had actually measured glacier motion. One of the earliest scientists to make formal study of glaciers was Johann Scheuchzer, who lived from 1672 to 1733. Scheuchzer (1723) recognised the fact of glacier motion and proposed that movement occurred by the mechanism of 'dilatation'. Dilatation supposedly involved the freezing of water that penetrated into cracks in the ice, and the thrusting forwards of the glacier by the expansion of that water upon freezing. This theory was still prominent a century later, when it was taken up and promoted by Jean de Charpentier. The theory was not finally laid to rest until the middle of the nineteenth century, when measurements by Louis Agassiz, James Forbes and others showed that movement did not cease in winter, as the dilatation theory implied it should. More realistic ideas of glacier flow and sliding were also emerging in these early years. The so-called 'gravitational theory', or sliding theory, was espoused by several eighteenth century glaciologists, primarily Altmann (1751), Gruner (1760) and de Saussure (1779–96). This theory developed the traditional belief of Alpine shepherds that glaciers moved downhill under the influence of gravity, and it incorporated the recognition that terrestrial heat would lead to melting of the base of the ice to assist sliding. At a similar time the idea of glacier flow by gradual deformation was also discussed. Bordier (1773) suggested that glacier ice was flexible and ductile like softened wax, directly prefiguring the theory of viscous flow that was to be the subject of dispute between Forbes and Tyndall a century later. De Saussure also distinguished between angular supraglacial debris and rounded boulders at the bed of the glacier, and recognised that ancient moraines represented former extensions of existing glaciers. On the basis of reading de Saussure's work, Hutton (1795) recognised that former expansions of Alpine glaciers were responsible for the disposition of distant erratics. Jean-Pierre Perraudin, an Alpine peasant and hunter born in 1767, recognised that striations in the proglacial zone were made by formerly more extensive glaciers (Forel, 1900); and several other scientists, including Esmarch (1827), Venetz (1833), and Bernhardi (1832), were developing ideas along similar lines. Thus, by the time in 1836 when Agassiz was converted to the glacier theory, it already had a substantial background, requiring only to be clearly exposited and championed.

After the glacier theory became widely known in the 1840s, the development of glaciology coincided with the growth of a wider culture of popular wilderness travel and romantic appreciation of nature. By the end of the nineteenth century it was by no means unusual for wealthy and educated Europeans to have visited glaciated areas and to have seen glaciers. Glaciology was expanded also by the growing interest of a broader scientific community. The problems raised by glaciology, especially those pertaining to the movement of glaciers, were of interest to the community of physicists, as well as to geologists. In the latter part of the nineteenth century, expanded programmes of field measurement were supplemented by laboratory investigations and theoretical analyses of the behaviour of glaciers and of ice. Franz Hugi had made measurements of glacier motion over several years from 1827 onwards, and Agassiz and Forbes in competition made detailed measurements in the early 1840s. Rendu's (1841) suggestion that differences between reported measurements of glacier motion might be due to differences in motion across the width of the Glacier was confirmed by measurements made by Forbes in 1842. Outside the traditional Alpine research areas, Rink (1877) reported how Helland in 1875 had measured velocity at several points across the 4500 m surface width of Jakobshavn Isbrae. He reported velocities of 0.02 m d^{-1} close by the lateral margin, 15 m d^{-1} 400–450 m from the margin, and 20 m d^{-1} at 1000 m from the margin. Detailed observations such as these led researchers such as Forbes, Rendu, Hopkins, Thompson, Tyndall, Huxley and Faraday to develop theories of glacier movement based on the dynamic properties of ice itself. By the latter part of the nineteenth century these involved concepts of sliding, viscous flow and regelation that were the direct precursors of modern glaciological understanding.

It is fascinating for a modern glaciologist to read texts from the end of the nineteenth century and to see that there has been little in the twentieth century to match the surge-like leaps of progress witnessed in the nineteenth. Of course, huge amounts of detail and elaboration of basic theory have been accomplished, and a number of key new ideas have been accepted. Even the basic geography of the Greenland and Antarctic ice sheets was barely known a century ago. However, a glaciologist from the end of the nineteenth century would, if transported to the present day, be able very quickly to catch up with the glaciological knowledge of his modern counterpart. The twentieth century has seen no conceptual revolution in the subject. The details of individual processes have been elaborated, and our understanding of the physical basis of the behaviour of ice has changed dramatically (Glen, 1987), but at the scale of glacier behaviour, only a very small number of new processes have been identified. Lliboutry (1994) pointed out that all of the processes of glacial erosion that are known today had already been put forward by 1900. Clarke (1987) suggested that much of the progress of the latter part of the twentieth century could have been accomplished a hundred years earlier if modern scientific tools had been available.

12.2.3 RECENT GLACIOLOGY

The understanding of glaciers that has emerged as a result of scientific investigations over the past century has formed the subject of the bulk of this book. Many of the principal concerns of glaciology have been the subject of constant research for over a hundred years, and the topics of conversation of modern glaciologists would be familiar to their intellectual great grandfathers. For example, the use of dye to trace water flow through glaciers was described by Forel (1898) and still featured prominently a century later in Hubbard and Nienow's (1997) review of the field, with numerous examples of its application in the intervening century. The same questions about debris entrainment into basal ice that were addressed by Chamberlin (1895) in the last decade of the nineteenth century were still being revisited by Hart (1995a), albeit from a contemporary standpoint in the throes of a deforming bed paradigm, in the last decade of the twentieth. When Deely and Parr (1914) identified the mechanisms of regelation and creep by which sliding ice overcame bed roughness, and introduced a sliding law where slip was proportional to stress and inversely proportional to friction, they were developing a theme that had been embarked upon 60 years earlier by Hopkins (1845), and were putting themselves at the head of a line of scientific enquiry that is still at the forefront of glaciology more than 80 years later. There is thus a great deal of continuity in the history of glaciology. However, there is also much in modern glaciology that is new. The history of glaciology during the twentieth century can be interpreted as an example of 'paradigm science', characterised by the occasional development of major new ideas or paradigms that dominate ways of thinking about the subject, and periods during which these new ideas are accommodated into the conventional wisdom, applied to existing scientific problems, and tested against new observations (Kuhn, 1962). A new idea forces its way into the glaciological consciousness and sets in motion a series of theoretical reshufflings as the implications and repercussions of the new idea rattle through the discipline, forcing changes in some older ideas, loosening previously sticky problems and generally shaking up the conceptual layout of the field. In recent glaciological history, one of the best known examples of a new paradigm impacting the discipline is the deforming bed theory.

The emergence of the deforming bed theory is typical of the emergence of new paradigms in modern glaciology (Boulton, 1986). Until the 1980s, modelling and theoretical analysis of glaciers and glacial processes assumed almost universally that glaciers rested on rigid bedrock. The glacier bed was referred to without question as the 'ice/rock interface' (Glen *et al.*, 1979). However, by the end of the 1970s a number of direct observations of glacier beds composed of deformable sediment had been made, and it had been recognised that the beds of former continental ice sheets in Europe and North America were characterised widely by sediment rather than bedrock. In the mid-1980s new geophysical techniques allowed greater knowledge to be gained about the beds of existing glaciers and it was observed that ice streams in Antarctica were underlain not by bedrock but by deforming till. Glaciology thus came in a relatively short period to accept a new glaciological reality (that glacier motion, and hence glacier morphology, could be controlled by substrate deformation): and the deforming bed model rose to prominence in the discipline. Every new paradigm is, for a while, treated as a panacea, and problems that could not be solved under the old theoretical framework are quickly tested against the new model. In the case of deforming beds, a range of issues including the origin of drumlins, glacier surges, and the potential response of the Antarctic ice sheet to changing climate were all shoved hard into the glass slipper of the deforming bed model. After the problem cases, even the bread-and-butter routine of a subject needs to accommodate new paradigms, and the last 20 years or so have been characterised by a permeation of deforming bed thinking through progressively more and more 'traditional' areas of glaciology. Models of subglacial deformation are now firmly embedded in glacier studies, and have facilitated advances in our understanding of observed patterns of glacier motion, ice-sheet profiles, the formation of debris-bearing basal ice, processes of glacier erosion and deposition, till depositional sequences, and the growth and decay of ice sheets. Any self-respecting ice-sheet model in the last decade of the twentieth century needs to accommodate deformable substrates.

In the recent history of glaciology, as in its early years, certain people and events are prominent. A glance through the reference list at the end of any textbook of glaciology will indicate the huge contributions of a few individuals to the discipline. The development of new ideas and the progress of the discipline are facilitated partly by the outstanding endeavour or insight of its practitioners and partly through the emergence over time of new technical capabilities. The personalities and technology of modern glaciology are central to its present status and its future potential.

12.3 THE ROLE OF TECHNOLOGY

Clarke (1987) suggested that technological advances, rather than revolutions in thinking, have often been the spurs to progress in the study of glaciers. Certainly the application of new ideas has often been limited or delayed by technological constraints, and the progress of the discipline has relied upon, and had its timing controlled by, technological innovation. Technological advances have permitted measurements that were made in the early years of glaciology to be made more reliably, frequently and over wider areas, and have made possible the measurement of new parameters that were not available to early glaciologists. Zwally (1987) reviewed some of the major technological advances that have facilitated progress in glaciology, and argued that continued technological progress would help to guide the evolution of the scientific questions considered by glaciologists.

It is clear that some major sub-fields of glaciology, such as ice coring, are based entirely on specific technologies. The use of glaciers in environmental reconstruction became possible only with the development of new ice-coring and analytical techniques in the second part of the twentieth century. The technology of ice coring was discussed in Chapter 11. The extraction of deep cores to bedrock through the major ice sheets was not possible in the early years of glaciology as neither the drilling technology nor the logistical capabilities of that period permitted it. However, not only the extraction of ice cores but also their analysis relies on technology that was not available until recently. This includes the chemical analysis of the ice, especially with regard to stable isotopes, which has proved to be a key element in the scientific value of ice cores. It is also true of ice-sheet modelling that while the questions might have existed in the minds of early glaciologists as to how ice sheets manoeuvred across landscapes, the development of modern ice-sheet models has depended on the development of modern computing capabilities. The gross and often crippling over-simplifications required by numerical modellers, which appal field-based glaciologists, are an inevitable consequence of computational limitations. Increased computer power has permitted the development of increasingly sophisticated models that can incorporate more and more of the complexity that field observation reveals. Clarke (1987) argued that there was no place in the traditional physicist's analysis for the delight in complexity that characterised the geographer's approach to glaciers, but observed that increasing computational power enabled new models to bring the two groups closer together.

It is also true that theoretical developments in areas of glaciology that are not in themselves technologically oriented have often depended on technological developments. For example, the emergence of deforming bed theory relied partly on the development of new geophysical techniques in glaciology that permitted observations of the glacier bed beneath thick ice. Likewise, development of the concept of ice-divide migration (Chapter 8) was facilitated by the availability of remotely sensed data on large-scale geomorphic lineations and glacier morphology. The exploration of the deep interiors of the large ice sheets and the discovery (for example) of features such as subglacial Lake Vostok have relied upon the emergence of geophysical and remote-sensing procedures that were not available to early glaciologists.

One area of glaciology that demonstrates very clearly the role of technology in the development of our understanding is the study of the debris-bearing basal layer of glaciers and ice sheets. The history of research on basal ice is interesting in itself as a case study of how evolving scientific technology allowed elucidation of competing theories that had developed over nearly a century. For many years, two main theories were disputed in the literature: the shearing theory and the freezing-on theory. Chamberlin (1895) and Salisbury (1896) were among the first to record basal debris penetrating upwards through the ice to the glacier surface, and both explained the phenomenon by the theory that sediment was entrained by shearing or thrusting from the bed along shear zones penetrating upwards into the ice. Shearing or thrusting mechanisms were proposed by many observers, including Lamplugh (1911), Goldthwait (1951,1971), Swinzow (1962) and Souchez (1967, 1971). Wakahama and Tusima (1981) reported that all the debris bands cropping out as moraine ridges on the ice surface near the margin of McCall Glacier in Alaska were located along thrust faults, and they successfully duplicated transport along shear planes in a laboratory model. An alternative theory was that debris was incorporated by the freezing of meltwater at the bed. Weertman (1961) argued that the shear hypothesis was theoretically unsound, and proposed instead that ice and debris were incorporated into the base of the glacier by freezing of subglacial water and sediment when the freezing isotherm moved downwards through the glacier onto the bed. Hooke (1973a), like Weertman, argued against the shear hypothesis on theoretical grounds, and proposed that the upwards transport of debris close to the margin could be explained by normal glacier flow coupled with ice movement over marginal accumulations of snow and ice. A likely compromise was that material was attached to the base by freezing, but could be transported into the ice along thrusts in the ice. Boulton (1967, 1970) suggested a freezing-on mechanism for sediment entrainment but noted that, once incorporated by freezing-on, debris seemed to be transported to different levels in the glacier along thrust planes. However, little evidence was available to confirm or refute either hypothesis with confidence. The majority of this work up until the mid-1970s was based on field observations of apparent deformation structures within the ice. What was lacking from analyses at this time was a definitive method for identifying ice frozen from meltwater at the bed, and for discriminating refrozen ice from meteorically derived glacier ice.

A major step forward in understanding the basal ice layer was made possible by the application of stable isotope techniques coupled with detailed structural and sedimentological observation. A seminal contribution was made by Lawson and Kulla (1978) and Lawson (1979) on the basis of detailed observations of the basal zone of the Matanuska Glacier in Alaska. They used $\delta^{18}O$ in ice to suggest that the debris-rich ice forming the lowest part of the glacier was refrozen from meltwater, and was therefore isotopically distinct from ice in the englacial zone. The development of co-isotopic analysis using both oxygen and hydrogen isotopes (Jouzel and Souchez, 1982; Souchez and Jouzel, 1984) permitted further developments, most importantly the possibility of distinguishing clearly between meteoric and refrozen ice – in other words, between glacier ice and regelation of congelation ice. Using co-isotopic analysis, Knight (1989) was able with confidence to falsify the freeze-on hypothesis for a thick layer of basal ice at part of the ice-sheet margin in West Greenland, as major parts of the sequence were, according to the co-isotopic signatures, composed of meteoric ice that had not undergone melting and refreezing.

Other analytical techniques based on the physical structure and chemical composition of the ice have allowed closer discrimination of processes of formation of basal ice, and identification of subglacial processes and conditions. Modern interpretations of basal ice are based on multi-analytical procedures in the field and in low-temperature laboratories involving chemical and physical analyses that have only recently become available. Detailed chemical and isotopic analyses now permit not only the differentiation of meteoric and refrozen ice, but reconstruction of rates of freezing, characteristics and provenance of source waters, and the nature of the hydrological and lithological system in which freezing occurred (e.g. Souchez and DeGroote, 1985; Hubbard and Sharp, 1993). The drilling of cores through to the basal layer in the interiors of the major ice sheets has also played an important role in basal ice research. Early information provided for example by Hansen and Langway (1966), Herron and Langway (1979) and Gow *et al.* (1968, 1979) and more recent data on basal ice retrieved from the GRIP and GISP2 cores at Summit, Greenland (e.g. Souchez *et al.*, 1994), have provided new insights into both the history of the ice sheets and the variety of processes affecting the characteristics of the basal ice. Hubbard and Sharp (1989), Souchez and Lorrain (1991) and Knight (1997) provided summaries of many of the key issues in the history of basal ice research. A combination of evolving techniques and an increasing range of study sites has resulted in the recognition of a wider range of basal ice types and processes of basal ice formation than hitherto envisaged. There is still a great degree of uncertainty and controversy in the field, but the amount of progress that has been made primarily as a result of the technical and theoretical advances of a small group of chemical glaciologists, inspiring conceptual advances among a much broader glaciological community, is remarkable.

Technological developments have thus been a key element in the development of glaciological thinking. Even the role of glaciers and glaciated environments in the human consciousness is strongly conditioned by the ability of technology to recreate those environments remotely: Dodds (1997) examined how different technologies of exploration and mapping, especially the use of aerial photography, transformed the perception of place and the human understanding of the Antarctic in the period 1918–1960 in the aftermath of the age of 'heroic' exploration. Many of the controversies that lingered for decades in the early years of glaciology, and many of the major misconceptions of early glaciologists, could have been overcome quickly with modern tools and resources. It remains true even to the present day that many areas of glaciology are limited not by the curiosity of the glaciologists but by their ability to acquire data from their subjects or to analyse and treat the samples and data that they have.

12.4 STYLES OF CONTEMPORARY RESEARCH, AND RESEARCHER

The study of glaciers is unusual in that although it draws expertise from a wide range of disciplines, the number of researchers who would explicitly call themselves glaciologists is relatively small. The International Glaciological Society, which is the principal professional organisation for glaciologists, had in 1997 a membership of only 750.

The study of glaciers has traditionally involved people who come from one of two different backgrounds. One set comprises those people with an interest in landforms and features of the physical environment. These might include geographers, geologists and geomorphologists. Another set comprises those with an interest in the physical or mathematical properties of phenomena, for whom glaciers sometimes represent little more than a case study. This second group might include physicists and mathematicians, along with metallurgists, crystallographers and ceramicists. Clarke (1987) suggested that the two great legacies in the tradition of glacier studies were those of classical physics and a romantic enthusiasm for nature. There is a clear and distinct route from childhood fascination with the great outdoors, mountains and wilderness into the study of the most quintessential of wilderness phenomena. There is another clear and distinct route from a fascination with physics and mathematics to the application of those arts to the understanding of materials. To those who are so inclined, the ice sheet represents one of the most simple and elegant expressions of physics amenable to mathematical treatment that the surface of this planet offers. To the romantic, there is a lure of glaciers that has its roots in the cultural outlook of many nations. Spufford (1997) discussed this lure in the context of the European fascination with polar exploration. There is something of the sublime that draws people of a certain bent towards glaciers, and, if those individuals are of an intellectual nature, towards glaciology.

There are thus different groups of people within glaciology who have approached their subject, wittingly or otherwise, from entirely different intellectual and aesthetic points of view. Furthermore, these people, by virtue of their different training routes in different academic disciplines, approach glaciology with entirely different scientific outlooks, speaking different scientific languages. Nevertheless, it is a tradition in glaciology that these different groups do interact. Inevitably in the early years of glaciology all glaciologists were scientists from some other branch of science. Agassiz was initially a biologist. Forbes had published in physics, astronomy, meteorology and geology before he entered the glaciological fray. The nature of glaciology, involving the behaviour of a common material in a geological environment, inevitably drew geologists and other natural scientists together. The understanding of the Earth surface processes of glaciation required an appreciation of the physical properties of materials. The very first volume of the *Journal of Glaciology* reported a joint meeting of the British Glaciological Society, the British Rheologists' Club and the Institute of Metals. Trying to distinguish the attitudes of physicists from geographers, Clarke (1987) traced the development of different styles of glaciology through the nature of presentations in the *Journal of Glaciology*. He noted that in the early years of the journal (1947–51) many papers included either equations (the 'Nye school') or maps (the 'Lewis school') but that maps and equations never appeared together in the same paper. By the 1980s, however, a third school (the 'Röthlisberger school') had become equally important, comprising papers that combined maps and equations. An outstanding feature of modern glaciology is collaborative effort coordinating skills from different backgrounds, and increasingly the modern glaciologist is trained and competent in approaches from more than one field of science.

In the past, schools of glaciologists swimming inquisitively through the conceptual ocean of glaciology were marked by their different goals or approaches. In modern glaciology they are marked more commonly by their group associations and by their geographical, as well as conceptual, hunting grounds. A striking feature of modern glaciology is the tendency for groups of collaborators to focus their attentions over prolonged periods at individual sites. A part of this trend in modern glaciological research programmes has been the ascendancy of what can be called 'siege glaciology'. This style of research involves comprehensive, long-term data acquisition on a range of topics at a specific site. Base camps are established, teams of researchers are coordinated over periods of several years to handle different aspects of the multi-analytical scientific and technical programmes, and long-term funding and management strategies are employed. A primitive version of this style of research was pioneered by early glaciologists like Agassiz, whose expeditions into the field involved long sojourns at often-revisited sites. In modern glaciology it is a style of research that can be attributed in part to the logistical requirements of collaborative, technology-based field research. The logistics of technological glaciology, with coring equipment, electronic monitors and other paraphernalia, favour long-term static encampment. The amount of information that we have the technical capability to retrieve is too great to be acquired, or interpreted, in a short period of time. Once a logistical and administrative infrastructure for research at a particular site is established, the use of the site becomes self-reinforcing. Once a glacier has been well studied, the data that has been collected there serves as a framework to encourage further study. Furthermore, there is a preference on the part of many glaciologists for multi-parameter research. Because so many aspects of the glacier system are closely interlinked, efforts to understand one part of the system are increasingly deemed to require data from other parts of the system. No self-respecting programme of study into the mechanisms of basal motion could proceed without an array of boreholes to monitor basal water pressure, or ploughmeters to test the substrate, or equipment to monitor glaci- seismic activity. Hence major research programmes grow up to involve large groups of researchers around a relatively small number of glaciers. There are many examples of this. Reading through the literature of recent years one quickly becomes familiar with a mere handful of glaciers: Storglaciären, Haut Glacier d'Arolla, Variegated Glacier, Jakobshavn Isbrae. In the ice-sheet literature one quickly gets to know the Siple Coast ice streams, the GRIP and GISP2 cores, Vostok, and a few other sites. Each of these, and a small number of others, have huge lists of publications attaching to them, while a huge number of other glaciers or ice-sheet locations are more or less unmentioned in the literature.

One consequence of siege glaciology is the development of siege teams: groups of people who, over a period of years, work in close association on a particular glacier. These teams commonly involve several principal investigators who work in various combinations over a period of time, along with junior team- members such as postgraduate students and postdoctoral assistants. Over time these teams evolve such that junior members mature and achieve senior status. It is striking in some instances how these upcomers inherit aspects of the main programme. It might be that they continue to work at a specific site after the initiating investigators have moved on, or it might be that they develop a related programme at a different site. Knight (1993) reviewed 6 years of glaciological research in Iceland, and identified a 'research chain' effect whereby field sites were transferred from senior to junior generations of researchers, often from a research supervisor to a former research student. Inheritance of sites from former student/supervisor associations was a major factor in the geographical distribution of research. An analysis of the research literature produced about individual glaciers demonstrates the importance of siege teams in the development of work in glaciology. For those glaciers that are well represented in the literature, the vast bulk of published material can be attributed to individuals associated with collaborative siege teams, often all with direct links to just one or two institutions, constituting an extended collaborative family of workers. For example, consider some of the recent research at Haut Glacier d'Arolla, Switzerland. An inexhaustive search identified 62 published papers about this glacier and its immediate proglacial area that were published or in press between 1990 and mid-1996. Of these, 60 include amongst their authors at least one member of one of the two major research groups that are working in that area. Thirty-one of the papers include one particular individual among the authors, and a representative of one pair of researchers (or their research students) features in the author-lists of 52 of the 62 papers identified. Similarly at Storglaciären, Sweden, the vast majority of the large volume of recent work that has appeared in the international literature can be attributed to members of a research 'family tree' that traces back to one principal investigator.

Siege programmes now constitute a substantial proportion of the research being carried out on glaciers. There are several types, including ice-coring sieges, ice-sheet sieges and valley-glacier sieges. Ice-coring sieges develop almost inevitably because the logistics of establishing ice-coring programmes are so prodigious that collaboration is essential and diversification of sites prohibitively expensive. The most recent and substantial of these ice-coring sieges has been the programme at Summit in central Greenland, described in Chapter 11. Ice-sheet siege programmes arise when field-based research on ice sheets is required because a substantial home base for the research team is essential, and severe logistical limitations apply. Ice-core programmes and ice-sheet field-research programmes tend to perpetuate themselves through diversification into non-field approaches, such as modelling the ice-core sites, or remote sensing the ice-sheet sites. It is easier and cheaper to use data that has already been collected from one perfectly adequate ice-sheet location than to establish a new programme at a new site to collect more. Valley-glacier siege sites become established either because of a particular phenomenon such as a predicted surge or recognised extreme fast flow, or because the gradual accumulation of work by a group at a site encourages further contributions. Some examples of these types of programme include huge bodies of research that have been accomplished at sites such as Storglaciären in Sweden, Haut Glacier d'Arolla in the Swiss Alps, Variegated Glacier in Alaska, the Siple Coast ice streams in Antarctica, and Jakobshavn Isbrae in West Greenland.

12.5 THE FUTURE

The sooner this book goes out of date, the better it will be for glaciology. It has been a recurrent motif throughout these chapters that our knowledge of glaciers goes only a certain way before devolving into speculation. It is to be hoped that the position of that boundary will advance over time. The aim of glaciology is a clear and detailed understanding of how glaciers work, that can be applied both to questions in glaciology and to unsolved problems in the linked global environment. However, there are many fundamental glacier properties and processes that we do not yet understand. Almost every aspect of glaciology that has been touched upon in this book remains the subject of ongoing research, and the list of unanswered questions in glaciology is huge. For example, at one extreme we don't know what controls the stability of ice sheets, and at another we don't know exactly how tiny debris inclusions influence the deformation of ice. There are unproven theories current in the literature, such as the idea of thermal convection in ice sheets (Hughes, 1985; Pappalardo *et al.*, 1998), the potential implications of which are enormous. We don't even know for sure whether the world's two ice sheets are currently growing or shrinking.

Our ignorance exists at two levels: first, the mechanics of individual processes or phenomena; second, the ways in which individual phenomena interact in controlling gross glacier behaviour. Contemporary glaciology fights on both these fronts. Many of its preoccupations are cast in the grand scale: ice-sheet responses to climate change; ice streams, mega-floods and ice-sheet stability; continent-scale subglacial hydrology and geomorphology. The components of these problems, however, can be couched in much narrower terms. Problems of ice-sheet dynamics and continental hydro-geomorphology can be reduced to assemblages of questions about the properties of ice, water and mineral material in different thermal and pressure environments.

A major part of modern research effort is dedicated to modelling glacier behaviour. A great ice-sheet model would be like a symphony, drawing together into a unified picture the disparate themes of the glacier story. Creating an ice-sheet model is like translating an epic natural poem into something understandable, and usable, at a human scale. Modelling either of whole ice sheets or of components of the glacier system aims to facilitate predictions and reconstructions of glacier behaviour. Adequate modelling for these purposes requires some understanding of the processes and parameters that control glacier behaviour. However, modelling in many aspects of glaciology has been forced to rely on uncertain input parameters. For example, Hooke *et al.* (1990) pointed out that virtually all of the important assumptions in Nye's (1953) theoretical analysis of the closure rate of subglacial channels are violated in the natural situation, but that calculation of tunnel closure rates nevertheless relies commonly upon his analysis. Röthlisberger's (1972) seminal discussion of flow in subglacial channels relied on a series of assumptions and adjustments to make the modelled situation compatible with reality. For the model to make accurate predictions of borehole water pressures, for example, basal ice would need to be as much as 150 times softer than normal. Several explanations to justify such high deformation rates have been put forward, but Hooke *et al.* (1990) showed that the problem is more easily solved by abandoning Röthlisberger's assumption of circular or semi-circular cross-section tunnels. Hooke et. al. replaced Röthlisberger's tunnel shapes with new shapes of their own, but were able to justify them only on the grounds of model consistency, not by recourse to any direct observation of actual subglacial tunnels. Developments in modelling procedures need to progress hand in hand with observational glaciological data. Modelling, like remote sensing, needs ground-truth data. Models, like engines, require fuel, and the fuel of a good model is good baseline data.

The early literature of ice-sheet modelling is abundantly stocked with imaginary glaciers that have little relation to reality. They have rigid horizontal beds, perfect parabolic profiles and debris-free ice/bed contacts, and they exist in a steady state for indefinite periods. However, we have moved forwards now into a period of growing awareness on the part of modellers that sanitised models in which simplifying assumptions make modelling possible but unrealistic are inadequate. The world really is a complicated place, and modelling it without its complexity is not always useful. Models of ice sheets that do not involve ice streams are unhelpful. Models of glaciers with rigid, smooth, horizontal beds are inadequate. The challenge for modellers at the turn of the twenty-first century is to model realistic glaciers and ice sheets. This will require careful assimilation of relevant field observations, but unfortunately, and perhaps surprisingly, relevant observations may not be abundant. For example, direct observations of basal processes on a variety of different substrates remain startlingly few. Even on the fundamental issue of the flow law of ice, models remain at the mercy of gross simplifications, and we are a very long way from achieving a relevant flow law for glaciers rather than merely a flow law for ice.

Notwithstanding the points made earlier in this chapter about the willingness of glaciologists to collaborate and to combine skills from different backgrounds, glaciology has yet to take full advantage of many possible avenues of interdisciplinary collaboration. There have been substantial contributions to glaciology from physicists, chemists and mathematicians, and there has been some interdisciplinary dialogue in fields such as fracture mechanics and creep (e.g. Lliboutry, 1987), but much potentially valuable interdisciplinary collaboration in the fields of material sciences has yet to be explored. Most of the glaciologists who study glaciers in the field have little contact with ceramicists, metallurgists or others who study equivalent phenomena in other materials. Cross-boundary collaborations on physical and chemical phenomena and in mathematical or technical procedures have not widely translated to the scale of whole glaciers or whole physical systems. The sort of collaboration that is possible has been demonstrated recently by collaboration between glaciologists, hydrologists and chemists. Recent progress in understanding glacier hydrology, which in turn has shed light on glacier dynamics, has come from the application of chemical principals to the study of glaciers. There are further clues to glacier behaviour hidden in the physical characteristics of the ice that are not yet being adequately interpreted. Crystallographers and structural glaciologists will in future throw as much new light on glacier behaviour as chemists recently have done through ice-core analysis and hydrochemical studies. Future collaborations are also likely between terrestrial glaciologists and those studying ice on other bodies in the solar system. Discoveries about the characteristics of extraterrestrial ice masses, such as the evidence for thermal convection or diapirism in the ice shell of the Jovian moon Europa (Pappalardo *et al.*, 1998), can cause us to look with new eyes at familiar problems.

One attractive characteristic of glaciology is the way in which all of its components clearly interrelate. Thermal regime, hydrology, substrate rheology, ice deformation, mass balance, geomorphology, glacier dynamics, and all the other topics that have occupied this book, combine transparently into a single system. Furthermore, glaciers play a key role in the global environment, both contributing to and serving as indicators of key environmental changes. One consequence of this is that glaciology is a field in which huge questions can be made explicit. These might relate to the role of glaciers and glaciation in the general global system, or to the interaction of components within the glacier system. There are several 'Grand Unsolved Problems' (Grosswald and Hughes, 1995) that reappear in the glaciological literature from time to time, and the prospect of some great unified and unifying, theory of glaciology that links material properties, dynamic processes, morphology and environment is alluring.

12.6 CONCLUSION

Glaciology is a small profession.

(Hughes, 1985, p. 39)

The boundaries of glaciology are a little fuzzy. From the inside, looking out, the limits are hard to define and the discipline is huge. From the outside, the fuzzy little blob that is glaciology, with tentacles wriggling out into neighbouring fields, is easier to delimit. I am often tempted to suggest to my students, usually in the context of an examination, that 'when you've seen one glacier, you've seen 'em all (discuss).' The goal of one sort of glaciological research seems to be to achieve a universally applicable set of principles by which the glacier condition can be defined. Each specific instance, a surge here, a jökulhlaup there, will then be recognisable as a variant on some established and understood theme. Glaciology is enriched, however, by a sense of variety and surprise in the natural phenomena that it considers. Partly this may be because glaciers themselves are places of variety and surprise, not to say wonder, and the study of them in the field is associated with a whole range of experiences beyond the purely scientific. Partly, however, the surprise element in the study of glaciers arises because there is still so much about them that we do not know.

The vastest things are those we may not learn
We are not taught to die, nor to be born
Nor how to burn
With love.
How pitiful is our enforced return to those small things we are the masters of.

Mervyn Peake: *The vastest things . . .*

REFERENCES

Adkins, J.F., Boyle, E.A., Keigwin, L. and Cortijo, E. (1997) Variability of the North Atlantic thermohaline circulation during the last interglacial period. *Nature* **390** (6656), 154–156.

Aellen, M. (1995) Glacier mass balance studies in the Swiss Alps. *Zeitschrift für Gletscherkunde und Glazialgeologie* **31**, 159–168.

Agassiz, L. (1840) *Études sur les Glaciers* 2 volumes. H. Nicolet, Neuchâtel.

Ageta, Y. and Fujita, K. (1996) Characteristics of mass balance of summer-accumulation type glaciers in the Himalayas and Tibetan Plateau. *Zeitschrift für Gletscherkunde und Glazialgeologie* **32**, (2), 61–65

Ageta, Y. and Higushi, K. (1984) Estimation of mass balance components of a summer-accumulation type glacier in the Nepal Himalaya. *Geografiska Annaler* **66A** (3), 249–255.

AGRISPINE (1983) *The Use of Landsat Data to Monitor Iceberg Production. Monitoring iceberg production from West Greenland tidewater glaciers using Landsat data. Results from the AGRISPINE experiment for the Jakobshavn Isbrae*. Macaulay Institute for Soil Research, Aberdeen and National Remote Sensing Centre, Farnborough.

Ahlmann, H.W. (1933) Scientific results of the Swedish-Norwegian Arctic expedition in the summer of 1931, part VIII. *Geografiska Annaler* **15**, 161–216, 269–295.

Ahlmann, H.W. (1935) Contribution to the physics of glaciers. *Geographical Journal* **86** (2), 97–113.

Ahlmann, H.W. (1948) *Glaciological research on the north Atlantic coasts*. (*Royal Geographical Society Research Series* 1.) Royal Geographical Society, London.

Alean, J. (1985) Ice avalanches: some empirical information about their formation and reach. *Journal of Glaciology* **31** (109), 324–325.

Alexander, D. (1993) *Natural Disasters*. UCL Press, London.

Alley, R.B. (1988a) Concerning the deposition and diagenesis of strata in polar firn. *Journal of Glaciology* **34** (118), 283–290

Alley, R.B. (1988b) Fabrics in polar ice sheets: development and prediction. *Science* **240**, 493–495.

Alley, R.B. (1989a) Water-pressure coupling of sliding and bed deformation: I. Water system. *Journal of Glaciology* **35** (119), 108–118.

Alley, R.B. (1989b) Water-pressure coupling of sliding and bed deformation: II. Velocity–depth profiles. *Journal of Glaciology* **35** (119), 119–129.

Alley, R.B. (1991) Deforming-bed origin for Southern Laurentide till sheets? *Journal of Glaciology* **37** (125), 67–76.

Alley, R.B. (1992a) Flow-law hypotheses for ice-sheet modelling. *Journal of Glaciology* **38** (129), 245–256.

Alley, R.B. (1992b) How can low-pressure channels and deforming tills coexist subglacially? *Journal of Glaciology* **38** (138), 200–207.

Alley, R.B. (1993) In search of ice stream sticky spots. *Journal of Glaciology* **39** (133), 447–454.

Alley, R.B. and Whillans, I.M. (1991) Changes in the West Antarctic ice sheet. *Science* **254** (5034), 959–963.

Alley, R.B., Perepezko, J.H. and Bentley, C.R. (1986a) Grain growth in polar ice: I. Theory. *Journal of Glaciology* **32** (112), 415–424.

Alley, R.B., Perepezko, J.H. and Bentley, C.R. (1986b) Grain growth in polar ice: II. Application. *Journal of Glaciology* **32** (112), 425–433.

Alley, R.B., Perepezko, J.H. and Bentley, C.R. (1988) Long term climate changes from crystal growth. *Nature* **332**, 592–593.

Alley, R.B., Meese, D.A., Shuman, C.A., Taylor, K.C., Grootes, P.M., White, J.W.C., Ram, M., Waddington, E.D., Mayewski, P.A. and Zielinski, G.A. (1993) Abrupt increase in Greenland snow accumulation at the end of the Younger Dryas event. *Nature* **362** (6420), 527–529.

Altmann, J.G. (1751) *Versuch Einer Historischen und Physischen Beschreibung der Helvetischen Eisgebirge*. Zurich.

Ambach, W., Elässer, M., Behrens, H. and Moser, H. (1974) Studie zum Schmelzwasser-abfluss aus dem Akkumulationsgebiet eines Alpengletschers (Hintereisferner, Ötztaler Alpen). *Zeitschrift für Gletscherkunde und Glazialgeologie* **10**, 181–187.

Ames, A. and Hastenrath, S. (1996) Mass balance and ice flow of the Uruashraju glacier, Cordillers Blanca, Peru. *Zeitschrift füur Gletscherkunde und Glazialgeologie* **32**, 83–89.

Anandakrishnan, S. and Alley, R.B. (1997) Stagnation of ice stream C, West Antarctica by water piracy. *Geophysical Research Letters* **24** (3), 265–268.

Anandakrishnan, S. and Bentley, C.R. (1993) Micro-earthquakes beneath Ice Streams B and C, West Antarctica: observations and implications. *Journal of Glaciology* **39** (133), 455–462.

Andrée, M., Beer, J., Loetscher, H.P., Moor, E., Oeschger, H., Bonani, G., Hofmann, H.J., Morenzonie, E., Nessi, M., Suter, M. and Wolfli, W. (1986) Dating polar ice by ^{14}C accelerator mass spectrometry. *Radiocarbon* **28** (N2A), 417–423.

Andrews, J.T. (1968) Postglacial rebound in arctic Canada; similarity and prediction of uplift curves. *Canadian Journal of Earth Sciences* **5**, 39–47.

Andrews, J.T. (1970) A geomorphological study of post-glacial uplift with particular reference to Arctic Canada. *Institute of British Geographers Special Publication Number 2*. Institute of British Geographers, London.

Andrews, J.T., Milliman, J.D., Jennings, A.E., Rynes, N. and Dwyer, J. (1994) Sediment thicknesses and Holocene glacial marine sedimentation rates in three East Greenland fjords (ca.68°N). *Journal of Geology* **102**, 669–683.

Anonymous (1969) Mass balance terms. *Journal of Glaciology* **8** (52), 3–7.

Aristarain, A.J. and Delmas, R.J. (1993) Firn-core study from the southern Patagonia ice cap, South America. *Journal of Glaciology* **39** (132), 249–254.

Armstrong, R.L. (1989) Mass balance history of Blue Glacier, Washington, U.S.A. In: Oerlemans, J. (ed.) (1989) *Glacier Fluctuations and Climatic Change*. Kluwer Academic Publishers, Dordrecht, 183–192.

Armstrong, T.E., Roberts, B. and Swithinbank, C. (1973) *Illustrated Glossary of Snow and Ice*, 2nd. edn. Scott Polar Research Institute, Cambridge.

Astakhov, V.I. and Isayeva, L.L. (1988) The 'ice hill': an example of retarded deglaciation. *Quaternary Science Reviews* **7**, 29–40.

Azuma, N. and Higashi, A. (1985) Formation processes of ice fabric pattern in ice sheets. *Annals of Glaciology* **6**, 130–134.

Azuma, N. and Mae, S. (1988) The effect of C-axis orientation fabric on the flow rate of ice; derivation from in situ observation of deformation behaviour of individual grains. In: Saeki, H. and Hirayama, K-I. (eds) *Proceedings, the 9th International Symposium on Ice, 23–27 August 1988, Sapporo, Japan. Volume I.* International Association for Hydraulic Research, Committee on ice problems, Delft, 67–76.

Bader, H. (1961) *The Greenland Ice Sheet*. U.S. Army Cold Regions Research and Engineering Laboratory (monograph 1-B2), Hanover, New Hampshire.

Bader, H. (1962) *Theory of Densification of Dry Snow on High Polar Glaciers*. U.S. Army Cold Regions Research and Engineering Laboratory (Research Report 108), Hanover, New Hampshire.

Baker, V.R. (1983) Large scale palaeohydrology. In: Gregory, K.J. (ed.) *Background to Palaeohydrology*. John Wiley and Sons, Chichester, 455–478.

Baker, V.R. and Bunker, R.C. (1985) Cataclysmic late Pleistocene flooding from glacial Lake Missoula: a review. *Quaternary Science Reviews* **4**, 1–41.

Balise, M.J. and Raymond, C.F. (1985) Transfer of basal sliding variations to the surface of a linearly viscous glacier. *Journal of Glaciology* **31** (109), 308–318.

Bamber, J.L. (1987) Internal reflecting horizons in Spitzbergen glaciers. *Annals of Glaciology* **9**, 5–10.

Bard, E., Hamelin, B., Fairbanks, R.G. and Zindler, A. (1990) Calibration of the C-14 timescale over the past 30,000 years using mass-spectrometer U-Th ages from Barbados corals. *Nature* **345** (6274), 405–410.

Bard, E., Hamelin, B., Arnold, M., Montaggione, L., Cabioch, G., Faure, G. and Rougerie, F. (1996) Deglacial sea-level record from Tahiti corals and the timing of global meltwater discharge. *Nature* **382** (6588) 241–244.

Barnes, P. and Robin, G. de Q. (1966) Implications for glaciology. *Nature* **210** (5039), 882–883.

Barnes, P., Tabor, D. and Walker, J.C.F. (1971) The friction and creep of polycrystalline ice. *Proceedings of the Royal Society of London, Series A* **324**, 127–155.

Barnola, J.M., Raynaud, D., Korotkevich, Y.S. and Lorius, C. (1987) Vostok ice core provides 160,000-year record of atmospheric CO_2. *Nature* **329** (6138), 408–414.

Barrie, J.V. (1980) Iceberg–seabed interaction (northern Labrador Sea). *Annals of Glaciology* **1**, 71–76.

Barron, E.J., Hay, W.W. and Thompson, S. (1989) The hydrologic cycle – a major variable during Earth history. *Global and Planetary Change* **75** (3), 157–174.

Barsch, D. (1988) Rockglaciers. In: Clark, M.J. (ed.) *Advances in Periglacial Geomorphology*. J. Wiley and Sons, Chichester, 69–90.

Battle, M., Bender, M., Sowers, T., Tans, P.P., Butler, J.H., Elkins, J.W., Ellis, J.T., Conway, T., Zhang, N., Lang, P. and Clarke, A.D. (1996) Atmospheric gas concentrations over the past century measured in air from firn at the South Pole. *Nature* **383** (6597), 231–235.

Bauer, A. (1955) The balance of the Greenland ice sheet. *Journal of Glaciology* **2** (17), 456–462.

Beget, J.E. (1986) Modelling the influence of till rheology on the flow and profile of the Lake Michigan lobe, southern Laurentide ice sheet, U.S.A. *Journal of Glaciology* **32** (111), 235–241.

Behrens, H., Löschhorn, U., Ambach, W. and Moser, H. (1976) Studie zum Schmelzwasser-abfluss aus dem Akkumulationsgebiet eines Alpengletschers (Hintereisferner, Ötztaler Alpen). *Zeitschrift fur Gletscherkunde und Glazialgeologie* **12**, 69–74.

Benn, D.I. and Evans, D.J.A. (1998) *Glaciers and Glaciation*. Edward Arnold, London.

Bennett, M.R. and Boulton, G.S. (1993a) A reinterpretation of Scottish 'hummocky moraine' and its significance for the deglaciation of the Scottish Highlands during the younger Dryas or Loch Lomond stadial. *Geological Magazine* **130**, 301–318.

Bennett, M.R. and Boulton, G.S. (1993b) Deglaciation of the younger Dryas of Loch Lomond stadial ice-field in the northern Highlands, Scotland. *Journal of Quaternary Science* **8**, 133–145.

Bennett, M.R. and Glasser, N.F. (1996) *Glacial Geology*. Wiley, London.

Benson, C.S. (1962) Stratigraphic studies in the snow and firn of the Greenland ice sheet. *Snow Ice and Permafrost Research Establishment (SIPRE) Research Report* **70**.

Bentley, C.R. (1987) Antarctic ice streams: a review. *Journal of Geophysical Research* **92** (B9), 8843–8858.

Bernhardi, H. (1832) Wie kammen die aus den norden stammenden felsbruchstucke und gescheibe, welche man in Norddeutschland und den benachbarten landern findet, an ihre gegenwartigen fundorte. *Jahrb. für Min., Geogn., Geol. und Petrefaktenkunde* **3**, 257–267.

Bezinge, A. (1987) Glacial meltwater streams, hydrology and sediment transport: the case of the Grande Dixence hydroelectricity scheme. In: Gurnell, A.M. and Clark, M.J. (eds) *Glacio-fluvial Sediment Transfer*. Wiley, Chichester, 473–498.

Bezinge, A., Clark, M.J., Gurnell, A.M. and Warburton, J. (1989) The management of sediment transported by glacial melt-water streams and its significance for the estimation of sediment yield. *Annals of Glaciology* **13**, 1–5.

Bindschadler, R.A. (1983) The importance of pressurized subglacial water in separation and sliding at the glacier bed. *Journal of Glaciology* **29** (101), 3–19.

Bindschadler, R.A., Harrison, W.D., Raymond, C.F. and Crosson, R. (1977) Geometry and dynamics of a surge-type glacier. *Journal of Glaciology* **18** (79), 181–194.

Bishop, B.C. (1957) Shear moraines in the Thule area, northwest Greenland. *SIPRE Research Report* **17**, 47 pp.

Bintanja, R. (1995) *The Antarctic Ice Sheet and Climate*. University of Utrecht, Utrecht.

Björnsson, H. (1992) Jökulhlaups in Iceland: prediction, characteristics and simulation. *Annals of Glaciology* **16**, 95–106.

Björnsson, H. (1996) Scales and rates of glacial sediment removal: a 20 km long, 300 m deep trench created beneath Breiðamerkurjökull during the little ice age. *Annals of Glaciology* **22**, 141–146.

Blake, E., Clarke, G.K.C. and Gérin, M.C. (1992) Tools for examining subglacial bed deformation. *Journal of Glaciology* **38** (130), 388–396.

Blake, E., Fischer, U.H. and Clarke, G.K.C. (1994) Direct measurement of sliding at the glacier bed. *Journal of Glaciology* **40** (136), 595–599.

Blanchon, P. and Shaw, J. (1995) Reef drowning during the last deglaciation: evidence for catastrophic sea-level rise and ice-sheet collapse. *Geology* **23** (1), 4–8.

Blankenship, D.D., Bentley, C.R., Rooney, S.T. and Alley, R.B. (1986) Seismic measurements reveal a saturated porous layer beneath an active Antarctic ice stream. *Nature* **322** (6074), 54–57.

Blatter, H. (1987) On the thermal regime of an arctic alley glacier: A study of White Glacier, Axel Heiberg Island, N.W.T., Canada. *Journal of Glaciology* **33** (114), 200–211

Bogen, J. (1989) Glacial sediment production and development of hydro-electric power in glacierized areas. *Annals of Glaciology* **13**, 6–11.

Bogen, J. (1996) Erosion rates and sediment yields of glaciers. *Annals of Glaciology* **22**, 48–52.

Böhm, R. (1995) Long-term changes of glaciers in the Sonnblick Region in the Austrian Alps. *Zeitschrift füur Gletscherkunde und Glazialgeologie* **31**, 169–170.

Bond, G., Heinrich, H., Broecker, W., Labeyrie, L., McManus, J., Andrews, J., Huon, S., Jantschik, R., Clasen, S., Simet, C., Tedesco, K., Klas, M., Bonani, G. and Ivy, S. (1992) Evidence for massive discharges of icebergs into the north-Atlantic ocean during the last glacial period. *Nature* **360** (6401), 245–249.

Bond, G., Broecker, W., Johnsen, S., McManus, J., Labeyrie, L., Jouzel, J. and Bonani, G. (1993) Correlations between climate records from north-Atlantic sediments and Greenland ice. *Nature* **365** (6442), 143–147.

Bordier, G. (1773) *Picturesque Journey to the Glaciers of Savoy*. Geneva.

Boulton, G.S. (1967) The development of a complex supraglacial moraine at the margin of Sorbreen, Ny Friesland, Vestspitsbergen. *Journal of Glaciology* **6** (47), 717–735.

Boulton, G.S. (1968) Flowtills and related deposits on some Vestspitsbergen Glaciers. *Journal of Glaciology* **7** (51), 391–412.

Boulton, G.S. (1970) On the origin and transport of englacial debris in Svalbard glaciers. *Journal of Glaciology* **9** (56), 213–229.

Boulton, G.S. (1972) The role of thermal regime in glacial sedimentation. In: Price, R.J. and Sugden, D.E. (eds) *Polar Geomorphology. Institute of British Geographers special publication number 4*. Institute of British Geographers, London, 1–19.

Boulton, G.S. (1974) Processes and patterns of glacial erosion. In: Coates, D.R. (ed.) *Glacial Geomorphology*. State University of New York, Binghampton, NY, 41–87.

Boulton, G.S. (1975a) Processes and patterns of subglacial sedimentation: a theoretical approach. In: Wright, A.E. and Moseley, F. (eds) *Ice Ages: Ancient and Modern*. Seel House Press, Liverpool, 7–42.

Boulton, G.S. (1975b) The genesis of glacial tills: a framework for geotechnical interpretation. In: Midlands Soil Mechanics and Foundation Engineering Society: *The Engineering Behaviour of Glacial Materials*, MSMFES, Birmingham, 52–59.

Boulton, G.S. (1978) Boulder shapes and grain size distribution of debris as indicators of transport paths through a glacier and till genesis. *Sedimentology* **25** (6), 773–799.

Boulton, G.S. (1982) Subglacial processes and the development of glacial bedforms. In: Davidson-Arnott, R., Nickling, W. and Fahey, B.D. (eds) *Research in Glacial, Glacio-fluvial, and Glacio-lacustrine Systems, Proc.6th Guelph Symposium on Geomorphology*, 1980, Geo Books, Norwich, 1–31.

Boulton, G.S. (1986) A paradigm shift in glaciology? *Nature* **322** (6074), 18.

Boulton, G.S. (1996) Theory of glacial erosion, transport and deposition as a consequence of subglacial sediment deformation. *Journal of Glaciology* **42** (140), 43–62.

Boulton, G.S. and Caban, P. (1995) Groundwater flow beneath ice sheets: part II – its impact on glacier tectonic structures and moraine formation. *Quaternary Science Reviews* **14** (6), 563–588.

Boulton, G.S. and Clarke, C.D. (1990) A highly mobile Laurentide ice sheet revealed by satellite images of glacial lineations. *Nature* **346** (6287), 813–817.

Boulton, G.S. and Dobbie, K.E. (1993) Consolidation of sediments by glaciers: relations between sediment geotechnics, soft-bed glacier dynamics and subglacial ground-water flow. *Journal of Glaciology* **39** (131), 26–44.

Boulton, G.S. and Hindmarsh, R.C.A. (1987) Sediment deformation beneath glaciers: rheology and geological consequences. *Journal of Geophysical Research* **92**, 9059–9082.

Boulton, G.S. and Jones, A.S. (1979) Stability of temperate ice caps and ice sheets resting on beds of deformable sediment. *Journal of Glaciology* **24** (90), 29–42.

Boulton, G.S. and Paul, M.A. (1976) The influence of genetic processes on some geotechnical properties of glacial tills. *Quarterly Journal of Engineering Geology* **9**, 159–194.

Boulton, G.S. and Spring, U. (1986) Isotopic fractionation at the base of polar and sub-polar glaciers. *Journal of Glaciology* **32** (112), 475–485.

Boulton, G.S., Dent, D.L. and Morris, E.M. (1974) Subglacial shearing and crushing, and the role of water pressures in tills from south-east Iceland. *Geografiska Annaler* **56A**, 135–145.

Boulton, G.S., Caban, P.E. and van Gijssel, K. (1995) Groundwater flow beneath ice sheets: part I – large scale patterns. *Quaternary Science Reviews* **14** (6), 545–562.

Boyce, J.L. and Eyles, N. (1991) Drumlins carved by deforming till streams beneath the Laurentide ice sheet. *Geology* **19**, 787–790.

Bozhinskiy, A.N., Krass, M.S. and Popovnin, V.V. (1986) Role of debris cover in thermal physics of glaciers. *Journal of Glaciology* **32** (111), 255–266.

Bradley, R.S. and Serreze, C. (1987) Mass balance of two high arctic plateau ice caps. *Journal of Glaciology* **33** (113), 123–128.

Braithwaite, R.J. (1984) Can the mass balance of a glacier be estimated from its equilibrium line altitude? *Journal of Glaciology* **30** (106), 364–368.

Braithwaite, R.J. (1995a) Mass balance of the Greenland ice sheet: a century of progress but there is still much to do. *Zeitschrift für Gletscherkunde und Glazialgeologie* **31**, 51–56.

Braithwaite, R.J. (1995b) Positive degree-day factors for ablation on the Greenland ice sheet studied by energy-balance modelling. *Journal of Glaciology* **41** (137), 153–160.

Braithwaite, R.J. and Olesen, O.B. (1988) Effect of glaciers on annual run-off, Johan Dahl Land, south Greenland. *Journal of Glaciology* **34**, (117), 200–207.

Braithwaite, R.J. and Olesen, O.B. (1989) Calculation of glacier ablation from air temperature, West Greenland. In: Oerlemans, J. (ed.) *Glacier fluctuations and climatic change*. Kluwer Academic Publishers, Dordrecht, 219–234.

Braithwaite, R.J. and Olesen, O.B. (1990a) Response of the energy balance on the margin of the Greenland ice sheet to temperature changes. *Journal of Glaciology* **36** (123), 217–221

Braithwaite, R.J. and Olesen, O.B. (1990b) A simple energy balance model to calculate ice ablation at the margin of the Greenland ice sheet. *Journal of Glaciology* **36** (123), 222–228.

Braithwaite, R.J. and Olesen, O.B. (1990c) Increased ablation at the margin of the Greenland ice sheet under a greenhouse-effect climate. *Annals of Glaciology* **14**, 20–22.

Braithwaite, R.J. and Olesen, O.B. (1993) Seasonal variation of ice ablation at the margin of the Greenland ice sheetand its sensitivity to climate change, Quamanârssûp sermia, West Greenland. *Journal of Glaciology* **39**, (132), 267–274.

Braithwaite, R.J., Laternser, M. and Pfeffer, W.T. (1994) Variations of near-surface firn density in the lower accumulation area of the Greenland ice sheet, Pâkitsoq, West Greenland. *Journal of Glaciology* **40** (136), 477–485.

Bretz, J.H. (1969) The Lake Missoula floods and the Channeled Scabland. *Journal of Geology* **77**, 505–543.

Broecker, W.S. and Denton, G.H. (1990) The role of ocean–atmosphere reorganizations in glacial cycles. *Quaternary Science Reviews* **9** (4), 305–341.

Broecker, W.S. (1994) Massive iceberg discharges as triggers for global climate change. *Nature* **372** (6505), 421–424.

Brown, G.H., Tranter, M. and Sharp, M.J. (1996) Experimental investigations of the weathering of suspended sediment by alpine glacial meltwater. *Hydrological Processes* **10** (4), 579–598.

Brown, N.E., Hallet, B. and Booth, D.B. (1987) Rapid soft bed sliding of the Puget glacial lobe. *Journal of Geophysical Research* **92-B**, 8985–8997.

Budd, W.F and Jacka T.H. (1989) A review of ice rheology for ice sheet modelling. *Cold Regions Science and Technology* **16** (2), 107–144.

Budd, W.F. and McInnes, B.J. (1979) Periodic surging of the Antarctic ice sheet – an assessment by modelling. *Hydrological Sciences Bulletin* **24**, 95–104.

Budd, W.F. and Young, N.W. (1983) Application of modelling techniques to measured profiles of temperatures and isotopes. In: Robin, G. de Q. (ed.) *The Climate Record in Polar Ice Sheets*. Cambridge University Press, Cambridge, 150–177.

Budd, W.F., Jenssen, D. and Radok, U. (1970) The extent of basal melting in Antarctica. *Polarforschung* **6**, 293–306.

Budd, W.F., Keage, P.L. and Blundy, N.A. (1979) Empirical studies of ice sliding. *Journal of Glaciology* **23** (89), 157–170.

Burbank, D.W. and Fort, M.B. (1985) Bedrock control on glacial limits: examples from the Ladakh and Zanskar ranges, noth-western Himalaya, India. *Journal of Glaciology* **31** (108), 143–149.

Burkimsher, M. (1983) Investigations of glacier hydrological systems using dye tracer techniques: Observations at Pasterzengletscher, Austria. *Journal of Glaciology* **29** (103), 403–416.

Cai, J. (1994) Sediment yields, lithofacies architecture and mudrock characteristics in glacimarine environments. Unpublished thesis. Northern Illinois University.

Cameron, D. (1965) Early discoverers. XXII. Goethe – discoverer of the Ice Age. *Journal of Glaciology* **5** (41), 751–754.

Chamberlin, T.C. (1895) Recent glaciological studies in Greenland. *Bulletin Geological Society of America* **6**, 199–220.

Charlesworth, J.K. (1955) The late-glacial history of the Highlands and Islands of Scotland. *Transactions of the Royal Society of Edinburgh* **62**, 103–929.

Chen, J. and Funk, M. (1990) Mass balance of Rhonegletscher during 1882/83 and 1986/87. *Journal of Glaciology* **36** (123), 199–209.

Chernova, L.P. (1981) Influence of mass balance and run-off on relief-forming activity of mountain glaciers. *Annals of Glaciology* **2**, 69–70.

Chinn, T.J.H. and Dillon, A. (1987) Observations on a debris-covered polar glacier 'Whiskey Glacier', James Ross Island, Antarctic Peninsula, Antarctica. *Journal of Glaciology* **33** (115), 300–310.

Chorley, R.J., Dunn, A.J. and Beckinsale, R.P. (1964) *The History of the Study of Landforms or the Development of Geomorphology*. Methuen, London.

Church, M. and Ryder, J.M. (1972) Paraglacial sedimentation: consideration of fluvial processes conditioned by glaciation. *Bulletin of the Geological Society of America* **83**, 3059–3072.

Clague, J.J. (1985) Deglaciation of the Prince Rupert – Kitimat area, British Columbia. *Canadian Journal of Earth Sciences* **22** (2), 256–265.

Clague, J.J. and Evans, S.G. (1993) Historic retreat of Grand Pacific and Melberne Glaciers, Saint Elias Mountains, Canada: an analogue for decay of the Cordilleran ice sheet at the end of the Pleistocene? *Journal of Glaciology* **39** (133), 619–624.

Clague, J.J. and Mathews, W.H. (1973) The magnitude of Jökulhlaups. *Journal of Glaciology* **12** (66), 501–504.

Clark, P.U. (1994) Unstable behaviour of the Laurentide ice-sheet over deforming sediments and its implications for climate change. *Quaternary Research* **41** (1), 19–25.

Clark, P.U., Licciardi, J.M., MacAyeal, D.R. and Jenson, J.W. (1996) Numerical reconstruction of a soft-bedded Laurentide Ice Sheet during the last glacial maximum. *Geology* **24** (8), 679–682.

Clarke, G.K.C. (1982) Glacier outburst floods from 'Hazard Lake', Yukon Territory, and the problem of flood magnitude prediction. *Journal of Glaciology* **28** (98), 3–21.

Clarke, G.K.C. (1987) A short history of scientific investigations on glaciers. *Journal of Glaciology* (special issue), 4–24.

Clarke, G.K.C. and Blake, E.W. (1991) Geometric and thermal evolution of a surge-type glacier in its quiescent state: Trapridge Glacier, Yukon Territory, Canada, 1969–89. *Journal of Glaciology* **37** (125), 158–169.

Clarke, G.K.C. and Matthews, W.H. (1981) Estimates of the magnitude of glacier outburst floods from Lake Donjek, Yukon Territory, Canada. *Canadian Journal of Earth Sciences* **18** (9), 1452–1463.

Clarke, G.K.C. and Waddington, E.D. (1991) A three-dimensional theory of wind pumping. *Journal of Glaciology* **37** (125), 89–96.

Clarke, G.K.C., Collins, S.G. and Thompson, D.E. (1984) Flow, thermal structure, and subglacial conditions of a surge-type glacier. *Canadian Journal of Earth Sciences* **21** (2), 232–240.

Clarke, G.K.C., Fisher, D.A. and Waddington, E.D. (1987) Wind pumping: a potentially significant heat source in ice sheets. In: *The Physical Basis of Ice Sheet Modelling, International Association of Hydrological Sciences Publication* **170**, 169–80.

Clarke, T.S. (1991) Glacier dynamics in the Susitna River basin, Alaska, U.S.A. *Journal of Glaciology* **37** (125), 97–106.

Clausen, H.C. and Hammer, C.U. (1988) The Laki and Tambora eruptions as revealed in Greenland ice cores from eleven locations. *Annals of Glaciology* **10**, 16–22.

Clayton, L., Teller, J.T. and Attig, J.W. (1985) Surging of the southwestern part of the Laurentide Ice Sheet. *Boreas* **14**, 235–241.

Colbeck, S.C. (1989) Air movement in snow due to windpumping. *Journal of Glaciology* **35** (120), 209–213.

Collins, D.N. (1978) Hydrology of an Alpine glacier as indicated by the chemical composition of meltwater. *Zeitschrift für Gletscherkunde und Glazialgeologie* **13**, 219–238.

Collins, D.N. (1979) Quantitative determination of the subglacial hydrology of two Alpine glaciers. *Journal of Glaciology* **23** (89), 347–362.

Collins, D.N., Lowe, A.T. and Boult, S. (1996) Solute fluxes in meltwaters draining from glacierised high mountain basins. In: Bottrell, S.H. (ed.) *Fourth International Symposium on the Geochemistry of the Earth's Surface*, University of Leeds, 728–732.

Collins, I.F. and McCrae, I.R. (1985) Creep buckling of ice shelves and the formation of pressure rollers. *Journal of Glaciology* **31** (109), 242–252.

Conklin, M.H., Sigg, A., Neftel, A. and Bales, R.C. (1993) Atmosphere-snow transfer function for H_2O_2: microphysical considerations. *Journal of Geophysical Research* **98** (D10), 18367–18376.

Conway, M. (1894) *Climbing and Exploration in the Karakoram-Himalaya*. Appleton and Company, New York.

Court, A. (1957) The classification of glaciers. *Journal of Glaciology* **3** (21), 2–7.

Craig, H. (1961) Isotopic variations in meteoric waters. *Science* **133** (3465), 1702–1703.

Crimes, T.P., Chester, D.K. and Thomas, G.S.P. (1992) Exploration of sand and gravel resources by geomorphological analysis in the glacial sediments of the eastern Lleyn-peninsula, Gwynedd, north Wales. *Engineering Geology* **32** (3), 137–156.

Croot, D.G. (1988) *Glaciotectonics: Forms and Processes*. A.A. Balkema, Rotterdam.

Cuffey, K.M. and Alley, R.B. (1996) Is erosion by deforming subglacial sediments significant? (Toward till continuity.) *Annals of Glaciology* **22**, 17–24.

Cuffey, K.M., Alley, R.B., Grootes, P.M., Bolzan, J.M. and Anandakrishnan, S. (1994) Calibration of the $\delta^{18}O$ isotopic palaeothermometer for central Greenland using borehole temperatures. *Journal of Glaciology* **40** (135), 341–349.

Cunningham, F. (1990) *James David Forbes, Pioneer Scottish Glaciologist*. Scottish Academic Press, Edinburgh.

Cunningham, J. and Waddington, E.D. (1990) Boudinage: a source of stratigraphic disturbance in glacial ice in central Greenland. *Journal of Glaciology* **36** (124), 269–272.

Dahl-Jensen, D. (1985) Determination of the flow properties at Dye 3, South Greenland, by bore-hole-tilting measurements and perturbation modelling. *Journal of Glaciology* **31** (108), 92–98.

Dahl-Jensen, D. (1989) Steady thermomechanical flow along two-dimensional flow lines in large grounded ice sheets. *Journal of Geophysical Research* **94**, 10355–10362.

Dahl-Jensen, D. and Johnsen, S.J. (1986) Palaeotemperatures still exist in the Greenland ice sheet. *Nature* **320** (6059), 250–252.

Dahl-Jensen, D., Gundestrup, N.S., Keller, K.R., Johnsen, S.J., Gogineni, S.P., Allen, C.T., Chuah, T.S., Miller, H., Kipstuhl, S. and Waddington, E.D. (1997) A search in north Greenland for a new ice-core drill site. *Journal of Glaciology* **43** (144), 300–306.

Daly, R.A. (1934) *The Changing World of the Ice Age*. Yale University Press, Newhaven.

Dansgaard, W. (1964) Stable isotopes in precipitation. *Tellus* **16**, 436–468.

Dansgaard, W., Johnsen, S.J., Clausen, H.B. and Gundestrup, N. (1973) Stable isotope glaciology. *Meddelelser om Grønland* **197** (2), 53 pp.

Dansgaard, W., Johnsen, S.J., Clausen, H.B., Dahl-Jensen, D., Gundestrup, N., Hammer, C.U., Hvidberg, C.S., Steffensen, J.P., Sveinbjörnsdottir, A.E., Jouzel, J. and Bond, G. (1993) Evidence for general instability of past climate from a 250-kyr ice-core record. *Nature* **364**, (6434), 218–220.

Davidson, C.I. (1989) Mechanism of wet and dry deposition of atmospheric contaminants to snow facies. In: Oeschger, H. and Langway, C.C. (eds) *The Environmental Record in Glaciers and Ice Sheets*. John Wiley and Sons, London. 29–52.

Dawson, A.G. (1992) *Ice age Earth*. Routledge, London.

De Marle, D.J. (1980) Design parameters for a South African iceberg power and water project. *Annals of Glaciology* **1**, 129–133.

Deely, R.M. and Parr, P.H. (1914) The Hintereis Glacier. *Philosophical Magazine* **6**, 153–176.

Demorest, M. (1941) Glacier flow and its bearing on the classification of glaciers. *Bulletin of the Geological Society of America* **52** (12, part 2), 2024–2025.

Ding Yongjian and Liu Jingshi (1992) Glacier lake outburst flood disasters in China. *Annals of Glaciology* **16**, 180–184.

Dodds, K.J. (1997) Antarctica and the modern geographical imagination (1918–1960). *Polar Record* **33** (184), 47–62.

Dolgushin, L.D., Lebedeva, I.M., Osipova, G.B. and Rototayeva, O.V. (1972) Vliyaniye eolovoy zapylennosti lednikov I poverkhnostoy moreny na tayaniye lednikov Sredney Azii (The influence of aeolian dusting of glaciers and superficial morain on glacier thawings in Central Asia). *Materialy Glyatsiologicheskikh Issledovaniy. Khronika. Obsuzhdeniya*, Vyp. **20**, 108–116.

Dowdeswell, J.A. (1989) On the nature of Svalbard icebergs. *Journal of Glaciology* **35** (120), 224–234.

Dowdeswell, J.A. and Scourse, J.D. (1990) *Glacimarine Environments: processes and sediments*. Geological Society Special Publication **53**, Geological Society, Bath.

Dowdeswell, J.A., Hambrey, M.J. and Wu, R. (1985) A comparison of clast fabric and shape in Late Pre-Cambrian and Modern glaciogenic sediments. *Journal of Sedimentary Petrology* **55** (5), 691–704.

Drewry, D.J. (1986) *Glacial Geologic Processes*. Edward Arnold, London.

Drewry, D.J. and Cooper, A.P.R. (1981) Processes and models of Antarctic glaciomarine sedimentation. *Annals of Glaciology* **2**, 117–122.

Dugmore, A.J. (1989) Tephrochronological studies of holocene glacier fluctuations in south Iceland. In: Oerlemans, J. (ed.) *Glacier Fluctuations and Climatic Change*. Kluwer Academic Publishers, Dordrecht, 37–55.

Dugmore, A.J. and Sugden, D.E. (1991) Do the anomolous fluctuations of Solheimajökull reflect ice-divide migration? *Boreas* **20** (2), 105–113.

Duval, P. (1977) The role of the water content on the creep rate of polycrystalline ice. *International Association of Hydrological Sciences* publication **118**, 29–33.

Duval, P. (1985) Grain growth and mechanical behaviour of polar ice. *Annals of Glaciology* **6**, 79–82.

Duval, P. and Lorius, C. (1980) Crystal size and climatic record down to the last ice age from Antarctic ice. *Earth and Planetary Science Letters* **48** (1), 59–64.

Dyugerov, M. (1996) Substitution of long-term mass balance data by measurements of one summer. *Zeitschrift für Gletscherkunde und Glazialgeologie* **32**, 177–184.

Echelmeyer, K. and Harrison, W.D. (1990) Jakobshavn Isbrae, West Greenland: seasonal variations in velocity – or lack thereof. *Journal of Glaciology* **36** (122), 82–88.

Echelmeyer, K. and Kamb, B. (1987) Glacier flow in a curving channel. *Journal of Glaciology* **33** (115), 281–292.

Echelmeyer, K. and Wang Zhongxiang (1987) Direct observation of basal sliding and deformation of basal drift at sub-freezing temperatures. *Journal of Glaciology* **33** (113), 83–98.

Echelmeyer, K., Butterfield, R. and Cuillard, D. (1987) Some observations on a recent surge of Peters Glacier, Alaska, U.S.A. *Journal of Glaciology* **33** (115), 341–345.

Echelmeyer, K., Clarke, T.S. and Harrison, W.D. (1991) Surficial glaciology of Jakobshavn Isbrae, West Greenland: Part I. Surface morphology. *Journal of Glaciology* **37** (127), 368–382.

Echelmeyer, K., Harrison, W.D., Clarke, T.S. and Benson, C. (1992) Surficial glaciology of Jakobshavn Isbrae, West Greenland: Part II. Ablation, accumulation and temperature. *Journal of Glaciology* **38** (128), 169–181.

Echelmeyer, K., Harrison, W.D., Larsen, C. and Mitchell, J.E. (1994) The role of the margins in the dynamics of an active ice stream. *Journal of Glaciology* **40** (136), 527–538.

Eicken, H., Oerter, H., Miller, H., Graf, W. and Kipfstuhl, J. (1994) Textural characteristics and impurity content of meteoric and marine ice in the Ronne Ice Shelf, Antarctica. *Journal of Glaciology* **40** (135), 386–398.

Elconin, R.F. and LaChapelle, E.R. (1997) Flow and internal structure of a rock glacier. *Journal of Glaciology* **43** (144), 238–244.

Ellis, M. (1982) *Ice and Icehouses through the Ages*. Southampton University Industrial Archaeology Group, Southampton University, UK.

Embleton, C. and King, C.A.M. (1968) *Glacial and Periglacial Geomorphology*. Edward Arnold, London.

Engelhardt, H. and Kamb, B. (1997) Basal hydraulic system of a West Antarctic ice stream: constraints from borehole observations. *Journal of Glaciology* **43** (144), 207–230.

Epstein, S. (1956) Variation of the O^{18}/O^{16} ratio of freshwater and ice. *Publication 400, US National Academy of Sciences*.

Epstein, S. and Sharp, R.P. (1959) Oxygen isotope variations in the Malaspina and Saskatchewan glaciers. *Journal of Geology* **67** (1), 88–102.

Esmarch, J. (1827) Remarks tending to explain the geological history of the Earth. *Edinburgh New Philosophical Journal* **2**, 107–121.

European Science Foundation (1995) *Greenland Ice Core Project: Final Report*. European Science Foundation, Strasbourg.

Evans, D.J.A. (1989) Apron entrainment at the margins of sub-polar glaciers, North-west Ellesmere Island, Canadian High Arctic. *Journal of Glaciology* **35** (121), 317–324.

Evenson, E.B., Pasquini, T.A., Stewart, R.A. and Stephens, G. (1979) Systematic provenance investigations in areas of alpine glaciation: applications to glacial geology and mineral exploration. In: Schlüchter, C. (ed.) *Moraines and Varves*. A.A. Balkema, Rotterdam, 25–42

Eyles, N. (ed.) (1983) *Glacial Geology*. Pergamon Press, Oxford.

Eyles, N. and Menzies, J (1983) The subglacial landsystem. In: Eyles, N. (ed.) *Glacial Geology*. Pergamon Press, Oxford.19–70.

Fairbanks, R.G. (1989) A 17,000-year glacio-eustatic sea level record: influence of glacial melting rates on the Younger Dryas event and deep-ocean circulation. *Nature* **342** (6250), 637–642.

Fairchild, I.J., Bradby, L. and Spiro, B. (1993) Carbonate diagenesis in ice. *Geology* **21** (10), 901–904.

Ferguson, R.I. (1973) Sinuosity of supra-glacial streams. *Geological Society of America Bulletin* **84**, 251–256.

Finsterwalder, R. (1959) Chamonix glaciers. *Journal of Glaciology* **3** (26), 547–548.

Firestone, J., Waddington, E.D. and Cunningham, J. (1990) The potential for basal melting under Summit, Greenland. *Journal of Glaciology* **36** (123), 163–168.

Fisher, D. (1973) Subglacial leakage of Summit Lake, British Columbia, by dye determinations. *International Association of Hydrological Sciences Publication* **95**, 111–116.

Fisher, D.A. and Koerner, R.M. (1986) On the special rheological properties of ancient microparticle-laden northern hemisphere ice as derived from bore-hole and core measurements. *Journal of Glaciology* **32** (112), 501–510.

Fitzgerald, D. and Rosen, P.S. (1987) *Glaciated Coasts*. Academic Press, San Diego.

Fitzsimons, S.J. (1990) Ice-marginal depositional processes in a polar maritime environment, Vestfold Hills, Antarctica. *Journal of Glaciology* **36** (124), 279–286.

Fleisher, P.J., Muller, E.H., Cadwell, D.H., Rosenfeld, C.L., Bailey, P.K., Pelton, J.M. and Puglisi, M.A. (1995) The surging advance of Bering Glacier, Alaska: a progress report. *Journal of Glaciology* **41** (137), 207–213.

Fletcher, N.H. (1970) *The Chemical Physics of Ice*. Cambridge University Press, Cambridge.

Fookes, P.G., Gordon, D.L. and Higginbottom, I.E. (1975) Glacial landforms, their deposits and engineering characteristics. In: *The Engineering Behaviour of Glacial Materials*. Midlands Soil Mechanics and Foundation Engineering Society, Birmingham, 18–51.

Forbes, J.D. (1842) The glacier theory. *Edinburgh Review* **75**, 49–105.

Forel, F.A. (1900) J.J. Perraudin, le précurseur glaciariste. *Ecolog. Geol. Helv.* **6**, 170.

Forel, F.A. (1898) Circulation des eaux dans la Glacier du Rhône. *Bulletin de la Société de Spéléologie (Spelunca)* **4** (16), 156–158.

Forster, R.R., Davis, C.H., Rand, T.W. and Moore, R.K. (1991) Snow-stratification investigation on an Antarctic ice stream with an X-band radar system. *Journal of Glaciology* **37** (127), 323–326.

Fortuin, J.P.F. and Oerlemans, J. (1990) Parameterization of the annual surface temperature and mass balance of Antarctica. *Annals of Glaciology* **14**, 78–84.

Fountain, A.G. (1994) Borehole water-level variations and implications for the subglacial hydraulics of South Cascade Glacier, Washington State, U.S.A. *Journal of Glaciology* **40** (135), 293–304.

Fowler, A.C. (1986) Sub-temperate basal sliding. *Journal of Glaciology* **32** (110), 3–5.

Fowler, A.C. (1987) Sliding with cavity formation. *Journal of Glaciology* **33** (115), 255–267.

Fowler, A.C. and Ng, F.S.L. (1996) The role of sediment transport in the mechanics of jökulhlaups. *Annals of Glaciology* **22**, 255–259.

Frolich, R.M., Vaughan, D.G. and Doake, C.S.M. (1989) Flow of Rutford ice stream and comparison with Carlson Inlet, Antarctica. *Annals of Glaciology* **12**, 51–56.

Fujita, K., Seko, K., Ageta, Y., Jianchen, P. and Tandong, Y. (1996) Superimposed ice in glacier mass balance on the Tibetan Plateau. *Journal of Glaciology* **42** (142), 454–460.

Funk, M., Echelmeyer, K. and Iken, A. (1994) Mechanisms of fast flow in Jakobshavns Isbrae, West Greenland: part II. Modelling of englacial temperatures. *Journal of Glaciology* **40** (136), 569–585.

Gagnon, R.E. and Gammon, P.H. (1995) Characterisation and flexural strength of iceberg and glacier ice. *Journal of Glaciology* **41** (137), 103–111.

Gagnon, R.E., Tilk, C. and Kiefte, H. (1994) Internal melt figures in ice by rapid adiabatic compression. *Journal of Glaciology* **40** (134), 132–134.

Gardner, J.S. and Hewitt, K. (1990) A surge of Bultar Glacier, Karakoram Range, Pakistan: A possible landslide trigger. *Journal of Glaciology* **36** (123), 159–162.

Giardino, J.R., Schroder, J.F. and Vitek, J.D. (eds) (1987) *Rock Glaciers*. Allen and Unwin, London.

Gibbs, M. and Kump, L. (1996) Global chemical weathering during glaciation. In: Bottrell, S.H. (ed.) *Fourth International Symposium on the Geochemistry of the Earth's Surface*. University of Leeds, 733–737.

Gillet, F. (1975) Steam, hot-water and electrical thermal drills for temperate glaciers. *Journal of Glaciology* **14**, (70), 171–179.

Giovinetto, M.B. and Zwally, H.J. (1995) An assessment of the mass budgets of Antarctica and Greenland using accumulation derived from remotely sensed data in areas of dry snow. *Zeitschrift für Gletscherkunde und Glazialgeologie* **31**, 25–37.

Glen, J.W. (1954) The stability of ice-dammed lakes and other water-filled holes in glaciers. *Journal of Glaciology* **2**, 316–318.

Glen, J.W. (1955) The creep of polycrystalline ice. *Proceedings of the Royal Society, series A*. **228** (1175), 519–38

Glen, J.W. (1958) The flow law of ice. A discussion of the assumptions made in glacier theory, their experimental foundations and consequences. *International Association of Hydrological Sciences* **47**, 171–183.

Glen, J.W. (1987) Fifty years of progress in ice physics. *Journal of Glaciology* (special issue), 52–59.

Glen, J.W., Adie, R.J., Johnson, D.M., Homer, D.R. and Macqueen, A.D. (eds) (1979) Symposium on glacier beds: the ice-rock interface. *Journal of Glaciology* **29** (89).

Goldthwait R.P. (1951) Development of end-moraines in east-central Baffin Island. *Journal of Geology* **59**, 567–577.

Goldthwait, R.P. (1971) Restudy of Red Rock ice cliff, Nunatarssuaq, Greenland. *CRREL Technical Report* **224**. U.S. Army Cold Regions Research and Engineering Laboratory, Hanover, New Hampshire.

Goldthwait, R.P. and Matsch, C. (1988) *Genetic classification of glaciogenic deposits*, Balkema. Rotterdam.

Goodman, D.J., King, G.C.P., Millar, D.H.M. and Robin, G. deQ. (1979) Pressure-melting effects in basal ice of temperate glaciers: laboratory studies and field observations under Glacier d'Argentière. *Journal of Glaciology* **23** (89), 259–272.

Goodman, D.J., Frost, H.J. and Ashby, M.F. (1981) The plasticity of polycrystalline ice. *Philosophical Magazine A* **43** (3), 665–695.

Goodwin, I.D. (1991) Snow-accumulation variability from seasonal surface observations and firn-core stratigraphy, eastern Wilkes Land, Antarctica. *Journal of Glaciology* **37** (127), 383–387.

Goodwin, I.D. (1993) Basal ice accretion and debris entrainment within the coastal ice margin, Law Dome, Antarctica. *Journal of Glaciology* **39** (131), 157–166.

Gordon, J.E., Darling, W.G., Whalley, W.B. and Gellatly, A.F. (1988) δD-$\delta^{18}O$ relationships and the thermal history of basal ice near the margins of two glaciers in Lyngen, North Norway. *Journal of Glaciology* **34** (118), 265–268.

Goto-Azuma, K., Koerner, R.M., Nakawo, M. and Kudo, A. (1997) Snow chemistry of Agassiz Ice Cap, Ellesmere Island, Northwest Territories, Canada. *Journal of Glaciology* **43** (144) 199–206.

Gow, A.J. (1965) Snow studies in Antarctica. *Cold Regions Research and Engineering Laboratory (CRREL) Research Report* **177.**

Gow, A.J., Ueda, H.T. and Garfield, D.E. (1968) Antarctic ice sheet: preliminary results of first core hole to bedrock. *Science* **161**, 1011–1013.

Gow, A.J., Epstein, S. and Sheehy, W. (1979) On the origin of stratified debris in ice cores from the bottom of the Antarctic ice sheet. *Journal of Glaciology* **23** (89), 185–192.

Gray, J.M. (1993) Quaternary geology and waste disposal in south Norfolk, England. *Quaternary Science Reviews* **12**, 899–912.

Greene, D. (1992) Topography and former Scottish tidewater glaciers. *Scottish Geographical Magazine* **108**, 164–171.

Gregory, J.M. and Oerlemans, J. (1998) Simulated future sea-level rise due to glacier melt based on regionally and seasonally resolved temperature changes. *Nature* **391** (6666), 474–476.

GRIP (Greenland Ice-core Project) members (1993) Climate instability during the last interglacial recorded in the GRIP ice core. *Nature* **364** (6434), 203–207.

Gripp, K. (1929) Glaciologische und geologische ergenisse der Hamburgischen Spitzbergen expedition 1927. *Naturwissenschaftlicher verein in Hamburg* **22** (2–4), 146–249.

Grootes, P., Stuiver, M., White, J., Johnsen, S. and Jouzel, J. (1993) Comparison of oxygen isotope records from the GISP2 and GRIP Greenland icecores. *Nature* **366** (6455), 552–554.

Grosswald, M.G. and Hughes, T.J. (1995) Paleoglaciology's grand unsolved problem. *Journal of Glaciology* **41** (138), 313–332.

Grove, J.M. (1987) Glacier fluctuations and hazards. *The Geographical Journal* **153** (3), 351–369

Grove, J.M. (1988) *The Little Ice Age*. Methuen, London.

Gruner, G.S. (1760) *Die eisgebirge des Schweizerlandes*, Bern.

Gudmundsson, G.H. (1997a) Basal flow characteristics of a linear medium sliding frictionless over small bedrock undulations. *Journal of Glaciology* **43** (143), 71–79.

Gudmundsson, G.H. (1997b) Basal flow characteristics of a non-linear flow sliding frictionless over strongly undulating bedrock. *Journal of Glaciology* **43** (143), 80–89.

Gudmundsson, M.T., Sigmundsson, F. and Björnsson, H. (1997) Ice–volcano interaction of the 1996 Gjálp subglacial eruption, Vatnajökull, Iceland. *Nature* **389**, 954–957.

Haeberli, W. (1983) Frequency and characteristics of glacier floods in the Swiss Alps. *Annals of Glaciology* **4**, 85–90.

Haeberli, W. (1985) Creep of mountain permafrost: internal structure and flow of Alpine rock glaciers. *Mitteilungen der Versuchsanstalt für Wasserbau, Hydrologie, und Glaziologie, ETH-Zurich* **77**, 142.

Haeberli, W. (1989) Glacier ice-cored rock glaciers in the Yukon Territory, Canada. *Journal of Glaciology* **35** (120), 294–295.

Haeberli, W., Müller, P., Alean, P. and Bösch, H. (1989) Glacier changes following the little ice age

– a survey of the international data basis and its perspectives. In: Oerlemans, J. (ed.) (1989) *Glacier Fluctuations and Climatic Change*. Kluwer Academic Publishers, Dordrecht, 77–102.

Haldorsen, S. (1981) Grain-size distribution of subglacial till and its relation to glacial crushing and abrasion. *Boreas* **10** (1), 91–105.

Hall, A. and Weston, K. (1993) The interaction between an ice sheet and its atmospheric boundary layer. *Journal of Glaciology* **39** (133), 601–608.

Hallet, B. (1976) Deposits formed by subglacial precipitation of $CaCO_3$. *Geological Society of America Bulletin* **87**, 1003–1015.

Hallet, B. (1979) A theoretical model of glacial abrasion. *Journal of Glaciology* **23** (89), 39–50.

Hallet, B. (1981) Glacial abrasion and sliding: their dependence on the debris concentration in basal ice. *Annals of Glaciology* **2**, 23–28.

Hallet, B. and Anderson, M. (1980) Detailed glacial geomorphology of a proglacial bedrock area at Castleguard glacier, Alberta, Canada. *Zeitschrift für Gletscherkunde und Glazialgeologie* **16**, 171–184.

Hallet, B., Hunter, C. and Bogen, J. (1996) Rates of erosion and sediment evacuation by glaciers: A review of field data and their implications. *Global and Planetary Change* **12** (1–4), 213–235.

Hallet, B., Lorrain, R.D. and Souchez, R.A. (1978) The composition of basal ice from a glacier sliding over limestones. *Geological Society of America Bulletin* **89** (2), 314–320

Hambrey, M.J. (1975) The origin of foliation in glaciers: evidence from some Norwegian examples. *Journal of Glaciology* **14** (70), 181–185.

Hambrey, M.J. (1994) *Glacial Environments*. UCL Press, London.

Hambrey, M.J., Milnes, A.G. and Siegenthaler, H. (1980) Dynamics and structure of Griesgletscher, Switzerland. *Journal of Glaciology* **25** (92), 215–228.

Hamley, T.C. and Budd, W.F. (1986) Antarctic iceberg distribution and dissolution. *Journal of Glaciology* **32** (111), 242–251.

Hamley, T.C., Smith, I.N. and Young, N.W. (1985) Mass-balance and ice-flow-law parameters for East Antarctica. *Journal of Glaciology* **31** (109), 334–339.

Hammer, C.U. (1977) Past volcanism revealed by Greenland ice sheet impurity. *Nature* **270**, 482–486.

Hammer, C.U., Clausen, H.B., Dansgaard, W., Gundestrup, N., Johnsen, S.J. and Reeh, N. (1978) Dating of Greenland ice cores by flow models, isotopes, volcanic debris and continental dust. *Journal of Glaciology* **20** (82), 3–26.

Hamran, S., Aarholt, E., Hagen, J.O. and Mo, P. (1996) Estimation of relative water content in a sub-polar glacier using surface-penetration radar. *Journal of Glaciology* **42** (142), 533–537.

Hansen, B.L. and Langway, C.C. Jr (1966) Deep core drilling in ice and core analysis at Camp Century, Greenland, 1961–1966. *Antarctic Journal of the United States* **1** (5), 207–208.

Hanson, B. (1995) A fully three-dimensional finite-element model applied to velocities on Storglaciären, Sweden. *Journal of Glaciology* **41** (137), 91–102.

Hanson, B. and Hooke, R.LeB. (1994) Short-term velocity variations and basal coupling near a bergschrund, Storglaciären, Sweden. *Journal of Glaciology* **40** (134), 67–74.

Harrison, S.P., Prentice, I.C. and Bartlein, P.J. (1992) Influence of insolation and glaciation on atmospheric circulation in the north-Atlantic sector – implications of general-circulation model experiments for the late Quaternary climatology of Europe. *Quaternary Science Reviews* **11** (3), 283–299.

Harrison, W.D. and Raymond, C.F. (1976) Impurities and their distribution in temperate glacier ice. *Journal of Glaciology* **16** (74), 173–180.

Harrison, W.D., Echelmeyer, K.A., Chacho, E.F., Raymond, C.F. and Benedict, R.J. (1994) The 1987–88 surge of West Fork Glacier, Susitna basin, Alaska, U.S.A. *Journal of Glaciology* **40** (135), 241–254.

Hart, J.K. (1995a) An investigation of the deforming layer/debris-rich basal-ice continuum, illustrated from three Alaskan glaciers. *Journal of Glaciology* **41** (139), 619–633.

Hart, J.K. (1995b) Subglacial erosion, deposition and deformation associated with deformable beds. *Progress in Physical Geography* **19** (2), 159–172.

Hart, J.K. (1997) The relationship between drumlins and other forms of subglacial glaciotectonic deformation. *Quaternary Science Reviews* **16** (1), 93–108.

Hart, J.K., Hindmarsh, R.C.A. and Boulton, G.S. (1990) Different styles of subglacial glaciotectonic deformation in the context of the Anglian ice sheet. *Earth Surface Processes and Landforms* **15**, 227–241.

Harvey, L.D.D. (1988) Climatic impact of ice-age aerosols. *Nature* **334** (6180), 333–336.

Hastenrath, S. (1984) *The Glaciers of Equatorial East Africa*. D. Reidel, Dordrecht.

Hastenrath, S. (1989) Ice flow and mass changes of Lewis Glacier, Mount Kenya, East Africa:

observations 1974–86, modelling and predictions to the year 2000 A.D. *Journal of Glaciology* **35** (121) 325–332.

Hastenrath, S. (1992) Ice-flow and mass changes of Lewis Glacier, Mount Kenya, East Africa, 1986–90: observations and modelling. *Journal of Glaciology* **38** (128), 36–42.

Hastenrath, S. and Ames, A. (1995a) Diagnosing the imbalance of Yanamarey Glacier in the Cordillera Blanca of Peru. *Journal of Geophysical Research – Atmospheres* **100**, D3 5105–5112.

Hastenrath, S. and Ames, A. (1995b) Recession of the Yanamarey Glacier in Peru's Cordillera Blanca during the 20th century. *Journal of Glaciology* **41** (137), 191–196.

Heinrich, H. (1988) Origin and consequences of cyclic ice rafting in the north-east Arctic ocean during the past 130,000 years. *Quaternary Research* **29**, 141–152.

Herron, S. and Langway, C.C. Jr (1979) The debris-laden ice at the bottom of the Greenland ice sheet. *Journal of Glaciology* **23** (89), 193–207.

Herron, S. and Langway, C.C. Jr (1987) Derivation of paleoelevations from total air content of two deep Greenland ice cores. *International Association of Hydrological Sciences Publication* **170**, 283–295.

Higuchi, K. (1969) On the possibility of artificial control of the water balance of perennial ice. *Symposium on the Hydrology of Glaciers, Cambridge. International Association of Hydrological Sciences publication* **95**, 207–212.

Hindmarsh, R.C.A. (1993) Modelling the dynamics of ice sheets. *Progress in Physical Geography* **17** (4), 391–412.

Hodge, S.M. (1974) Variations in the sliding of a temperate glacier. *Journal of Glaciology* **13** (69), 349–369.

Hodge, S.M. (1976) Direct measurement of basal water pressures: a pilot study. *Journal of Glaciology* **16** (74), 205–218.

Hodge, S.M. (1979) Direct measurement of basal water pressures: progress and problems. *Journal of Glaciology* **23** (89), 309–319.

Hodgkins, R. and Dowdeswell, J. (1994) Tectonic processes in Svalbard tide-water glacier surges: evidence from structural glaciology. *Journal of Glaciology* **40** (136), 553–560.

Hoinkes, H. (1967) Glaciology in the International Hydrological Decade. IUGG General Assembly, Bern, IASH Commission on snow and ice. *IASH Publication* **79**, 112–126.

Hollin, J.T. (1965) Wilson's theory of ice ages. *Nature* **208** (5005), 12–16.

Hollin, J.T. (1972) Interglacial climates and Antarctic ice surges. *Quaternary Research* **2**, 401–408.

Hollin, J.T. (1980) Climate and sea level in isotope stage 5: an East Antarctic ice surge at 95,000 BP? *Nature* **283**, 629–633.

Holmlund, P. (1988) Internal geometry and evolution of moulins, Storglaciären, Sweden. *Journal of Glaciology* **34** (117), 242–248.

Holmlund, P. and Hooke, R.LeB. (1983) High water-pressure events in moulins, Storglaciären, Sweden. *Geografiska Annaler* **65(A)** (1–2), 19–25.

Holmlund, P., Karlén, W. and Grudd, H. (1996a) Fifty years of mass balance and glacier front observations at the Tarfala research station. *Geografiska Annaler* **78A** (2–3), 105–114.

Holmlund, P., Burman, H. and Rost, T. (1996b) Sediment-mass exchange between turbid meltwater streams and proglacial deposits of Storglaciären, northern Sweden. *Annals of Glaciology* **22**, 63–67.

Hooke, R.L. (1969) Crystal shape in polar glaciers and the philosophy of ice-fabric diagrams. *Journal of Glaciology* **28** (98), 35–42.

Hooke, R.L. (1973a) Flow near the margin of the Barnes ice cap and the development of ice-cored moraines. *Geological Society of America Bulletin* **84** (3), 3929–2948.

Hooke, R.L. (1973b) Structure and flow in the margin of the Barnes ice cap, Baffin Island, N.W.T., Canada. *Journal of Glaciology* **12** (66), 423–438.

Hooke, R.L. (1981) Flow law for polycrystalline ice in glaciers: comparison of theoretical predictions, laboratory data, and field measurements. *Reviews of Geophysics and Space Physics* **19** (4), 664–672.

Hooke, R.L. (1984) On the role of mechanical energy in maintaining subglacial water conduits at atmospheric pressure. *Journal of Glaciology* **30** (105), 180–187.

Hooke, R.L. and Clausen, H.B. (1982) Wisconsin and Holocene $\delta^{18}O$ variations, Barnes Ice Cap, Canada. *Geological Society of America Bulletin* **93** (8), 784–789.

Hooke, R.L. and Hock, R. (1993) Evolution of the internal drainage system in the lower part of the ablation area of Storglaciären, Sweden. *Geological Society of America Bulletin* **105** (4) 537–546

Hooke, R.L. and Hudleston, P.J. (1978) Origin of foliation in glaciers. *Journal of Glaciology* **20** (83), 285–299.

Hooke, R.L., Dahlin, B.B. and Kauper, M.T. (1972) Creep of ice containing dispersed fine sand. *Journal of Glaciology* **11** (63), 327–336.

Hooke, R.L., Raymond, C.F., Hotchkiss, R.L. and Gustafson, R.J. (1979) Calculations of velocity and temperature in a polar glacier using the finite-element method. *Journal of Glaciology* **24** (90), 131–146

Hooke, R.L., Holmlund, P. and Iverson, N.R. (1987) Extrusion flow demonstrated by bore-hole deformation measurements over a riegel, Storglaciären, Sweden. *Journal of Glaciology* **33** (113), 72–78.

Hooke, R.L., Calla, P., Holmlund, P., Nilsson, M. and Stroeven, A. (1989) A 3 year record of seasonal variations in surface velocity, Storglaciären, Sweden. *Journal of Glaciology* **35** (120), 235–247.

Hooke, R.L., Laumann, T. and Kohler, J. (1990) Subglacial water pressures and the shape of subglacial conduits. *Journal of Glaciology* **36** (122), 67–71.

Hooke, R.L., Pohjola, V.A., Jansson, P. and Kohler, J. (1992) Intra-seasonal changes in deformation profiles revealed by borehole studies, Storglaciären, Sweden. *Journal of Glaciology* **38** (130), 348–358.

Hooke, R.L., Hanson, B., Iverson, N.R., Jansson, P. and Fischer, U.H. (1997) Rheology of till beneath Storglaciären, Sweden. *Journal of Glaciology* **43** (143), 172–179.

Hopkins, W.A. (1845) On the motion of glaciers. *Philosophical Magazine* **26**, 250.

Hoppe, G. (1959) Glacial morphology and inland ice recession in northern Sweden. *Geografiska Annaler* **41** (4), 193–212.

Hubbard, B. (1991) Freezing-rate effects on the physical characteristics of basal ice formed by net adfreezing. *Journal of Glaciology* **37** (127), 339–347.

Hubbard, B. and Nienow, P. (1997) Alpine subglacial hydrology: research status and implications for glacial geology. *Quaternary Science Reviews* **16**, 939–955.

Hubbard, B. and Sharp, M. (1989) Basal ice formation and deformation: a review. *Progress in Physical Geography* **13** (4), 529–558.

Hubbard, B. and Sharp, M. (1993) Weertman regelation, multiple refreezing events and the isotopic evolution of the basal ice layer. *Journal of Glaciology* **39** (132), 275–291.

Hubbard, B. and Sharp, M. (1995) Basal ice facies and their formation in the western Alps. *Arctic and Alpine Research* **27** (4), 301–310.

Hubbard, B.Y., Sharp, M.J., Willis, I.C., Nielsen, M.K. and Smart, C.C. (1995) Borehole water-level variations and the structure of the subglacial hydrological system of Haut Glacier d'Arolla, Valais, Switzerland. *Journal of Glaciology* **41 (139), 572–584.**

Hudleston, P.J. and Hooke, R.L. (1980) Cumulative deformation in the Barnes Ice Cap and implications for the development of foliation. *Tectonophysics* **66**, 127–146.

Hughes, T.J. (1973) Is the West Antarctic ice sheet disintegrating? *Journal of Geophysical Research* **78**, 7884–7910.

Hughes, T.J. (1975) The west Antarctic ice sheet: instability, disintegration and initiation of ice ages. *Reviews of Geophysics and Space Physics* **13**, 502–526.

Hughes, T.J. (1985) Thermal convection in ice sheets: we look but do not see. *Journal of Glaciology* **31** (107), 39–48.

Hughes, T.J. (1986) The Jakobshavns effect. *Geophysical Research Letters* **13**, 46–48.

Hughes, T.J. (1989) Bending shear: the rate-controlling mechanism for calving ice walls. *Journal of Glaciology* **35** (120), 260–266.

Hughes, T.J. (1992a) On the pulling power of ice streams. *Journal of Glaciology* **38** (128), 125–151.

Hughes, T.J. (1992b) Theoretical calving rates from glaciers along ice walls grounded in water of variable depths. *Journal of Glaciology* **38** (129), 282–294.

Hugi, F.J. (1831) Observations on the Glaciers of the Alps. *Edinburgh New Philosophical Journal* **11**, 74–81.

Humlum, O. (1985) Changes in texture and fabric of particles in glacial traction with distance from source, Myrdalsjökull, Iceland. *Journal of Glaciology* **31** (108), 150–156.

Humphrey, N.F. and Raymond, C.F. (1994) Hydrology, erosion and sediment production in a surging glacier: Variegated Glacier, Alaska, 1982–83. *Journal of Glaciology* **40** (136), 539–552.

Humphrey, N.F., Raymond, C.F. and Harrison, W. (1986) Discharges of turbid water during mini-surges of Variegated Glacier, Alaska, U.S.A. *Journal of Glaciology* **32** (111), 195–207.

Hunter, L.E. (1994) Grounding-line systems of modern temperate glaciers and their effects on glacier stability. Thesis, North Illinois University.

Hunter, L.E., Powell, R.D. and Lawson, D.E. (1996) Flux of debris transported by ice at three Alaskan tidewater glaciers. *Journal of Glaciology* **42** (140), 123–135.

Hutton, J. (1795) *Theory of the Earth, with Proofs and Illustrations*. William Creech, Edinburgh.

Huybrechts, P., Letréguilly, A. and Reeh, N. (1991) The Greenland ice sheet and greenhouse warming. *Palaeogeography, Palaeoclimate, Palaeoecology (Global and Planetary Change)* **89**, 399–412.

IAHS (1990) *International Classification of Seasonal Snow on the Ground.* International Commission for snow and ice, International Association of Hydrological Sciences, Wallingford, Oxon.

IHD (International Hydrological Decade) (1970) Coordinating Council, 7th session, Paris, 1971, Part 1.4.1.1 of report SC/IHD/VII/45 Rev.

Iken, A. (1981) The effect of the subglacial water pressure on the sliding velocity of a glacier in an idealized numerical model. *Journal of Glaciology* **27** (97), 407–421.

Iken, A. and Bindschadler, R.A. (1986) Combined measurements of subglacial water pressure and surface velocity of Findelengletscher, Switzerland: conclusions about drainage system and sliding mechanism. *Journal of Glaciology* **32** (110), 101–119.

Iken, A. and Truffer, M. (1997) The relationship between subglacial water pressure and velocity of Findelengletscher, Switzerland, during its advance and retreat. *Journal of Glaciology* **43** (144), 328–338.

Iken, A., Röthlisberger, H., Flotron, A. and Haeberli, W. (1983) The uplift of Unteraargletscher at the beginning of the melt season – a consequence of water storage at the bed? *Journal of Glaciology* **29** (101), 28–47.

Iken, A., Echelmeyer, K., Harrison, W. and Funk, M. (1993) Mechanisms of fast flow in Jakobshavns Isbrae, West Greenland: part I. Measurement of temperature and water level in deep boreholes. *Journal of Glaciology* **39** (131), 15–25.

Isaksson, E. (1992) The western Barents Sea and the Svalbard archipelago 18,000 years ago – a finite-difference computer model reconstruction. *Journal of Glaciology* **38** (129), 295–301.

Iverson, N.R. (1990) Laboratory simulations of glacial abrasion: comparison with theory. *Journal of Glaciology* **36** (124), 304–314.

Iverson, N.R. (1991) Potential effects of subglacial water-pressure fluctuations on quarrying. *Journal of Glaciology* **37** (125), 27–36.

Iverson, N.R. (1993) Regelation of ice through debris at glacier beds – implications for sediment transport. *Geology* **21** (6), 559–562.

Iverson, N.R. and Semmens, D.J. (1995) Intrusion of ice into porous media by regelation: A mechanism of sediment entrainment by glaciers. *Journal of Geophysical Research* **100** (B6), 10219–10230.

Iverson, N.R., Hanson, B., Hooke, R.LeB. and Jansson, P. (1995) Flow mechanism of glaciers on soft beds. *Science* **267** (5195), 80–81.

Jacobel, R.W., Gades, A.M., Gottschling, D.L., Hodge, S.M. and Wright, D.L. (1993) Interpretation of radar-detected internal layer folding in West Antarctic ice streams. *Journal of Glaciology* **39** (133), 528–537.

Jacobs, J.D., Simms, E.L. and Simms, A. (1997) Recession of the southern part of barnes Ice Cap, Baffin Island, Canada, between 1961 and 1993, determined from digital mapping of Landsat TM. *Journal of Glaciology* **43** (143), 98–102.

Jacobs, S.S., MacAyeal, D.R. and Ardai, J.L. Jr (1986) The recent advance of the Ross ice shelf, Antarctica. *Journal of Glaciology* **32** (112), 464–474.

Jacobs, S.S., Helmer, H.H., Doake, C.S.M., Jenkins, A. and Frolich, R.M. (1992) Melting of ice shelves and the mass balance of Antarctica. *Journal of Glaciology* **38** (130), 375–387.

Jansson, P. (1995) Water pressure and basal sliding on Storglaciären, northern Sweden. *Journal of Glaciology* **41** (138), 232–240.

Jayaraman, K.S. (1996) 'Ban tourists' call to save glacier source of the Ganges. *Nature* **384** (6610), 602.

Jenkins, A., Vaughan, D.G., Jacobs, S.S., Hellmer, H.H. and Keys, J.R. (1997) Glaciological and oceanographic evidence of high melt rates beneath Pine Island Glacier, West Antarctica. *Journal of Glaciology* **43** (143), 114–121.

Jenson, J.W., MacAyeal, D.R., Clark, P.U., Ho, C.L. and Vela, J.C. (1996) Numerical modeling of subglacial sediment deformation – implications for the behavior of the Lake-Michigan lobe, Laurentide ice-sheet. *Journal of Geophysical Research* **101** (B4), 8717–8787.

Jóhannesson, T. (1997) The response of two Icelandic glaciers to climatic warming computed with a degree-day glacier mass-balance model coupled to a dynamic glacier model. *Journal of Glaciology* **43** (143), 321–327.

Jóhannesson, T., Raymond, C. and Waddington, E. (1989a) A simple method for determining the response time of glaciers. In: Oerlemans, J. (ed.) *Glacier Fluctuations and Climatic Change.* Kluwer Academic Publishers, Dordrecht, 343–352.

Jóhannesson, T., Raymond, C. and Waddington, E. (1989b) Time-scale for adjustment of glaciers to changes in mass balance. *Journal of Glaciology* **35** (121), 355–369.

Johnsen, S.J., Dahl-Jensen, D., Dansgaard, W. and Gundestrup, N. (1995) Greenland palaeotemperatures derived from GRIP bore hole temperature and core isotope profiles. *Tellus* **47B**, 624–629.

Johnson, P.G. and Lacasse, D. (1988) Rock glaciers of the Dalton Range, Kluane Ranges, south-west Yukon Territory, Canada. *Journal of Glaciology* **34** (118), 327–332.

Johnson, P.H. (1995) The formation of clotted ice, a basal ice facies of some glaciers. In: McLelland, S.J., Skellern, A.R. and Porter, P.R. (eds) *Postgraduate Research in Geomorphology, Selected papers from the 17th BGRG Postgraduate Symposium*. British Geomorphological Research Group.

Jones, S.J. and Glen, J.W. (1969) The effect of dissolved impurities on the mechanical properties of ice crystals. *Philosophical Magazine* 8th series **19** (157), 13–24.

Jouzel, J. and Souchez, R.A. (1982) Melting-refreezing at the glacier sole and the isotopic composition of the ice. *Journal of Glaciology* **29** (98), 35–42.

Kamb, B. (1970) Sliding motion of glaciers: theory and observation. *Reviews of Geophysics and Space Physics* **8** (4), 673–728.

Kamb, B. (1972) Structure of the ices. In: Horne, R.A. (ed.) *Water and Aqueous Solutions*. Wiley-Interscience, New York.

Kamb, B. (1987) Glacier surge mechanism based on linked cavity configuration of the basal water conduit system. *Journal of Geophysical Research* **92** (B9), 9083–9100.

Kamb, B. (1991) Rheological nonlinearity and flow instability in the deforming-bed mechanism of ice-stream motion. *Journal of Geophysical Research* **96** (B10), 16585–16595.

Kamb, B. and Echelmeyer, K.A. (1986) Stress-gradient coupling in glacier flow: I. Longitudinal averaging of the influence of ice thickness and surface slope. *Journal of Glaciology* **32** (111), 267–284.

Kamb, B. and Engelhardt, H. (1987) Waves of accelerated motion in a glacier approaching surge: the mini-surges of Variegated Glacier, Alaska, U.S.A. *Journal of Glaciology* **33**, (113), 27–46.

Kamb, B. and LaChapelle, E. (1964) Direct observations of the mechanism of glacier sliding over bedrock. *Journal of Glaciology* **5** (38), 159–172.

Kamb, B., Raymond, C.F., Harrison, W.D., Engelhardt, H., Echelmeyer, K.A., Humphrey, N., Brugman, M.M. and Pfeffer, T. (1985) Glacier surge mechanism: 1982–83 surge of Variegated Glacier, Alaska. *Science* **227** (4686), 469–479.

Kamb, W.B. (1962) Refraction corrections for universal stage measurements. I. Uniaxial crystals. *American Mineralogist* **47** (3), 227–245.

Kapitsa, A.P., Ridley, J.K., Robin, G.de Q., Siegert, M.J. and Zotikov, I.A. (1996) A large deep freshwater lake beneath the ice of central East Antarctica. *Nature* **381** (6584), 684–686.

Kaser, G. and Noggler, B. (1996) Glacier fluctuations in the Ruwenzori Range (East Africa) during the 20th century. A preliminary report. *Zeitschrift für Gletscherkunde und Glazialgeologie* **32**, 109–117.

Kaser, G., Hastenrath, S. and Ames, A. (1996) Mass balance profiles on tropical glaciers. *Zeitschrift für Gletscherkunde und Glazialgeologie* **32**, 75–81.

Kidson, C. (1982) Sea level changes of the Holocene. *Quaternary Science Reviews* **1**, 121–151.

Kirkbride, M.P. (1993) The temporal significance of transitions from melting to calving termini at glaciers in the central Southern Alps of New Zealand. *Holocene* **3**, 232-–0.

Kleman, J. and Borgström, I. (1994) Glacial landforms indicative of a partly frozen bed. *Journal of Glaciology* **40** (135), 255–264.

Kleman, J., Hätterstrand, C., Borgström, I. and Stroeven, A. (1997) Fennoscandian palaeoglaciology reconstructed using a glacial geological inversion model. *Journal of Glaciology* **43** (144), 283–299.

Knight, P.G. (1987) Observations at the edge of the Greenland ice sheet; boundary condition implications for modellers. *International Association of Hydrological Sciences Publication* **170**, 359–366.

Knight, P.G. (1988) The basal ice and debris sequence at the margin of an equatorial ice cap; El Cotopaxi, Ecuador. *Geografiska Annaler* **70A** (1), 9–13.

Knight, P.G. (1989) Stacking of basal debris layers without bulk freezing-on: isotopic evidence from West Greenland. *Journal of Glaciology* **35** (120), 214–216.

Knight, P.G. (1990) On the origin of debris-bearing basal ice; West Greenland. PhD Thesis, University of Aberdeen.

Knight, P.G. (1992) Ice deformation very close to the ice-sheet margin in West Greenland. *Journal of Glaciology* **38** (128), 3–8.

Knight, P.G. (1993) The geography of field research in Iceland. *Scottish Geographical Magazine* **109** (3), 180–186.

Knight, P.G. (1994) Two-facies interpretation of the basal layer of the Greenland ice sheet contributes to a unified model of basal ice formation. *Geology* **22** (11), 971–974.

Knight, P.G. (1997) The basal ice layer of glaciers and ice sheets. *Quaternary Science Reviews* **16**, 1–19.

Knight, P.G. and Knight, D.A. (1994) Glacier sliding, regelation water flow, and development of basal ice. *Journal of Glaciology* **40** (136), 600–601.

Knight, P.G. and Russell, A.J. (1993) Most recent observations of the drainage of an ice-dammed lake at Russell Glacier, West Greenland, and a new hypothesis regarding mechanisms of drainage. *Journal of Glaciology* **39** (133), 701–703

Knight, P.G. and Thompson, D.A. (1992) Big savings at the ice margin. *Geographical Magazine* **64** (1), 22–26

Knight, P.G. and Tweed, F.S. (1991) Periodic drainage of ice-dammed lakes as a result of variations in glacier velocity. *Hydrological Processes* **5** (2), 175–184.

Knight, P.G., Sugden, D.E. and Minty, C. (1994). Ice flow around large obstacles as indicated by basal ice exposed at the margin of the Greenland ice sheet. *Journal of Glaciology* **40** (135), 359–367.

Koerner, R.M. (1970) The mass balance of the Devon Island ice cap, Northwest Territories, Canada, 1961–66. *Journal of Glaciology* **9** (57), 325–336.

Koerner, R.M. (1997) Some comments on climatic reconstructions from ice cores drilled in areas of high melt. *Journal of Glaciology* **43** (143), 90–97.

Kohler, J. (1995) Determining the extent of pressurised flow beneath Storglaciären, Sweden, using results of tracer experiments and of input and output discharge. *Journal of Glaciology* **41** (138), 217–231.

Kotlyakov, V.M. (1970) Land glaciation part in the Earth's water balance. *International Association of Hydrological Sciences Publication* **92** (1), 54–57.

Krass, M.S. (1984) Ice on planets of the solar system. *Journal of Glaciology* **30** (106), 259–274.

Krimmel, R.M. (1989) Mass balance and volume of South Cascade Glacier, Washington, 1958–1985. In: Oerlemans, J. (ed.) *Glacier Fluctuations and Climatic Change*. Kluwer Academic Publishers, Dordrecht, 193–206.

Kuhn, B.F. (1787) Versuch uber den mechanismus der Gletscher. *A. Hopfner's Magazin für die Naturkunde Helvetiens (Zurich)* **1**, 119–136.

Kuhn, M. (1981) Climate and Glaciers. *International Association of Hydrological Sciences Publication* **131**, 3–20.

Kuhn, M. (1994) Der Mieminger Schneeferner, ein Beispiel eines lawinenernährten Kargletschers. *Zeitschrift fur Gletscherkunde und Glazialgeologie* **29** (2), 153–171.

Kuhn, M. (1995) The mass balance of very small glaciers. *Zeitschrift für Gletscherkunde und Glazialgeologie* **31**, 171–179.

Kuhn, T.S. (1962) *The Structure of Scientific Revolutions*. University of Chicago Press, Chicago.

Kulkarni, A.V. (1992) Mass balance of Himalayan glaciers using AAR and ELA methods. *Journal of Glaciology* **38** (128), 101–104.

Kump, L.R. and Alley, R.B. (1994) Global chemical weathering on glacial time scales. In: National Research Council *Material Fluxes on the Surface of the Earth*. National Academy Press, Washington DC, 46–60

LaChapelle, E. (1968) Stress-generated ice crystals in a nearly isothermal two-phase system. *Journal of Glaciology* **7** (50), 183--98.

LaChapelle, E.R. (1992) *Field Guide to Snow Crystals* (Reprint Edition). International Glaciological Society, Cambridge.

Lagally, M. (1932) Zur thermodynamik der gletscher. *Zeitschrift für Gletscherkunde* **20**, 199–214.

Lamplugh, G.W. (1911) On the shelly moraine of the Sefstrom glacier, and other Spitzbergen phenomena illustrative of British glacial conditions. *Proceedings of the Yorkshire Geological Society* **17**, 216–241.

Langway, C.C. (1958) Ice fabrics and the universal stage. *U.S. Snow Ice and Permafrost Research Establishment Technical Report* **62**.

Langway, C.C. (1967) Stratigraphic analysis of a deep core from Greenland. *U.S. Army Cold Regions Research and Engineering Laboratory Research Report* **77.**

Langway, C.C. (1970) Stratigraphic analysis of a deep ice core from Greenland. *Geological Society of America Special Paper* **125**, 186.

Lawson, D.E. (1979) Sedimentological analysis of the Western terminus of the Matanuska Glacier, Alaska. *U.S. Army Cold Regions Research and Engineering Laboratory Report* **79–9**.

Lawson, D.E. and Kulla, J.B. (1978) An oxygen isotope investigation of the origin of the basal zone of the Matanuska Glacier, Alaska. *Journal of Geology* **86** (6), 673–685.

Lawson, W. (1996) The relative strengths of debris-laden basal ice and clean glacier ice: some evidence from Taylor Glacier, Antarctica. *Annals of Glaciology* **23**, 270–276.

Ledroit, M., Remy, F. and Minster, J.-F. (1993) Observations of the Antarctic ice sheet with the

Seasat scatterometer: relationship to katabatic-innd intensity and direction. *Journal of Glaciology* **39** (132), 385–396.

Lefauconnier, B., Hagen, J.O., Pinglot, J.F. and Pourchet, M. (1994) Mass balance estimates on the glacier complex Kongsvegan and Sveabreen, Spitsbergen, Svalbard, using radioactive layers. *Journal of Glaciology* **40** (135), 368–376.

Legrand, M.R. and Delmas, R.J. (1984) The ionic balance of Antarctic snow: a 10-year detailed record. *Atmospheric. Environment* **18** (9), 1867–1874.

Lehman, S.J., Jones, G.A., Keigwin, L.D., Andersen, E.S., Butenko, G. and Østmo, S.-R. (1991) Initiation of Fennoscandian ice-sheet retreat during the last deglaciation. *Nature* **349** (6309), 513–516.

Lehmann, M. and Siegenthaler, U. (1991) Equilibrium oxygen- and hydrogen-isotope fractionation between ice and water. *Journal of Glaciology* **37** (125), 23–26.

Leopold, L.B. and Wolman, M.G. (1960) River meanders. *Geological Society of America Bulletin* **71**, 769–794.

Letréguilly, A. (1988) Relation between the mass balance of Canadian mountain glaciers and meteorological data. *Journal of Glaciology* **34** (116), 11–18.

Letréguilly, A., Huybrechts, P. and Reeh, N. (1991) Steady-state characteristics of the Greenland ice sheet under different climates. *Journal of Glaciology* **37** (125), 149–157.

Lewis, E.L. and Perkin, R.G. (1986) Ice pumps and their rates. *Journal of Geophysical Research* **91** (C10), 11,756–11,762.

Liestøl, O. (1955) Glacier-dammed lakes in Norway. *Norsk Geografisk Tidsskrift* **15**, 122–149.

Lighthill, M.J. and Whitham, B.G. (1955) On kinematic waves. *Proceedings of the Royal Society of London, Series A* **229**, 281–345.

Lingle, C.S., Hughes, T.J. and Kollmeyer, R.C. (1981) Tidal flexure of Jakobshavns Glacier, West Greenland. *Journal of Geophysical Research* **86** (B5), 3960–3968

Lingle, C.S., Post, A., Herzfeld, U.C., Molnia, B.F., Krimmel, R.M. and Roush, J.J. (1993) Bering glacier surge and iceberg-calving mechaism at Vitus Lake, Alaska, U.S.A. *Journal of Glaciology* **39** (133), 722–727.

Linkletter, G.O. and Warburton, J.A. (1976) A note on the contribution of rime and surface hoar to the accumulation of the Ross Ice Shelf, Antarctica. *Journal of Glaciology* **17** (76), 351–354.

Lipenkov, V., Barkov, N.I., Duval, P. and Pimienta, P. (1989) Crystalline texture of the 2083 m ice core at Vostok Station, Antarctica. *Journal of Glaciology* **35** (121), 392–398

Lipenkov, V., Chandaudap, F., Ravoire, J., Dulac, E. and Raynaud, D. (1995) A new device for the measurement of air content in polar ice. *Journal of Glaciology* **41** (138), 423–429.

Lliboutry, L. (1954) The origin of penitents. *Journal of Glaciology* **2** (15), 331–338.

Lliboutry, L. (1958) La dynamique de la Mer de Glace et la vague de 1891–95 d'après les mesures de Joseph Vallot. *International Association of Hydrological Sciences Publication* **47**, 125–138.

Lliboutry, L. (1964) *Traité de Glaciologie* (2 volumes). Masson, Paris.

Lliboutry, L. (1968) General theory of subglacial cavitation and sliding of temperate glaciers. *Journal of Glaciology* **7** (49), 21–58.

Lliboutry, L. (1969) Contribution à la théorie des ondes glaciaires. *Canadian Journal of Earth Sciences* **6**, 943–953.

Lliboutry, L. (1971) Permeability, brine content and temperature of temperate ice. *Journal of Glaciology* **10** (58), 15–29.

Lliboutry, L. (1975) Loi de glissement d'un glacier sans cavitation. *Annales Géophysiqes* **31**, 207–226.

Lliboutry, L. (1978) Glissement d'un glacier sur un plan parsemé d'obstacle hémispheriques. *Annales Géophysiqes* **34**, 147–162.

Lliboutry, L. (1979) Local friction laws for glaciers: a critical review and new openings. *Journal of Glaciology* **23** (89), 67–95.

Lliboutry, L. (1983) Modifications to the theory of intraglacial waterways for the case of subglacial ones. *Journal of Glaciology* **29** (102), 216–226.

Lliboutry, L. (1986) A discussion of Robin's 'Heat Pump' effect by extending Nye's model for the sliding of a temperate glacier. *Hydraulic effects at the bed and related phenomena Mitteilungen Nr.90 der Versuchsanstalt fur Wasserbau, Hydrologie, und Glaziologie, ETH-Zurich* 74–77.

Lliboutry, L. (1987) *Very Slow Flows of Solids. Basics of Modelling in Geodynamics and Glaciology.* Martinus Nijhoff Publishers, Dordrecht.

Lliboutry, L. (1990) About the origin of rock glaciers. *Journal of Glaciology* **36** (122), 125.

Lliboutry, L. (1993) Internal melting and ice accretion at the bottom of temperate glaciers. *Journal of Glaciology* **39** (131), 50–64.

Lliboutry, L. (1994) Monolithologic erosion of hard beds by temperate glaciers. *Journal of Glaciology* **40** (136), 433–450.

Lliboutry, L., Arnao, B.M., Pautre, A. and Schneider, B. (1977) Glaciological problems set by the control of dangerous lakes in the Cordillera Blanca, Peru. I. Historical failures of morainic dams, their causes and prevention. *Journal of Glaciology* **18** (79), 239–254.

Long, D.G. and Drinkwater, M.R. (1994) Greenland ice-sheet surface properties observed by the Seasat-A scatterometer at enhanced resolution. *Journal of Glaciology* **40** (135), 213–230.

Lorrain, R.D. and Demeur, P. (1985) Isotopic evidence for relic pleistocene glacier ice on Victoria Island, Canadian Arctic Archipelago. *Arctic and Alpine Research* **17** (1), 89–98.

Lorrain, R.D., Souchez, R.A. and Tison, J.-L. (1981) Characteristics of basal ice from two outlet glaciers in the Canadian Arctic – implications for glacier erosion. *Current Research, Geological Survey of Canada* paper **81–1B**, 137–144.

Loutre, M.F. (1995) Greenland ice-sheet over the next 5000 years *Geophysical Research Letters* **22** (7), 783–786

Lvovitch, M.I. (1970) World water balance (general report). *IASH/Unesco Symposium on world water balance. International Association of Hydrological Sciences Publication* **93** (2), 401–403.

MacAyeal, D.R. (1989) Ice-shelf response to ice-stream discharge fluctuations: III. The effects of ice-stream imbalance on the Ross Ice Shelf, Antarctica. *Journal of Glaciology* **35** (119), 38–42.

MacAyeal, D.R. (1993) Binge/purge oscillations of the Laurentide ice sheet as a cause of the North Atlantic Heinrich events. *Paleoceanography* **8** (6), 775–784.

MacAyeal, D.R. and Barcilon, V. (1988) Ice-shelf response to ice-stream discharge fluctuations: I. Unconfined ice tongues. *Journal of Glaciology* **34** (116), 121–127.

MacAyeal, D.R. and Lange, M.A. (1988) Ice-shelf response to ice-stream discharge fluctuations: II. Ideal rectangular ice shelf. *Journal of Glaciology* **34** (116), 128–135.

MacAyeal, D.R., Bindschadler, R.A., Shabtaie, S., Stephenson, S. and Bentley, C.R. (1987) Force, mass and energy budgets of the Crary Ice Rise complex, Antarctica. *Journal of Glaciology* **33** (114), 218–230.

MacAyeal, D.R., Bindschadler, R.A. and Scambos, T.A. (1995) Basal friction of ice stream E, West Antarctica. *Journal of Glaciology* **41** (138), 247–262.

Macheret, Yu.Ya., Moskalevsky, M.Yu. and Vasilenko, E.V. (1993) Velocity of radio waves in glaciers as an indicator of their hydrothermal state, structure and regime. *Journal of Glaciology* **39** (132), 373–384.

Mader, H.M. (1992a) Observations of the water-vein system in polycrystalline ice. *Journal of Glaciology* **38** (130), 333–347.

Mader, H.M. (1992b) The thermal behaviour of the water-vein system in polycrystalline ice. *Journal of Glaciology* **38** (130), 359–374.

Mair, R. and Kuhn, M. (1994) Temperature and movement measurements at a bergschrund. *Journal of Glaciology* **40** (136), 561–565.

Maizels, J.K. (1979) Proglacial aggradation and changes in braided channel patterns during a period of glacier advance: an Alpine example. *Geografiska Annaler* **61A** (1–2), 87–101.

Maizels, J.K. (1992) Boulder ring structures produced during jökulhlaup flows: origin and hydraulic significance. *Geografiska Annaler* **74A**, 21–33.

Maizels, J.K. and Russell, A.J. (1992) Quaternary perspectives on jökulhlaup prediction. *Quaternary Proceedings* **2**, 133–152.

Mann, D.H. (1986) Reliability of a fjord glacier's fluctuations for palaeoclimatic reconstructions. *Quaternary Research* **25**, 10–24.

Marko, J.R., Fissel, D.B. and Miller, J.D. (1988) Iceberg movement prediction off the Canadian east coast. In: El Sabh, M.I. and Murty, T.S. (eds) *Natural and Man-made Hazards*. D. Reidel, Dordrecht, 435–462.

Martinec, J. (1976) Snow and ice. In: Rodda, J.C. (ed.) *Facets of Hydrology*. John Wiley and Sons, London. 85–118.

Martinerie, P., Reynaud, D., Etheridge, D.M., Barnola, J.-M. and Mazaudier, D. (1992) Physical and climatic parameters which influence the air content in polar ice. *Earth and Planetary Science Letters* **112**, 1–13.

Matthes, F.E. (1934) Ablation of snow-fields at high altitudes by radiant solar heat. *Transactions of the American Geophysical Union* **15** (2), 380–385.

Matthes, F.E. (1939) Report of committee on glaciers, April 1939. *Transactions of the American Geophysical Union* **20**, 518–23.

Matthews, W.H. (1974) Surface profiles of the Laurentide Ice Sheet in its marginal areas. *Journal of Glaciology* **13** (67), 37–43.

Maurette, M., Jehanno, C., Robin, E. and Hammer, C. (1987) Characteristics and mass-distribution of extraterrestrial dust from the Greenland ice cap. *Nature* **328** (6132), 699–702.

Mayo, L.R. and March, R.S. (1990) Air temperature and precipitation at Wolverine Glacier, Alaska: glacier growth in a warmer, wetter climate. *Annals of Glaciology* **14**, 191–194.

Mayo, L.R., Meier, M.F. and Tangborn, W.V. (1972) A system to combine stratigraphic and annual mass-balance systems: a contribution to the International Hydrological Decade. *Journal of Glaciology* **11** (61), 3–14.

McClung, D.M. and Armstrong, R.L. (1993) Temperate glacier time response from field data. *Journal of Glaciology* **39** (132), 323–326.

McDonald, J. and Whillans, I.M. (1992) Search for temporal changes in the velocity of Ice Stream B, West Antarctica. *Journal of Glaciology* **38** (128), 157–161.

McIntyre, N.F. (1985) The dynamics of ice-sheet outlets. *Journal of Glaciology* **31** (108), 99–107.

McKenzie, G.D. and Goodwin, R.G. (1987) Development of collapsed glacial topography in the Adams Inlet area, Alaska, U.S.A. *Journal of Glaciology* **33** (113), 55–59.

McMeeking, R.M. and Johnson, R.E. (1985) On the analysis of longitudinal stress in glaciers. *Journal of Glaciology* **31** (109), 293–302.

McMeeking, R.M. and Johnson, R.E. (1986) On the mechanics of surging glaciers. *Journal of Glaciology* **32** (110), 120–132.

Meier, M.F. (1962) Proposed definitions for glacier mass budget terms. *Journal of Glaciology* **4** (33), 252–263.

Meier, M.F. (1964) Ice and glaciers. In: Chow V.T. (ed.) *Handbook of Applied Hydrology*. McGraw-Hill, New York, 16.1–16.32.

Meier, M.F. and Johnson, A. (1962) The kinematic wave on Nisqually Glacier, Washington. *Journal of Geophysical Research* **67**, 886.

Meier, M.F. and Post, A.S. (1962) Recent variations in net mass budgets of glaciers in western North America. *International Association of Scientific Hydrology Publication* **58**, 63–77.

Meier, M.F. and Post, A.S. (1969) What are glacier surges? *Canadian Journal of Earth Sciences* **6** (4, part 2), 807–817.

Meier, M.F. and Post, A.S. (1987) Fast tidewater glaciers. *Journal of Geophysical Research* **92** (B9), 9051–9058.

Meier, M.F. and Post, A. (1991) *Glaciers: a Water Resource*. United States Department of the Interior, US Geological Survey, Denver.

Mellor, M. (1960) Correspondence. Antarctic ice terminology: ice dolines. *Polar Record* **10** (64), 30–34.

Menzies, J. (ed.) (1995) *Modern Glacial Environments*. Butterworth-Heinemann, Oxford.

Mercer, J.H. (1961) The response of fjord glaciers to changes in the firn limit. *Journal of Glaciology* **3** (29), 850–858.

Mercer, J.H. (1971) Cold glaciers in the central Transantarctic Mountains, Antarctica: dry ablation areas and subglacial erosion. *Journal of Glaciology* **10** (59), 319–321.

Mercer, J.H. (1978) West Antarctic ice sheet and CO_2 greenhouse effect: a threat of disaster. *Nature* **271**, 321–325.

Merrand, Y. and Hallet, B. (1996) Water and sediment discharge from a large surging glacier: Bering Glacier, Alaska, U.S.A., summer 1994. *Annals of Glaciology* **22**, 233–240.

Merritt, J.W. (1992) A critical review of methods used in the appraisal of onshore sand and gravel resources in Britain. *Engineering Geology* **32** (1–2), 1–9.

Merry, C.J. and Whillans, I.M. (1993) Ice-flow features on Ice Stream B, Antarctica, revealed by SPOT HRV imagery. *Journal of Glaciology* **39** (133), 515–527.

Miller, M.M. (1951) Englacial investigations related to core drilling on the upper Taku Glacier, Alaska. *Journal of Glaciology* **1** (10), 578–580.

Miller, M.M. (1952) Preliminary notes concerning certain glacier structures and glacial lakes on the Juneau Ice Field. *American Geographical Society JIRP Report* **6**, 49–86.

Miller, S.L. (1969) Clathrate hydrates of air in Antarctic ice. *Science* **165**, 489–490.

Molnar, P. and England, P. (1990) Late Cenozoic uplift of mountain ranges and global climate change: chicken or egg? *Nature* **346** (6279), 29–34.

Mooers, H.D. (1989a) Drumlin formation: a time transgressive model. *Boreas* **18** (2), 99–107

Mooers, H.D. (1989b) On the formation of the tunnel valleys of the Superior Lobe, central Minnesota. *Quaternary Research* **32** (1), 24–35.

Mooers, H.D. (1990) A glacial process model: the role of spatial and temporal variations in glacier thermal regime. *Geological Society of America Bulletin* **102**, 243–251.

Morgan, V.I., Goodwin, I.D., Etheridge, D.M. and Wookey, C.W. (1991) Evidence from Antarctic ice cores for recent increases in snow accumulation. *Nature* **354** (6348), 58–60.

Morgan, V.I., Wookey, C.W., Li, J., van Ommen, T.D., Skinner, W. and Fitzpatrick, M.F. (1997) Site information and initial results from deep ice drilling on Law Dome, Antarctica. *Journal of Glaciology* **43** (143), 3–10.

Morland, L.W. and Morris, E.M. (1977) Stress in an elastic bedrock hump due to glacier flow. *Journal of Glaciology* **18** (78), 67–75.

Mote, T.L. and Anderson, M.R. (1995) Variations in snowpack melt on the Greenland ice sheet based on passive-microwave measurements. *Journal of Glaciology* **41** (137), 51–60.

Motyka, R.J. and Begét, J.E. (1996) Taku Glacier, Southeast Alaska, U.S.A.: Late Holocene history of a tidewater glacier. *Arctic and Alpine Research* **28** (1), 42–51.

Müller, F. (1980) Present and late Pleistocene equilibrium line altitudes in the Mt. Everest region: an application of the glacier inventory. *International Association of Hydrological Sciences Publication* **126**, 75–94.

Müller, F. and Keeler, C.M. (1969) Errors in short-term ablation measurements on melting ice surfaces. *Journal of Glaciology* **8** (52), 91–105.

Mulvaney, R. and Peel, D.A. (1988) A high resolution anion profile of an ice core from Dolleman Island, Antarctic Peninsula. *Annals of Glaciology* **11**, 204–205.

Mulvaney, R., Wolff, E.W. and Oates, K. (1988) Sulphuric acid at grain boundaries in Antarctic ice. *Nature* **331** (6153), 247–249.

Murray, T. and Clarke, G.K.C. (1995) Black-box modelling of the subglacial water system. *Journal of Geophysical Research* **100** (B7), 10231–10245.

Naranjo, J.L., Sigurdsson, H., Carey, S.N. and Fritz, W. (1986) Eruption of the Nevado del Ruiz Volcano, Colombia, on 13 November 1985: tephra fall and lahars. *Science* **233**, 961–963.

Naruse, R., Fukami, H. and Aniya, M. (1992) Short-term variations in flow velocity of Glaciar Soler, Patagonia, Chile. *Journal of Glaciology* **38** (128), 152–156.

Näslund, J.O. and Hassinen, S. (1996) Supraglacial sediment accumulations and large englacial water conduits at high elevations in Myrdalsjökull, Iceland. *Journal of Glaciology* **42** (140), 190–192.

Nickling, W.G. and Bennett, L. (1984) The shear strength characteristics of frozen coarse granular debris. *Journal of Glaciology* **30** (106), 348–357.

Nienow, P., Sharp, M. and Willis, I. (1996) Temporal switching between englacial and subglacial drainage pathways – dye tracer evidence from the Haut-Glacier-d'Arolla, Switzerland. *Geografiska Annaler* **78A** (1), 51–60.

Nijampurkar, V.N., Bhandari, N., Borole, D.V. and Bhattacharya, U. (1985) Radiometric chronology of Ghangme-Khangpu glacier, Sikkim. *Journal of Glaciology* **31** (107), 28–33.

Nye, J.F. (1952a) The mechanics of glacier flow. *Journal of Glaciology* **2** (12), 82–93.

Nye, J.F. (1952b) A comparison between the theoretical and the measured long profile of the Unteraar glacier. *Journal of Glaciology* **2** (12), 103–107.

Nye, J.F. (1953) The flow law of ice from measurements in glacier tunnels, laboratory experiments and the Jungfraufirn borehole experiment. *Proceedings of the Royal Society of London, Series A* **219**, 477–489.

Nye, J.F. (1958) A theory of wave formation on glaciers. *International Association of Hydrological Sciences Publication* **47**, 139–154.

Nye, J.F. (1960) The response of glaciers and ice-sheets to seasonal and climatic changes. *Proceedings of the Royal Society of London*, to seasonal and climatic changes. *Proceedings of the Royal Society of London, Series A* **256**, 559–584.

Nye, J.F. (1963) On the theory of the advance and retreat of glaciers. *Geophysical Journal of the Royal Astronomical Society* **7**, 431–456.

Nye, J.F. (1969a) A calculation on the sliding of ice over a wavy surface using a Newtonian viscous approximation. *Proceedings of the Royal Society of London, Series A* **311**, 445–467

Nye, J.F. (1969b) The effect of longitudinal stress on the shear stress at the base of an ice sheet. *Journal of Glaciology* **8** (53), 207–213.

Nye, J.F. (1970) Glacier sliding without cavitation in a linear viscous approximation. *Proceedings of the Royal Society of London, Series A* **311**, 445–467.

Nye, J.F. (1973) Water at the bed of a glacier. *International Association of Hydrological Sciences Publication* **95**, 189–194.

Nye, J.F. (1976) Water flow in glaciers: jökulhlaups, tunnels and veins. *Journal of Glaciology* **17** (76), 181–207.

Nye, J.F. (1989) The geometry of water veins and nodes in polycrystalline ice. *Journal of Glaciology* **35** (119), 17–22.

Nye, J.F. (1991) Thermal behaviour of glacier and laboratory ice. *Journal of Glaciology* **37** (127), 401–413.

Nye, J.F. and Frank, F.C. (1973) Hydrology of the intergranular veins in a temperate glacier. *International Association of Hydrological Sciences Publication* **95**, 157–161

Oerlemans, J. (ed.) (1989) *Glacier Fluctuations and Climatic Change.* Kluwer Academic Publishers, Dordrecht.

Oerlemans, J. (1992) Climate sensitivity of glaciers in southern Norway: application of an energy-balance model to Nigardsbreen, Hellstugubreen and Alfotbreen. *Journal of Glaciology* **38** (129), 223–232.

Oerlemans, J. (1993) Evaluating the role of climate cooling in iceberg production and the Heinrich events. *Nature* **364** (6440), 783–786.

Oerlemans, J. and Hoogendoorn, N.C. (1989) Mass-balance gradients and climatic change. *Journal of Glaciology* **35** (121), 399–405.

Oerlemans, J. and van der Veen, C.J. (1984) *Ice Sheets and Climate*. D. Reidel, Dordrecht

Oerter, H., Kipfstuhl, J., Determann, J., Miller, H., Wagenbach, D., Minikin, A. and Graf, W. (1992) Evidence for basal marine ice in the Filchner-Ronne ice shelf. *Nature* **358** (6385), 399–401.

Ogilvie, I.H. (1904) The effect of superglacial debris on the advance and retreat of some Canadian glaciers. *Journal of Geology* **12**, 722–743.

Ohmura, A. and Reeh, N. (1991) New precipitation and accumulation maps for Greenland. *Journal of Glaciology* **37** (125), 140–148.

Ohmura, A., Kasser, P. and Funk, M. (1992) Climate at the equilibrium line of glaciers. *Journal of Glaciology* **38** (130), 397–411.

Ohno, H., Ohata, T. and Higuchi, K. (1992) The influence of humidity on the ablation of continental-type glaciers. *Annals of Glaciology* **16**, 107–114.

Orheim, O. and Lucchitta, B. (1990) Investigating climate change by digital analysis of blue ice extent on satellite images of Antarctica. *Annals of Glaciology* **14,** 211–215.

Orowan, E. (1949) Remarks of joint meeting of the British Glaciological Society, the British Rheologists Club and the Institute of Metals. *Journal of Glaciology* **1** (5), 231–236.

Östling, M. and Hooke, R.Le B., (1986) Water storage in Storglaciären, Kebnekaise, Sweden. *Geografiska Annaler* **68A** (4), 279–290.

Østrem, G. (1959) Ice melting under a thin layer of morain, and the existence of ice cores in morain ridges. *Geografiska Annaler* **41** (4) 228–230.

Østrem, G. (1975) ERTS data in glaciology – an effort to monitor glacier mass balance from satellite imagery. *Journal of Glaciology* **15** (73), 403–415.

Oswald, G.K.A. and Robin, G.de Q. (1973) Lakes beneath the Antarctic ice sheet. *Nature* **245**, 251–254.

Overpeck, J., Rind, D., Lacis, D. and Healy, R. (1996) Possible role of dust induced regional warming in abrupt climate change during the last glacial period. *Nature* **384**, 447–449.

Paillard, D. (1995) The hierarchical structure of glacial climatic oscillations – interactions between ice-sheet dynamics and climate. *Climate Dynamics* **11** (3), 162–177.

Pappalardo, R.T., Head, J.W., Greeley, R., Sullivan, R.J., Pilcher, C., Schubert, G., Moore, W.B., Carr, M.H., Moore, J.M., Belton, M.J.S. and Goldsby, D.L. (1998) Geological evidence for solid-state convection in Europa's ice shell. *Nature* **391** (6665), 365–368.

Paterson, W.S.B. (1981) *The Physics of Glaciers* (2nd edition). Pergamon, Oxford.

Paterson, W.S.B. (1983) Techniques for measuring temperatures in glaciers and ice sheets. In: Robin, G.deQ. (ed.) *The Climatic Record in Polar Ice Sheets*. Cambridge University Press, Cambridge, 63–65.

Paterson, W.S.B. (1994) *The Physics of Glaciers* (3rd edition). Pergamon, Oxford.

Patterson, C.J. and Hooke, R.LeB. (1995) Physical environment of drumlin formation. *Journal of Glaciology* **41** (137), 30–38.

Patterson, E.A. (1984) A mathematical model for perched block formation. *Journal of Glaciology* **30** (106), 296–301.

Patzelt, G. (1985) The period of glacier advances in the Alps, 1965 to 1980. *Zeitschrift fur Gletscherkunde und Glazialgeologie* **21**, 403–407.

Pearman, G.I., Etheridge, D.De Silva, F. and Fraser, P.J. (1986) Evidence of changing concentrations of atmospheric CO_2, N_2O and CH_4 from bubbles in Antarctic ice. *Nature* **320** (6059), 248–250.

Pedley, M., Paren, J.G. and Potter, J.R. (1988) Localized basal freezing within George VI ice shelf, Antarctica. *Journal of Glaciology* **34** (116), 71–77.

Pelfini, M. and Smiraglia, C. (1992) Recent fluctuations of glaciers in Valtellina (Italian Alps) and climatic variations. *Journal of Glaciology* **38** (129), 309–313.

Peltier, W.R. (1987) Mechanisms of relative sea-level change and the geophysical response to ice-water loading. In: Devoy, R.J.N. (ed.) *Sea Surface Studies: A Global View*. Croom-Helm, London, 415–463.

Peltier, W.R. (1994) Ice age paleotopography. *Science* **265**, 195–201.

Pelto, M.S. (1988) The annual balance of North Cascade glaciers, Washington, U.S.A., measured and predicted using an activity-index method. *Journal of Glaciology* **34** (117), 194–199.

Pelto, M.S., Hughes, T.J. and Brecher, H. (1989) Equilibrium state of Jakobshavn Isbrae, West Greenland. *Annals of Glaciology* **12**, 127–131.

Petit, J.-R., Briat, M. and Rayer, A. (1981) Ice age aerosol content from East Antarctic ice cores samples and past wind strength. *Nature* **293** (5831), 391–394.

Petit, J.-R., Duval, P. and Lorius, C. (1987) Long term climatic changes indicated by crystal growth in polar ice. *Nature* **326** (6108), 62–64.

Pfeffer, W.T. (1992) Stress-induced foliation in the terminus of Variegated Glacier, Alaska, U.S.A., formed during the 1982–1983 surge. *Journal of Glaciology* **38** (129), 213–221.

Philberth, K. (1977) The disposal of radioactive waste in ice sheets. *Journal of Glaciology* **19** (81), 607–618.

Picciotto, E., Crozaz, G., Ambach, W. and Eisner, H. (1967) Lead-210 and strontium-90 in an alpine glacier. *Earth and Planetary Science Letters* **3**, 237–242.

Piotrowski, J.A. (1997) Subglacial hydrology in north-western Germany during the last glaciation: groundwater flow, tunnel valleys, and hydrological cycles. *Quaternary Science Reviews* **16** (2), 169–185.

Pohjola, V.A. (1993) TV-video observations of bed and basal sliding on Storglaciären, Sweden. *Journal of Glaciology* **39** (131), 111–118.

Pohjola, V.A. (1994) TV-video observations of englacial voids in Storglaciären, Sweden. *Journal of Glaciology* **40** (135), 231–240.

Poirier, J.-P. (1982) Rheology of ices: a key to the tectonics of the icy moons of Jupiter and Saturn. *Nature* **299**, 683–688.

Porter, S.C. (1970) Quaternary glacial record in Swat Kohistan, West Pakistan. *Bulletin of the Geological Society of America* **81**, 1421–1446

Porter, S.C. (1975) Glaciation limit in New Zealand's southern alps. *Arctic and Alpine Research* **7** (1), 33–37.

Post, A. (1969) Distribution of surging glaciers in western North America. *Journal of Glaciology* **8** (53), 229–240.

Post, A. and LaChapelle, E.R. (1971) *Glacier Ice.* University of Washington Press, Seattle.

Pourchet, M., Pinglot, J.F., Reynaud, L. and Holdsworth, G. (1988) Identification of Chernobyl fall-out as a new reference level in Northern Hemisphere glaciers. *Journal of Glaciology* **34** (117), 183–187.

Powell, R.D. (1981) A model for sedimentation by tidewater glaciers. *Annals of Glaciology* **2**, 129–134.

Powell, R.D. and Molnia, B.F. (1989) Glacimarine sedimentary processes, facies and morphology of the south-southeast Alaska shelf and fjords. *Marine Geology* **85**, 359–390.

Prest, V.K. (1990) Laurentide ice-flow patterns – a historical review, and implications of the dispersal of Belcher-island erratics. *Geographie Physique et Quaternaire* **44** (2), 113–136.

Proctor, D.V. (ed.) (1981) Ice carrying trade at sea. *Maritime Monographs and Reports* **49** (Proceedings of a symposium held at the National Maritime Museum, September 1979.)

Rado, C., Girard, C. and Perrin, J. (1987) Electrochaude: a self-flushing hot-water drilling apparatus for glaciers with debris. *Journal of Glaciology* **33** (144), 236–238.

Raymo, M.E. and Ruddiman, W.F. (1992) Tectonic forcing of late Cenozoic climate. *Nature* **359** (6391), 117–122.

Raymo, M.E., Ruddiman, W.F. and Froelich, P.N. (1988) Influence of late Cenozoic mountain building on ocean geochemical cycles. *Geology* **16**, 649–653.

Raymond, C.F. (1987) How do glaciers surge? A review. *Journal of Geophysical Research* **92** (B9), 9121–9134.

Raymond, C.F. and Harrison, W.D. (1975) Some observations on the behaviour of the liquid and gas phases in temperate glacier ice. *Journal of Glaciology* **14** (71), 213–233.

Raymond, C.F. and Harrison, W.D. (1988) Evolution of Variegated Glacier, Alaska, U.S.A., prior to its surge. *Journal of Glaciology* **34** (117), 154–169.

Raymond, C.F. and Malone, S. (1986) Propagating strain anomalies during mini-surges of variegated Glacier, Alaska, U.S.A. *Journal of Glaciology* **32** (111), 178–191.

Raymond, C.F., Benedict, R.J., Harrison, W.D., Echelmeyer, K.A. and Sturm, M. (1995) Hydrological discharges and motion of Fels and Black Rapids Glaciers, Alaska, U.S.A.: implications for the structure of their drainage systems. *Journal of Glaciology* **41** (138), 290–304.

Raynaud, D., Jouzel, J., Barnola, J.M., Chappellaz, J., Delmas, R.J. and Lorius, C. (1993) The ice core record of greenhouse gases. *Science* **259**, 926–934.

Rea, B.R. (1996) A note on the experimental production of a mechanically polished surface within striations. *Glacial Geology and Geomorphology*, http://boris.qub.ac.uk/ggg

Reeh, N. (1968) On the calving of ice from floating glaciers and ice shelves. *Journal of Glaciology* **7** (50), 215–232.

Reeh, N. (1988) A flow-line model for calculating the surface profile and the velocity, strain-rate, and stress fields in an ice sheet. *Journal of Glaciology* **34** (116), 46–54.

Reeh, N. (1989) The age-depth profile in the upper part of a steady-state ice sheet. *Journal of Glaciology* **35** (121), 406–417.

Reeh, N. (1994) Calving from Greenland glaciers: observations, balance estimates of calving rates, calving laws. In: Reeh, N. (ed.) *Workshop on the Calving Rate of West Greenland Glaciers in Response to Climate Change, Sept 13–15 1993*. Danish Polar Centre, Copenhagen.

Reeh, N. and Paterson, W.S.B. (1988) Application of a flow model to the ice-divide region of Devon Island ice cap, Canada. *Journal of Glaciology* **34** (116), 55–63.

Reeh, N., Thomsen, H. and Clausen, H.B. (1987) The Greenland ice-sheet margin – a mine of ice for paleo-environmental studies. *Palaeogeography, Palaeoclimatology, Palaeoecology* **58**, 229–237.

Reid, H.F. (1896) The mechanics of glaciers. *Journal of Geology* **4**, 912–928.

Rendu, L. (1841) Théorie des glaciers de la Savoie. *Mem. Soc. R. Acad. Savoie* **10**, 39–159.

Repenning, C.A. (1990) Of mice and ice in the elate Pliocene of North America. *Arctic* **43** (4), 314–323.

Retzlaff, R. and Bentley, C.R. (1993) Timing of stagnation of Ice Stream C, West Antarctica, from short-pulse radar studies of buried surface crevasses. *Journal of Glaciology* **39** (133), 553–561.

Reynaud, L. (1977) Glacier fluctuations in the Mont Blanc area (French Alps). *Zeitschrift für Gletscherkunde und Glazialgeologie* **13**, 155–166.

Reynaud, L. (1987) The November 1986 survey of the Grand Moulin on the Mer de Glace, Mont Blanc Massif, France. *Journal of Glaciology* **33** (113), 130–131.

Rhodes, J.J., Armstrong, R.L. and Warren, S.G. (1987) Mode of formation of 'ablation hollows' controlled by dirt content of snow. *Journal of Glaciology* **33** (114), 135–139.

Ridley, J.K., Cudlip, W. and Laxon, S.W. (1993) Identification of subglacial lakes using ERS-1 radar altimeter. *Journal of Glaciology* **39** (133), 625–634.

Rignot, E. (1996) Tidal motion, ice velocity and melt rate of Petermann Gletscher, Greenland, measured from radar interferometry. *Journal of Glaciology* **42** (142), 476–485.

Rigsby, G.P. (1960) Crystal orientation in glacier and in experimentally deformed ice. *Journal of Glaciology* **3** (27), 589–606.

Rink, H. (1877) *Danish Greenland*. Henry S. King and Company, London.

Robin, G. de Q. (1955) Ice movement and temperature distribution in glaciers and ice sheets. *Journal of Glaciology* **2** (18), 523–532.

Robin, G. de Q. (1976) Is the basal ice of a temperate glacier at the pressure melting point? *Journal of Glaciology* **16** (74), 183–196.

Robin, G. de Q. (ed.) (1983) *The Climate Record in Polar Ice Sheets*. Cambridge University Press, Cambridge.

Robinson, J.G. (1984) Ice dynamics and thermal regime at Taylor Glacier, south Victoria Land, Antarctica. *Journal of Glaciology* **30** (105), 153–160.

Rogerson, R.J. (1985) Measured re-advance of a debris-covered glacier terminus in the President Range, Yoho national Park, British Columbia, Canada. *Journal of Glaciology* **31** (107), 13–17.

Rogerson, R.J. (1986) Mass balance of four cirque glaciers in the Torngat Mountains of Northern Labrador, Canada. *Journal of Glaciology* **32** (111), 208–218.

Rogerson, R.J., Olson, M.E. and Branson, D. (1986) Medial moraines and surface melt on glaciers of the Torngat Mountains, northern Labrador, Canada. *Journal of Glaciology* **32** (112), 350–354.

Rose, K.E. (1979) Characteristics of ice flow in Marie Byrd Land, Antarctica. *Journal of Glaciology* **24** (90), 63–75.

Röthlisberger, H. (1972) Water pressure in intra- and subglacial channels. *Journal of Glaciology* **11** (62), 177–203.

Röthlisberger, H. (1978) Eislawinen und Ausbrüche von Gletscherseen. *Jahrbuch der Scweizerischen Naturforschenden Gesellschaft. Wissenschaftlicher Teil*, 170–212.

Röthlisberger, H. and Iken, A. (1981) Plucking as an effect of water pressure variations at the glacier bed. *Annals of Glaciology* **2**, 57–62.

Röthlisberger, H. and Lang, H. (1987) Glacial hydrology. In: Gurnell, A.M. and Clark, M.J. (eds.) *Glacio-fluvial Sediment Transfer*. John Wiley and Sons, Chichester, 207–284.

Röthlisberger, H., Iken, A. and Spring, U. (1979) Piezometric observations of water pressure at the bed of Swiss glaciers (abstract). *Journal of Glaciology* **23** (89), 429.

Rott, H., Skvarca, P. and Nagler, T. (1996) Rapid collapse of Northern Larsen ice shelf, Antarctica. *Science* **271**, 788–792.

Russell, A.J. (1989) A comparison of two recent jökulhlaups from an ice-dammed lake, Søndre Strømfjord, West Greenland. *Journal of Glaciology* **35** (120), 157–162.

Russell, A.J. (1990) Extraordinary melt-water run-off near Søndre Strømfjord, West Greenland. *Journal of Glaciology* **36** (124), 353.

Russell, A.J. (1991) The geomorphological and sedimentological effects of jökulhlaups. PhD Thesis, University of Aberdeen.

Russell, A.J. (1993a) Supraglacial lake drainage near Søndre Strømfjord, Greenland. *Journal of Glaciology* **39** (132), 431–433.

Russell, A.J. (1993b) Obstacle marks produced by flow around stranded ice blocks during a glacier outburst flood (Jökulhlaup) in west Greenland. *Sedimentology* **40**, 1091–1111.

Salisbury, R.D. (1896) Salient points regarding the glacial geology of north Greenland. *Journal of Geology* **4**, 769–810.

Saussure, H.B. de (1779–96) *Voyages dans les Alpes, précédés d'un essai sur l'histoire naturelle des environs de Gèneve*. 4 volumes, Neuchâtel and Geneva.

Savage, J.C. and Paterson, W.S.B. (1963) Borehole measurements in the Athabasca Glacier. *Journal of Geophysical Research* **68**, 4521–4536.

Scheuchzer, J.J. (1723) *Itinera per Helvetiae Alpinas Regiones facta annis 1702–11*. Leyden.

Schlüchter, C. (ed.) (1979) *Moraines and Varves*. A.A. Balkema, Rotterdam.

Schmeits, M.J. and Oerlemans, J. (1997) Simulation of the historical variations in length of Unterer Grindelwaldgletscher, Switzerland. *Journal of Glaciology* **43** (143), 152–164.

Schumskiy, P.A. (1946) *The Energy of Glaciation and the Life of Glaciers*. United State Publishing House. Moscow. (Translated, U.S. Snow, Ice and Permafrost Research Establishment, Translation 7, 1950).

Schumskiy, P.A. (1978) *Dynamic Glaciology*. Amerind Publishing Company, New Delhi.

Schwander, J. and Stauffer, B. (1984) Age difference between polar ice and the air trapped in its bubbles. *Nature* **311,** 45–47.

Schweizer, J. and Iken, A. (1992) The role of bed separation and friction in sliding over an undeformable bed. *Journal of Glaciology* **38** (128), 77–92.

Scott, R.F. (1905) *The Voyage of the Discovery*. Smith, Elder, London.

Seaberg, S.Z., Seaberg, J.Z., Hooke, R.L. and Wiberg, D.W. (1988) Character of the englacial and subglacial drainage system in the lower part of the ablation area of Storglaciären, Sweden, as revealed by dye-trace studies. *Journal of Glaciology* **34** (117), 217–227.

Seidov, D., Sarnthein, M., Stattegger, K., Prien, R. and Weinelt, M. (1996) North-Atlantic ocean circulation during the last glacial maximum and subsequent meltwater event – a numerical model. *Journal of Geophysical Research – Oceans* **101** (C7), 16305–16332.

Seligman, G. (1936) *Snow Structures and Ski Fields*. Macmillan, London.

Severinghaus, J.P., Sowers, T., Brook, E.J., Alley, R.B. and Bender, R.L. (1998) Timing of abrupt climate change at the end of the Younger Dryas interval from thermally fractionated gases in polar ice. *Nature* **391** (6663), 141–146.

Sharp, M.J. (1988) Surging glaciers: behaviour and mechanisms. *Progress in Physical Geography* **12**, 349–370.

Sharp, M.J. (1996) Weathering pathways in glacial environments: hydrological and lithological controls. In: Bottrell, S.H. (ed.) *Fourth International Symposium on the Geochemistry of the Earth's Surface*. University of Leeds, 652–655.

Sharp, M.J., Lawson, W. and Anderson, R.S. (1988) Tectonic processes in a surge-type glacier. *Journal of Structural Geology* **10** (5), 499–515.

Sharp, M.J., Gemmell, J.C. and Tison, J.-L. (1989) Structure and stability of the former subglacial drainage system of the Glacier de Tsanfleuron, Switzerland. *Earth Surface Processes and Landforms* **14**, 119–134.

Sharp, M.J., Richards, K., Willis, I., Arnold, N., Nienow, P., Lawson, W. and Tison, J.-L. (1993) Geometry, bed topography and drainage system structure of the Haut Glacier d'Arolla, Switzerland. *Earth Surface Processes and Landforms* **18**, 557–572.

Sharp, M.J., Jouzel, J., Hubbard, B. and Lawson, W. (1994) The character, structure and origin of the basal ice layer of a surge-type glacier. *Journal of Glaciology* **40** (135), 327–340.

Sharp, M.J., Tranter, M., Brown, G.H. and Skidmore, M. (1995) Rates of chemical denudation and CO_2 drawdown in a glacier-covered catchment. *Geology* **23**, 61–64.

Shaw, J. (1977a) Till body morphology and structure related to glacier flow. *Boreas* **6** (2), 189–201.

Shaw, J. (1977b) Tills deposited in arid polar environments. *Canadian Journal of Earth Sciences* **14** (6), 1239–1245.

Shaw, J. (1988) Sublimation till. In: Goldthwait, R.P. and Matsch, C. (eds) *Genetic Classification of Glaciogenic Deposits*. Balkema, Rotterdam, 141–142.

Shaw, J. (1994) A qualitative view of sub-ice-sheet landscape evolution. *Progress in Physical Geography* **18** (2), 159–184.

Shilts, W.W. (1976) Glacial till and mineral exploration. *Royal Society of Canada Special Publication* **12**, 205–224.

Shoemaker, E.M. (1986) Debris-influenced sliding laws and basal debris balance. *Journal of Glaciology* **32** (111), 224–231.

Shoemaker, E.M. (1990) A subglacial boundary-layer regelation mechanism. *Journal of Glaciology* **36** (124), 263–268.

Shoemaker, E.M. (1992) Subglacial floods and the origin of low-relief ice-sheet lobes. *Journal of Glaciology* **38** (128), 105–112.

Shoji, H. and Langway, C.C. (1988) Flow-law parameters of the Dye 3, Greenland, deep ice core. *Annals of Glaciology* **10**, 146–150.

Shoji, H. and Langway, C.C. (1989) Physical property reference horizons. In: Oeschger, H. and Langway, C.C. (eds) *The Environmental Record in Glaciers and Ice Sheets*. Wiley, New York, 161–165.

Shreve, R.L. (1966) Sherman landslide, Alaska. *Science* **154** (3757), 1639–1643.

Shreve, R.L. (1972) Movement of water in glaciers. *Journal of Glaciology* **11** (62), 205–214.

Shreve, R.L. (1984) Glacier sliding at subfreezing temperatures. *Journal of Glaciology* **30** (106), 341–347.

Small, R.J. (1983) Lateral moraines of Glacier de Tsidjiore Nouve: form, development and implications. *Journal of Glaciology* **29** (102), 250–259.

Small, R.J. (1987) Moraine sediment budgets. In: Gurnell, A.M. and Clark, M.J. (eds) *Glacio-fluvial Sediment Transfer*. John Wiley and Sons, Chichester, 165–197.

Small, R.J., Beecroft, I.R. and Stirling, D.M. (1984) Rates of deposition on lateral moraine embankments, Glacier de Tsidjiore Nouve, Valais, Switzerland. *Journal of Glaciology* **30** (106), 275–281.

Smith, A.M. (1997a) Basal conditions on Rutford Ice Stream, West Antarctica, from seismic observations. *Journal of Geophysical Research* **102** (B1), 543–552.

Smith, A.M. (1997b) Variations in basal conditions on Rutford Ice Stream, West Antarctica. *Journal of Glaciology* **43** (144), 245–255.

Smith, R.A. (1989) A historical perspective of the nineteenth century ice trade. In: Oerlemans, J. (ed.) *Glacier Fluctuations and Climatic Change*. Kluwer Academic Publishers, Dordrecht, 173–182.

Sollid, J.L., Etzelmüller, B., Vatne, G. and Ødegård, T.S. (1994) Glacial dynamics, material transfer and sedimentation of Erikbreen and Hannabreen, Liefdefjorden, northern Spitsbergen. *Zeitschrift für Geomorphologie* **97**, 123–144.

Sorge, E. (1935) Glaciologische Untersuchungen in Eismitte. In: Brockcamp, B., Jülg, H., Loewe, F. and Sorge, E. *Wissenschaftliche Ergebnisse der Deutschen Grönland-Expedition Alfred Wegener 1929 und 1930/31. Band III. Glaziologie*. Brochhaus, Leipzig, 67–270.

Souchez, R.A. (1967) The formation of shear moraines: an example from south Victoria Land, Antarctica. *Journal of Glaciology* **6** (48), 837–843.

Souchez, R.A. (1971) Ice cored moraines in south west Ellesmere Island. *Journal of Glaciology* **10** (59), 245–254.

Souchez, R.A. (1987) Isotopic fractionation at the base of polar and sub-polar glaciers. *Journal of Glaciology* **33** (114), 246.

Souchez, R.A. and DeGroote, J.M. (1985) δD-δ18O relationships in ice formed by subglacial freezing: palaeoclimatic implications. *Journal of Glaciology* **31** (109), 229–232.

Souchez, R.A. and Jouzel, J. (1984) On the isotopic composition in δD and $\delta^{18}O$ of water and ice during freezing. *Journal of Glaciology* **30** (106), 369–372.

Souchez, R.A. and Lemmens, M. (1985) Subglacial carbonate deposition: an isotopic study of a present-day case. *Palaeogeography, Palaeoclimatology, Palaeoecology* **51**, 357–364.

Souchez, R.A. and Lemmens, M. (1987) Solutes. In: Gurnell, A.M. and Clark, M.J. (eds) *Glacio-fluvial Sediment Transfer*. John Wiley and Sons, Chichester, 207–284.

Souchez, R.A. and Lorrain, R.D. (1975) Chemical sorting effect at the base of an alpine glacier. *Journal of Glaciology* **14** (71), 261–265.

Souchez, R.A. and Lorrain, R.D. (1978) Origin of the basal ice layer from alpine glaciers indicated by its chemistry. *Journal of Glaciology* **20** (83), 319–328.

Souchez, R.A. and Lorrain, R.D. (1991) *Ice Composition and Glacier Dynamics*. Springer-Verlag, Berlin.

Souchez, R.A and Tison, J.-L. (1981) Basal freezing of squeezed water: its influence on glacier erosion. *Annals of Glaciology* **2**, 63–66.

Souchez, R.A. Lorrain, R.D. and Lemmens, M. (1973) Refreezing of interstitial water in a subglacial cavity of an alpine glacier as indicated by the chemical composition of ice. *Journal of Glaciology* **12** (66), 453–459.

Souchez, R.A., Lemmens, M., Lorrain, R., Tison, J.-L., Jouzel, J. and Sugden, D.E. (1990) Influence of hydroxyl-bearing minerals on the isotopic composition of ice from the basal zone of an ice sheet. *Nature* **345** (6272), 244–246.

Souchez, R.A., Lemmens, M., Tison, J.-L., Lorrain, R. and Janssens, L. (1993) Reconstruction of basal boundary conditions at the Greenland Ice Sheet margin from gas composition of the ice. *Earth and Planetary Science Letters* **118**, 327–333.

Souchez, R.A., Tison, J.-L., Lorrain, R., Lemmens, M., Janssens, L., Stievenard, M., Jouzel, J., Sveinbjörnsdottir, A. and Johnsen, S.J. (1994) Stable isotopes in the basal silty ice preserved in the Greenland Ice Sheet at Summit; environmental implications. *Geophysical Research Letters* **21** (8), 693–696.

Souchez, R.A., Janssens, L., Lemmens, M. and Stauffer, B. (1995a) Very low oxygen concentration in basal ice from Summit, Central Greenland. *Geophysical Research Letters* **22** (15), 2001–2004.

Souchez, R.A., Lemmens, M. and Chappellaz, J. (1995b) Flow-induced mixing in the GRIP basal ice deduced from CO_2 and CH_4 records. *Geophysical Research Letters* **22** (1), 41–44.

Splettstoesser, J.F. (ed.) (1976) *Ice-core Drilling*. University of Nebraska Press, Lincoln, Nebraska, and London.

Spufford, F. (1997) *I May Be Some Time*. Faber and Faber, London.

Stacey, J.S. (1960) A prototype hotpoint for thermal boring on the Athabasca Glacier. *Journal of Glaciology* **3** (28), 783–786.

Stauffer, B.R. (1989) Dating of ice by radioactive isotopes. In: Oeschger, H. and Langway, C.C. (eds) *The Environmental Record in Glaciers and Ice Sheets*. John Wiley and Sons, New York, 123–139.

Stenborg, T. (1969) Studies of the internal drainage of glaciers. *Geografiska Annaler* **51(A)** 13–41.

Stenborg, T. (1970) Delay of run-off from a glacier basin. *Geografiska Annaler* **52(A)** 1–30.

Stolle, D.F.E. and Killeavy, M.S. (1986) Determination of particle paths using the finite-element method. *Journal of Glaciology* **32** (111), 219–223.

Strasser, J.C., Lawson, D.E., Larson, G.J., Evenson, E.B. and Alley, R.B. (1996) Preliminary results of tritium analyses in basal ice, Matanuska Glacier, Alaska, U.S.A.: evidence for subglacial ice accretion. *Annals of Glaciology* **22**, 126–133.

Sturm, M. and Benson, C. (1985) A history of Jökulhlaups from Strandline Lake, Alaska, U.S.A. *Journal of Glaciology* **31** (109), 272–280.

Sturm, M., Benson, C. and MacKeith, P. (1986) Effects of the 1966–68 eruptions of Mount Redoubt on the flow of Drift Glacier, Alaska, U.S.A. *Journal of Glaciology* **32** (112), 355–362.

Sturm, M., Hall, D.K., Benson, C.S. and Field, W.O. (1991) Non-climatic control of glacier-terminus fluctuations in the Wrangell and Chugach Mountains, Alaska, U.S.A. *Journal of Glaciology* **37** (127), 348–356.

Sugden, D.E. (1968) The selectivity of glacial erosion in the Cairngorm mountains, Scotland. *Transactions of the Institute of British Geographers* **45**, 79–92.

Sugden, D.E. (1974) Landscapes of glacial erosion in Greenland and their relationship to ice, topographic and bedrock conditions. *Institute of British Geographers Special Publication* **7**, 177–195.

Sugden, D.E. (1977) Reconstruction of the morphology, dynamics, and thermal regime of the Laurentide ice sheet at its maximum. *Arctic and Alpine Research* **9** (1), 21–47.

Sugden, D.E. (1978) Glacial erosion by the Laurentide ice sheet. *Journal of Glaciology* **20** (83), 367–392.

Sugden, D.E. (1991) The stepped response of ice sheets to climatic change. In: Harris, C. and Stonehouse, B. (eds) *Antarctica and Global Change*. Belhaven, London, 107–114.

Sugden, D.E. and John, B.S. (1976) *Glaciers and Landscape*. Edward Arnold, London.

Sugden, D.E., Clapperton, C.M. and Knight, P.G. (1985) A jökulhlaup near Sondre Stromfjord, West Greenland, and some effects on the ice-sheet margin. *Journal of Glaciology* **31** (109), 366–368.

Sugden, D.E., Knight, P.G., Livesey, N., Souchez, R., Lorrain, R., Tison, J.-L. and Jouzel, J. 1987(a) Evidence for two zones of debris entrainment beneath the Greenland ice sheet. *Nature* **328** (6127), 238–241.

Sugden, D.E., Clapperton, C.M., Gemmell, J.C. and Knight, P.G. (1987b) Stable isotopes and debris in basal glacier ice, South Georgia, Southern Ocean. *Journal of Glaciology* **33** (115), 1–6.

Sugden, D.E., Glasser, N. and Clapperton, C.M. (1992) Evolution of large roches moutonées. *Geografiska Annaler* **74A**, 253–264.

Sugden, D.E., Marchant, D.R., Potter Jr, N., Souchez, R.A., Denton, G.H., Swisher III, C.C. and Tison, J.-L. (1995) Preservation of Miocene glacier ice in East Antarctica. *Nature* **376** (6539), 412–414.

Summerfield, M.A. (1991) *Global Geomorphology*. Longman, Harlow.

Swinzow, G.K. (1962) Investigation of shear zones in the ice sheet margin, Thule area, Greenland. *Journal of Glaciology* **4** (32), 215–229.

Syvitski, J.P.M. (1989) On the deposition of sediment within glacier-influenced fjords. *Marine Geology* **85**, 301–329.

Takahachi, S., Fujii, T. and Ishida, T. (1973) Origin and development of polygonal ablation hollows on a snow surface. *Low Temperature Science* **31** (Series A), 191–207.

Takahachi, S., Naruse, R., Nakawo, M. and Mae, S. (1988) A bare ice field in east Queen Maud Land, Antarctica, caused by horizontal divergence of drifting snow. *Annals of Glaciology* **11**, 156–160.

Tangborn, W.V., Krimmel, R.M. and Meier, M.F. (1975) A comparison of glacier mass balance by glaciological, hydrological and mapping methods, South Cascade Glacier, Washington. *International Association of Hydrological Sciences, Publication*, 104, 185–196.

Tarr, R.S. and Martin, L. (1914) *Alaskan Glacier Studies*. National Geographical Society, Washington D.C.

Taylor, K.C., Hammer, C.U., Alley, R.B., Clausen, H.B., Dahl-Jensen, D., Gow, A.J., Gundestrup, N.S., Kipfstuhl, J., Moore, J.C. and Waddington, E.D. (1993a) Comparison of oxygen isotope records from the GISP2 and GRIP Greenland ice cores. *Nature* **366**, 549–552.

Taylor, K.C., Lomorey, G.W., Doyle, G.A., Alley, R.B., Grootes, P.M., Mayewski, P.A., White, J.W.C. and Barlow, L.K. (1993b) The 'flickering switch' of late Pleistocene climate change. *Nature* **361**, 432–436.

Teller, J.T. and Clayton, L. (eds) (1983) *Glacial Lake Agassiz*. Geological Association of Canada Special Paper **26**.

Theakstone, W.H. (1967) Basal sliding and movement near the margin of the glacier Østerdalsisen, Norway. *Journal of Glaciology* **6** (43), 805–816.

Theakstone, W.H. (1979) Observations within cavities at the bed of the Glacier Osterdalsisen, Norway. *Journal of Glaciology* **23** (89), 273–282.

Thompson, A. (1988) Historical development of the proglacial landforms of Svinafellsjökull and Skaftafellsjökull, southeast Iceland. *Jökull* **38**, 17–30.

Thompson, L.G. and Mosley-Thompson, E. (1981) Microparticle concentration variations linked with climate change: evidence from polar ice cores. *Science* **212**, 812–815.

Thomsen, H.H., Thorning, L. and Braithwaite, R.J. (1988) *Glacier-hydrological Conditions on the Inland Ice North-east of Jakobshavn/Ilulisat, West Greenland*. Grønlands Geologiske Undersøgelse. Report no. **138**.

Thorarinsson, S. (1953) Some new aspects of the Grimsvötn problem. *Journal of Glaciology* **2** (14), 267–275.

Thorarinsson, S. (1957) The jökulhlaup from the Katla area in 1955 compared with other jökulhlaups in Iceland. *Jökull* **7**, 21–25.

Thorpe, P.W. (1991) Surface profiles and basal shear stresses of outlet glaciers from a Late-glacial mountain ice field in western Scotland. *Journal of Glaciology* **37** (125), 77–88.

Tison, J.-L. and Lorrain, R.D. (1987) A mechanism of basal ice-layer formation involving major ice-fabric changes. *Journal of Glaciology* **33** (113), 47–50.

Tison, J.-L., Petit, J.-R., Barnola, J.-M and Mahaney, M.C. (1993) Debris entrainment at the ice-bedrock interface in sub-freezing temperature conditions (Terre Adelie, Antarctica). *Journal of Glaciology* **39** (132), 303–315.

Tison, J.-L., Thorsteinsson, T., Lorrain, R.D. and Kipfstuhl, J. (1994) Origin and development of textures and fabrics in basal ice at Summit, Central Greenland. *Earth and Planetary Science Letters* **125**, 421–437.

Trabant, D.C. and Meyer, D.F. (1992) Flood generation and destruction of 'Drift' Glacier by the 1989–90 eruption of Redoubt Volcano, Alaska. *Annals of Glaciology* **16**, 33–38.

Tranter, M. (1996) Glacial runoff as a sink for atmospheric CO_2 during the last glacial–interglacial transition. In: Bottrell, S.H. (ed.) *Fourth International Symposium on the Geochemistry of the Earth's Surface*. University of Leeds, 709–713.

Tranter, M. and Raiswell, R. (1991) The composition of the englacial and subglacial component in bulk meltwaters draining the Gornergletscher, Switzerland. *Journal of Glaciology* **37** (125), 59–66.

Tranter, M., Brown, G., Raiswell, R., Sharp, M. and Gurnell, A. (1993) A conceptual model of solute acquisition by Alpine glacial meltwaters. *Journal of Glaciology* **39** (133), 573–581.

Tufnell, L. (1984) *Glacier Hazards*. Longman, Harlow.

Tyndall, J. (1858) On some physical properties of ice. *Proceedings of the Royal Society of London, Series A* **9**, 76–80.

Tyndall, J. (1872) *The Forms of Water in Clouds and Rivers, Ice and Glaciers*. London.

UNESCO (1970) *Seasonal Snow Cover*. UNESCO/IASH/WMO Technical Papers in Hydrology, no. **2**.

US Department of Energy (1985) *Glaciers, ice sheets and sea level: effect of a CO_2 induced climatic change.* Department of Energy, Washington D.C.

van der Meer, J.J.M. (ed.) (1987) *Tills and Glaciotectonics.* A.A. Balkema, Rotterdam and Boston.

van der Veen, C.J. and Whillans, I.M. (1989) Force budget: I. Theory and numerical methods. *Journal of Glaciology* **35** (119), 53–60.

van der Veen C.J. and Whillans I.M. (1992) Determination of a flow center on an ice cap. *Journal of Glaciology* **38** (130), 412–416.

van der Veen, C.J. and Whillans, I.M. (1994) Development of fabric in ice. *Cold Regions Science and Technology* **22** (2), 171–195.

van der Wal, R.S.W., Oerlemans, J. and van der Hage, J.C. (1992) A study of ablation variations on the tongue of Hintereisferner, Austrian Alps. *Journal of Glaciology* **38** (130), 319–324.

van der Wateren, F.M. (1995) Processes of Glaciotectonism: In: Menzies, J. (ed.) *Modern Glacial Environments.* Butterworth-Heinemann, Oxford, 309–336.

Vaughan, D.G. (1993a) Chasing the rogue icebergs. *New Scientist* **137** (1855), 24–27.

Vaughan, D.G. (1993b) Relating the occurrence of crevasses to surface strain rates. *Journal of Glaciology* **39** (132), 255–266.

Venetz, I. (1833) Memoire sur les variations de la temperature dans les Alpes de la Suisse. *Denkschriften der allg. Schweiz. Gesellsch. Gesammten Naturwissenschaften* **1** (2), 1–38.

Vere, D.M. and Benn, D.I. (1989) Structure and debris characteristics of medial moraines in Jotunheimen, Norway: implications for moraine classification. *Journal of Glaciology* **35** (120), 276–280.

Vincent, C. and Vallon, M. (1997) Meteorological controls on glacier mass balance: empirical relations suggested by measurements on glacier de Sarennes, France. *Journal of Glaciology* **43** (143), 131–137.

Vinje, T.E. (1980) Some satellite-tracked iceberg drifts in the Antarctic. *Annals of Glaciology* **1**, 83–88.

Vivian, R. (1977) Tourism, summer skiing, hydroelectricity and protection of the public in the French Alpine glacial area: the development of an applied glaciology. *Journal of Glaciology* **19** (81), 639–642.

Vivian, R. and Bocquet, G. (1973) Subglacial cavitation phenomena under the Glacier d'Argentière, Mont Blanc, France. *Journal of Glaciology* **12** (66), 439–451.

Vornberger, P.L. and Whillans, I.M. (1990) Crevasse deformation and examples from Ice Stream B, Antarctica. *Journal of Glaciology* **36** (122), 3–10.

Waddington, B.S. and Clarke, G.K.C. (1995) Hydraulic properties of subglacial sediment determined from the mechanical response of water-filled boreholes. *Journal of Glaciology* **41** (137), 112–124.

Waddington, E.D. (1986) Wave ogives. *Journal of Glaciology* **32** (112), 325–334.

Waddington, E.D. (1987) Geothermal heat flux beneath ice sheets. *International Association of Hydrological Sciences Publication* **170**, 217–226.

Wakahama, K. and Tusima, M. (1981) Observations of inner moraines near the terminus of McCall Glacier in Arctic Alaska and Laboratory experiments on the mechanism of picking up moraines into a glacier body (abstract only). *Annals of Glaciology* **2**, 116.

Walder, J.S. (1982) Stability of sheet flow of water beneath temperate glaciers and implications for glacier surging. *Journal of Glaciology* **28** (99), 273–293.

Walder, J.S. (1986) Hydraulics of subglacial cavities. *Journal of Glaciology* **32** (112), 439–445.

Walder, J.S. and Costa, J.E. (1996) Outburst floods from glacier-dammed lakes: the effect of mode of lake drainage on flood magnitude. *Earth Surface Processes and Landforms* **21**, 701–723.

Walder, J.S. and Driedger, C.L. (1995) Frequent outburst floods from South Tahoma Glacier, Mount Rainier, U.S.A.: relation to debris flows, meteorological origin and implications for subglacial hydrology. *Journal of Glaciology* **41** (137), 1–10.

Walder, J.S. and Fowler, A. (1994) Channelized subglacial discharge over a deformable bed. *Journal of Glaciology* **40** (134), 3–15.

Walder, J.S. and Hallet, B. (1979) Geometry of former subglacial water channels and cavities. *Journal of Glaciology* **23** (89), 335–346.

Walker, J.C.F. and Waddington, E.D. (1988) Early discoverers XXXV: Descent of glaciers: some early speculations on glacier flow and ice physics. *Journal of Glaciology* **34** (118), 342–348.

Warburton, J. (1990) An alpine proglacial sediment budget. *Geografiska Annaler* **72A** (3–4), 261–272.

Warren, C.R. (1991) Terminal environment, topographic control and fluctuation of west Greenland glaciers. *Boreas* **20** (1), 1–45.

Warren, C.R. (1992) Iceberg calving and the glacioclimatic record. *Progress in Physical Geography* **16**, 253–282.

Warren, C.R., Glasser, N.F., Harrison, S., Winchester, V., Kerr, A.R. and Rivera, A. (1995) Characteristics of tide-water calving at Glaciar San Rafael, Chile. *Journal of Glaciology* **41** (138), 273–289.

Weeks, W.F. (1980) Iceberg water: an assessment. *Annals of Glaciology* **1**, 5–10.

Weeks, W.F. and Campbell, W.J. (1973) Icebergs as a fresh-water source, an appraisal. *Journal of Glaciology* **12** (65), 207–234.

Weertman, J. (1957) On the sliding of glaciers. *Journal of Glaciology* **3** (21), 33–38.

Weertman, J. (1961) Mechanism for the formation of inner moraines found near the edge of cold ice caps and ice sheets. *Journal of Glaciology*, **3** (30), 965–978.

Weertman, J. (1968) Diffusion law for the dispersion of hard particles in an ice matrix that undergoes simple shear deformation. *Journal of Glaciology* **7** (50), 161–165.

Weertman, J. (1972) General theory of water flow at the base of a glacier or ice sheet. *Reviews of Geophysics and Space Physics* **10** (1), 287–333.

Weertman, J. (1973a) Position of ice divides and ice centres on ice sheets. *Journal of Glaciology* **12** (66), 353–360.

Weertman, J. (1973b) Creep of ice. In: Whalley, E., Jones, S.J. and Gold, L.W. (eds) *Physics and Chemistry of Ice*. Royal Society of Canada, Ottawa, 320–337.

Weertman, J. (1983) Migration of centres of ice domes and of ice-shelf–inland-ice boundaries. In: Robin, G. de Q. (ed.) *The Climatic Record in Polar Ice Sheets*. Cambridge University Press, 26–28.

Weertman, J. (1986) Basal water and high-pressure basal ice. *Journal of Glaciology* **32** (112), 455–463.

Weertman, J. and Birchfield, G.E. (1983) Stability of sheet water flow under a glacier. *Journal of Glaciology* **29** (103), 374–382.

Weertman, J., Sibert, J., Weeks, W.F. and Sternig, J. (1974) Radioactive wastes on ice. *Science Public Affairs* **30** (1).

Weidick, A. (1975) Estimates on the mass balance changes of the Inland Ice since Wisconsin-Weichsel. *Grønlands Geologiske Undersøgelse Rapport* **148**, 87–91.

Weiss, R.F., Bucher, P., Oeschger, H. and Craig, H. (1972) Compositional variations of gasses in temperate glaciers. *Earth and Planetary Science Letters* **16**, 178–184.

Wendler, G. (1989) On the blowing snow in Adelie Land, Eastern Antarctica. In: Oerlemans, J. (ed.) *Glacier Fluctuations and Climatic Change*. Kluwer Academic Publishers, Dordrecht, 261–280.

Wendler, G. and Kelley, J. (1988) On the albedo of snow in Antarctica: A contribution to I.A.G.O. *Journal of Glaciology* **34** (116), 19–25.

Whalley, W.B. (1979) The relationship of glacier ice and rock glacier at Grubengletscher, Kanton Wallis, Switzerland. *Geografiska Annaler* **61A** (1–2), 49–61.

Whalley, W.B. and Langway, C.C. Jr (1980) A scanning electron microscope examination of subglacial quartz grains from Camp Century core, Greenland – a preliminary study. *Journal of Glaciology* **25** (91), 125–131.

Whillans, I.M. and van der Veen, C.J. (1993) Patterns of calculated basal drag on Ice Streams B and C, Antarctica. *Journal of Glaciology* **39** (133), 437–446.

Whillans, I.M. and van der Veen, C.J. (1997) The role of lateral drag in the dynamics of Ice Stream B, Antarctica. *Journal of Glaciology* **43** (144), 231–237.

Whillans, I.M., Jackson, M. and Tseng, Y.-H. (1993) Velocity pattern in a transect across Ice Stream B, Antarctica. *Journal of Glaciology* **39** (133), 562–572.

Whiteman, C.A. (1995) Processes of Terrestrial Deposition. In: Menzies, J. (ed.) *Modern Glacial Environments*. Butterworth-Heinemann, Oxford, 293–308.

Wilkinson, D.S. (1988) A pressure sintering model for the densification of polar firn and glacier ice. *Journal of Glaciology* **34** (116), 40–45.

Williams, R.S. and Ferrigno, J.G. (eds) (1988) Satellite image atlas of glaciers of the world. *United States Geological Survey Professional Paper* **1386** (11 volumes).

Williams, R.S. Jr, Hall, D.K. and Benson, C.S. (1991) Analysis of glacier facies using satellite techniques. *Journal of Glaciology* **37** (125), 120–128.

Williams, S.N. (ed.) (1989–1990) The November 13, 1985, eruption of Nevado del Ruiz Volcano, Colombia. *Journal of Volcanology and Geothermal Research* **40–41**.

Willis, I.C. (1995) Intra-annual variations in glacier motion: a review. *Progress in Physical Geography* **19** (1), 61–106.

Willis, I.C., Sharp, M.J. and Richards, K.S. (1990) Configuration of the drainage system of Midtdalsbreen, Norway, as indicated by dye-tracing experiments. *Journal of Glaciology* **36** (122), 89–101.

Wilson, A.T. (1964) Origin of ice ages: an ice shelf theory for pleistocene glaciation. *Nature* **201**, 147–149.

Wilson, A.T. (1969) The climatic effects of large-scale surges of ice sheets. *Canadian Journal of Earth Sciences* **6**, 911–918.

Wintges, T. (1985) Studies on crescentic fractures and crescentic gouges with the help of close-range photogrammetry. *Journal of Glaciology* **31** (109), 340–349.

Wold, B. and Østrem, G. (1979) Subglacial constructions and investigations at Bondhusbreen, Norway. *Journal of Glaciology* **23** (89), 363–379.

Wolfe, P.M. and English, M.C. (1995) Mass balance of a small valley glacier in the Canadian high Arctic, Ellesmere Island, Northwest Territories. *Zeitschrift für Gletscherkunde und Glazialgeologie* **31**, 93–103.

Yao, T.D., Petit, J.R., Jouzel, J., Lorius, C. and Duval, P. (1990) Climatic record from an ice margin in East Antarctica. *Annals of Glaciology* **14**, 323–327.

Yoshida, M., Yamamoto, K., Higuchi, K., Iida, H., Ohata, T. and Nakamura, T. (1990) First discovery of fossil ice of 1000–1700 year B.P. in Japan. *Journal of Glaciology* **36** (123), 258–259.

Zeller, E., Saunders, D.F. and Angono, E.E. (1973) Putting radioactive wastes on ice: a proposal for an international radionuclide depository in Antarctica. *Science Public Affairs* **29** (1).

Zotikov, I.A. (1986) *The Thermophysics of Glaciers*. D. Reidel Publishing Company, Dordrecht.

Zwally, H.J. (1987) Technology in the advancement of glaciology. *Journal of Glaciology*, Special Edition, 66–77.

Zwally, H.J. and Fiegles, S. (1994) Extent and duration of Antarctic surface melt. *Journal of Glaciology* **40** (136), 463–476.

Zwally, H.J. (1990) Ice sheet elevation change. *Annals of Glaciology* **14**, 366.

Zwally, H.J., Brenner, A.C., Major, J.A., Bindschadler, R.A. and Marsh, J.G. (1989) Growth of Greenland ice sheet: measurement. *Science* **246**, 1587–1589.

INDEX